Keywords in Radical Geography: *Antipode* at 50

Edited by the *Antipode* Editorial Collective:

Tariq Jazeel, Andy Kent, Katherine McKittrick, Nik Theodore,
Sharad Chari, Paul Chatterton, Vinay Gidwani, Nik Heynen,
Wendy Larner, Jamie Peck, Jenny Pickerill,
Marion Werner and Melissa W. Wright

**Celebrating 50 years of publishing some of the
best and most provocative radical geography**

1969-2019

WILEY Blackwell

This edition first published 2019
© 2019 Antipode Foundation Ltd.

Registered Office(s)
John Wiley & Sons, Inc., 111 River Street, Hoboken, NJ 07030, USA
John Wiley & Sons Ltd, The Atrium, Southern Gate, Chichester, West Sussex, PO19 8SQ, UK

Editorial Office
111 River Street, Hoboken, NJ 07030, USA
9600 Garsington Road, Oxford, OX4 2DQ, UK

For details of our global editorial offices, customer services, and more information about Wiley products visit us at www.wiley.com.

Wiley also publishes its books in a variety of electronic formats and by print-on-demand. Some content that appears in standard print versions of this book may not be available in other formats.

Library of Congress Cataloging-in-Publication Data

9781119558156 (paperback);
This book is also available and free-to-view online via www.onlinelibrary.wiley.com (search for ISBN 9781119558071)

Cover design by Wiley

Set in size of font and font name by Scientific Publishing Services (P) Ltd.

Printed in the UK

Contents

CONTENTS

Series Editors' Preface

The *Antipode Book Series* explores radical geography 'antipodally,' in opposition, from various margins, limits or borderlands.

Antipode books provide insight 'from elsewhere', across boundaries rarely transgressed, with internationalist ambition and located insight; they diagnose grounded critique emerging from particular contradictory social relations in order to sharpen the stakes and broaden public awareness. An *Antipode* book might revise scholarly debates by pushing at disciplinary boundaries, or by showing what happens to a problem as it moves or changes. It might investigate entanglements of power and struggle in particular sites, but with lessons that travel with surprising echoes elsewhere.

Antipode books will be theoretically bold and empirically rich, written in lively, accessible prose that does not sacrifice clarity at the altar of sophistication. We seek books from within and beyond the discipline of geography that deploy geographical critique in order to understand and transform our fractured world.

Vinay Gidwani
University of Minnesota, USA

Sharad Chari
University of California, Berkeley, USA

Antipode Book Series Editors

Keywords in Radical Geography: An Introduction

Nik Theodore

Department of Urban Planning and Policy, University of Illinois at Chicago, Chicago, IL, USA

Tariq Jazeel

Department of Geography, University College London, London, UK

Andy Kent

Editorial Office, Antipode: A Radical Journal of Geography, Cardiff, UK;
antipode@live.co.uk

Katherine McKittrick

Department of Gender Studies, Queen's University, Kingston, ON, Canada

Le mot juste: "Radical" Geography

"It appears that *Antipode* will survive, at least for a while, as a radical journal of geography." Thus began Richard Peet's (1972:iv) modest statement outlining *Antipode*'s first editorial policy—quite modest, in fact, considering radical geography's decidedly immodest goals to "transform the scope of a conventional discipline criticized as irrelevant to the great issues of the time" (Peet 2000:951). The journal itself was to directly contribute to this transformation by fostering "the search for organizational models for promoting social change" (Peet 1977:244) and providing a medium "for the dissemination of non-traditional ideas" (Stea 1969:1). Rather than being (just) "another Geography journal", *Antipode*'s "aspiration [then and now] was to produce geographical knowledge that might connect to a larger project for the transformation of economy, society and environment" (Castree and Wright 2009:2).

As has been well chronicled, *Antipode* was founded in 1969 as an intellectual and political intervention in the discipline of geography; in the wider social sciences; and in a world riven by war, racism, sexism, colonialism, and injustice. "The key to *Antipode*'s origin is the term 'radical'", recalled *Antipode*'s first editor, Ben Wisner (n.d.). "We were groping for root causes of ... problems, contradictions, inconsistencies, and hypocrisies", spurred by the promises of the Civil Rights movement in the United States as well as the spectre of nuclear war; the ravages of poverty and famine; and the violence of racism, colonialism, inequality, and uneven development. "Early radical geography was anarchic and exuberant, naïve yet nuanced" (Peet 2000:951) and, for better or worse, the pages of *Antipode* reflected these qualities. The journal has, from its inception, moved with and across wider debates in radical geography. For this reason, we write with an understanding that radical geography and *Antipode* are in conversation, co-constitutive and relational rather than distinct intellectual projects.

Early issues of the journal were eclectic, exploring, for example, the imperialist underpinnings of geographic thought, the militarisation of remote sensing technologies, and white supremacy and racism in geographical texts and urban policy (Anderson 1969; Blaut 1969, 1970; Donaldson 1969; Earickson 1971; Stewart 1969). Articles debated radical methodologies and the merits of advocacy planning, as well as the possibilities for revolutionary social change versus more incremental institutional reform (Amaral and Wisner 1970; Breitbart 1972; Corey 1972; Morrill 1969; Peet 1969; Stea 1970). Volume 1 included an interview with community organiser Ruby Jarrett on what today would be termed "spatial stigmatisation" and "racial-territorial enclosure" (Jarrett and Wisner 1969).

In their survey of the development of radical geography in North America, Linda Peake and Eric Sheppard characterize the late 1960s to mid 1970s as a time of politicised discovery in the face of the unquestioned whiteness of "a segregated and institutionally racist discipline" (2014:315):

> The ... [period] saw a flourishing of different voices in *Antipode*, *Transition*, and the USG [Union of Socialist Geographers] newsletters; socialist, feminist, anti-racist, anarchist and environmentalist approaches to studying social problems and advocating social change were all evident. This reflected the multivalent, intersecting protest and social movements unleashed by a 1960s politics of radicalism, anti-racism, sexual liberation and emancipation, in which various protagonists were involved in multiple ways, and the complex linkages between these and academic trajectories. (2014:314)[1]

The vitality of these ideas and movements energised a growing cadre of geographers who were searching for a vehicle to challenge conservative disciplinary structures. Channelling such "anarchic" energies can be an exacting endeavour, however, and attempts to archive radical thought without a shared theoretical framework struck some as incoherent. If *Antipode*'s early volumes were a politicised, though inchoate, foray into some of geography's many misrepresentations, silences and other misdeeds, David Harvey's (1972:6) clarion call, published in volume 4, issue 2, to overthrow prevailing paradigms would soon make plain the stakes for the discipline:

> The objective social conditions demand that we say something sensible and coherent or else forever (through lack of credibility or, even worse, through the further deterioration of the objective social conditions) remain silent. It is the emerging objective social conditions and our patent inability to cope with them which essentially explains the necessity for a revolution in geographic thought.

A proliferation of Marxist scholarship followed, published in the pages of *Antipode* and, to a much lesser extent, in other, more mainstream, geography journals. Thus, the emergence of a sub-discipline that could be named *as* radical geography was, through the early to mid 1970s, emerging *pari passu* with the dominance of Marxist critiques across the social sciences and humanities. However, while Marxism provided coherence, as an emergent orthodoxy it began to crowd out other nascent strands of radical geographical thought. Fortunately, by the early 1980s, radical geography once again opened out to grasp the heterodox strands of critical thought that were initially incubated in the early pages of the journal and had continued to develop elsewhere within the sub-discipline.

Unlike some other widely read reflections on this period, Peake and Sheppard's (2014) account veers away from celebrating the various achievements made during this time, preferring instead to evaluate the evolution of radical geographical scholarship against the ideals, tenets and demands of radical thought and praxis. As Marxism came to represent a new orthodoxy among radical scholars, some of the openness and creativity that characterised the emergence of radicalism in the discipline was, for a period at least, lost. With few outlets within mainstream geography for the publication of radical texts, *Antipode*'s privileging of articles centring on Marxist theory (O'Keefe 1979) inadvertently narrowed the scope for the publication of other forms of radical research. The impacts of this *de facto* closure were widely felt as early as the mid 1970s across the sub-discipline of radical geography. The field was slow to embrace a number of radical politico-intellectual currents, notably feminist theorising focusing on the situatedness and social construction of knowledge (Christopherson 1989; McDowell 1992a, 1992b). As far as *Antipode* is concerned, aside from a few noteworthy exceptions, early feminist scholarship primarily developed outside the pages of the journal.[2] In addition, radical geography more broadly retreated from the study of race and racism (Peake and Sheppard 2014; cf. Mahtani 2014), something that today is being remedied through the growing subfields of Black geographies and Indigenous geographies.

Although the development of non-Marxist viewpoints and analysis proved challenging, the journal was nonetheless the site of some of the earliest efforts in Anglophone geography to publish "theory from the South", something that continues to energise debates concerning the contested geographies of knowledge production. The call for papers for a special issue on "underdevelopment and domination/dependence" edited by Milton Santos (1975:91), for example, singled out prevailing North American and Eurocentric biases that rendered much of what had been written in geography of little interest to readers in the global South. A rejection of "the empirical and atomized formulations which have been imposed as theories on the Third World ... [was to be] questioned", in part through the affirmative inclusion of scholars from "underdeveloped countries" (ibid.).[3] However, even this laudable effort exposed a disjuncture, still in evidence today, between cultures and norms of academic writing as it is practiced in various parts of the world. Ultimately, fewer scholars from the global South were included in the special issue than originally had been envisioned, something that, as Ferretti and Pedrosa (2018) argue, was deeply troubling for Santos.

In the late 1980s, Susan Christopherson's feminist indictment of the discipline encapsulated a critique that had been building in many quarters, calling into question the very knowledge-making and pedagogical practices that constitute radical geography:

> For even among those who advocate political action and theoretical transformation, educational and institutional practice has remained profoundly conservative. Despite

the lip-service given to the integrity of individual experience, little attention is given to how to teach women, minorities and working-class people ... to translate their experiences into theory. (1989:87)

Flatly unconvinced that geography was prepared to confront its class, race and gender biases, Christopherson offered little hope for a "transformed geography" willing to fundamentally interrogate the exercise and basis of authority, "including our own" (ibid.). In case there might be any misunderstanding, Christopherson's response to the rhetorical question of whether she expects such a transformation was shorn of optimism: "No." The reasons for this terse judgment lay in what Christopherson identified as growing elitism within the discipline and a propensity to engage in intellectually moribund citational practices. Although Christopherson was writing 30 years ago, her appraisal of radical geography's epistemology *and contradictions* remains every bit as vital and compelling today. "[S]ome people are quickly out of fashion", she writes (1989:88), while the "[j]ustifiable fear" "of being left out" or "of being labeled" leads many to gravitate to "the 'in' subject" of the moment. Given the debates that are swirling across radical geography in 2019 concerning the field's knowledge-making and theory-building practices, readers today will no doubt have an immediate reaction to Christopherson's pointed and perceptive critique. Likely this will take the form of knowing assent, the assumption being that this critique is directed at someone other than the reader her/himself.[4]

This is not the place for us to weigh in on the nature of the discipline's contemporary debates about appropriate modes of knowledge production. We will do so anyway. One of the generative aspects of changing currents within geography has been the heightened awareness of longstanding biases regarding which authors are read, debated and cited. The prevailing winds determining which theories, concepts, methodologies, *and scholars* are regarded as being "in" or "out" of fashion periodically shift. For this reason, the politics of citation do not tack in a single direction, though the dominant course unquestionably was set long ago. Geography's institutionalised racism, the discipline's class and gender biases, and its underlying heterosexism (Chouinard and Grant 1995; Nast 2002) are among the deeply ingrained biases that have contributed to the field's damaging silences and exclusions along the axes of race, gender and sexuality (Katz 1996). That knowledge production has suffered as a direct result of these silences and exclusions is without question. Scholars must continue to struggle against structures and practices that discriminate and exclude; the transformation of geography, the prospects of which merited such profound scepticism by Susan Christopherson, must continue to be the goal.

At the same time, we must guard against a dangerous crosscurrent that threatens to stifle debate even as new corridors are opening for an expanded and robust discussion of epistemology and consideration of a more expansive set of subjects, methodologies, analyses, and critiques. Christopherson's observation that "some people are quickly out of fashion" points to a troubling characteristic of the discipline, one that is every bit as present today as when she penned these words three decades ago. By now it should be clear that stifling scholarly debate, which can prematurely rob us of key observations and insights, is antithetical to radicalism within geography and to the transformations that are long overdue. That the field has failed to live up to its own mandates of openness and inclusiveness provides no warrant for attempts to shut down areas of scholarship and to shut out scholars in the process. We need more engagement, not less; deeper engagement, not the summary dismissal of theories and ideas rendered unfashionable. Calls for increased political relevance and greater clarity in writing are well received (Mitchell 2006). But at the same time let us not lose sight of the importance of different modalities of thought and writing, including those that are theoretically dense and not immediately accessible. As problematic as the "race for theory" (Christian 1987) can be, a "race from theory" entails its own dangers and has its own costs. If geography, as a transforming discipline, is always in the process of becoming, we will need theory, in its manifold forms and modalities, to assist us in navigating this transformation and in producing knowledge that challenges injustice.

Yet this is to raise another important question concerning what constitutes theory, or what distinguishes a theoretical text from data, narrative, or the "mere empirics" of disciplinary inquiry. To begin to push at this question is to also render transparent the political economy of radical geographical knowledge production insofar as it is to turn the lens back on to Christopherson's observation that "some people are quickly out of fashion". For fashion has a market, and the market for radical knowledge bears upon the intellectual life of the sub-discipline. Herein lies a tension that *Antipode* has had to, and must continue to, navigate: it is a journal that plays a significant role in channelling the currents of radical knowledge production, yet its mandate has always been to be something of a counterpoint to prevailing orthodoxies. That much is central to its radical mandate. To give space to theorisation that may not (yet) be legible *as* theory has to be part of the journal's task, as long as that theorisation is committed to complementing, building, and extending radical ways of knowing while also seeking social change. Since the journal became a commercial affair in 1986, this tension between market and radicality is one that all its editors have had to negotiate.

Resisting the impulse to quiet different voices was essential to the evolution of Anglophone radical geography in the last decades of the 20th century. The discipline as a whole was fortunate that, throughout the 1970s and well into the 1990s, scholars continued to develop radical thought and practice, though often outside, or in complicated relationships with, the early radical geographical project. According to Peake and Sheppard (2014:315), "[t]his period was especially important for the establishment of the emerging fields of geographies of race and racism, feminist (Marxist, liberal and other variants) geography, and (although slightly less so) for geographies of sexualities", though the fact that these emergent approaches and traditions "existed within or alongside radical geography and in other times and places apart from it" (2014:314) can be seen as "indicative not only of the transversal and unpredictable intellectual and spatial paths of the evolution of Anglophone North American critical geography, but also of the impossibility of attempts to explore its evolution through a core (Clark [University], SFU [Simon Fraser University]) versus periphery (everywhere else) model of knowledge dissemination" (2014:314–315). The radical tradition within geography, in other words, has—since its very beginnings in the 1960s—been a much broader and more polyvalent undertaking than it is sometimes characterised.

So, what should we make of radical geography's achievements over the course of its first 50 years? "In 1969 when the first issue of *Antipode* was published the very notion that there could be a radical approach to geographical questions was an anathema to much of the profession", write Eric Sheppard and Joe Doherty (1986:1): "Geography had never had a significant critical, radical tradition." Seen in this light, the evolution and enduring relevance of radical geographical thought is notable. At the same time, Peake and Sheppard's appraisal of radical geography's trajectory and achievements is also likely tempered by the further realisation, frequently remarked upon in editorials published in *Antipode* and other journals, that in the 1970s, 1980s and early 1990s radical geography continued to face sustained resistance, both within and beyond the academy. "A series of subtle oppressions" (Peet 1985:3) experienced by radical geographers left many feeling "beleaguered" (Blomley 2006:87), "battered and bruised" (Chouinard 1994:2), "diminished, dispirited and divided" (Walker and McDowell 1993:2).

In retrospect, now more than five decades into Anglo-American geography's radical turn, these characterisations might strike some readers as surprising. The number of geographers who self-identify as "radical" or (more commonly) "critical" has soared, as have the citation counts of many of the field's leading figures; the days in which Left geographers could be regarded "an embattled minority" (Walker 1989:81) within the discipline having long past (Castree 2000; Johnston 2000; Peake and Sheppard 2014). So too there has been an embrace of radical geography beyond the Anglo-American academy (Belina et al. 2009; Finn and Hanson 2017). For its part,

> *Antipode* has played a crucial role in recasting and transforming the discipline of geography, the forms of geographical theory, and the practice of geographical research. Hostility to inequality, intolerance, and injustice are now at the core of the discipline and the plea for relevancy has been heard. (Peck and Wills 2000:3)

Though falling well short of the transfiguration called for by Christopherson, that progressive change has occurred within the discipline is undeniable. Along with *Environment and Planning D: Society and Space*; *Gender, Place and Culture*; *ACME: An International E-Journal for Critical Geographies*; *Human Geography: A New Journal of Radical Geography*; and now-defunct outlets including *Transition* and the newsletter of the Union of Socialist Geographers, all of which share a commitment to advancing radical geographical analysis and praxis, *Antipode* helped catalyse and propel a transformation of the discipline such that "by the late 1990s radical/critical geography ... [had] become the new canon, the new mainstream" (Peake and Sheppard 2014:318). Admittedly, though, for some radical/critical geographers (see Blomley 2006; Castree 2000; Chouinard 1994; Peet 2000; Waterstone 2002) this has been a dubious achievement as the pressures of academic professionalisation have made unwanted incursions that have come at the expense of activism, action and engagement with the world in which radical scholarship seeks to intentionally intervene.

A "Living Archive" of Radical Geography

How might one assess the achievements of *Antipode* and its transition from an upstart, countercultural journal whose founders were committed to disciplinary and societal transformation, though a protracted period in which the journal's very survival was in doubt (see Heynen et al. 2017:8–12), and on to its position as one of the mainstays of the discipline? For Peake and Sheppard (2014:309), "*Antipode*'s emergence was the relational effect of multiple conditions of possibility", occurring as it did during a period of social upheaval in the 1960s that laid bare many of the dangers and contradictions of the times. Key to its significance, certainly far more than citation counts, impact factors and other such metrics, is the simple fact that "it created visibility, and a place, for radical geography by dint of being a concrete and recognized academic object (a journal)" (Peake and Sheppard 2014:309). Here Stuart Hall's inquiry into the formation and consolidation of the archive is instructive to the task of historical assessment. As was the case with *Antipode*, "No archive arises out of thin air", Hall (2001:89) writes; "Each archive has a 'pre-history', in the sense of prior conditions of existence". Peake and Sheppard's illuminating contribution to the historical record documenting the emergence of radical geography in North America—one that invites further elaboration[5]—reveals that a number of geographers, both those who have been widely acknowledged and those who have been wrongly overlooked, played important roles in shaping emergent geographical thought, methodologies, and praxis, contributing to what ultimately would coalesce into geography's radical edge (see also Peake 2015).

"Constituting an archive represents a significant moment, on which we need to reflect with care. It occurs at that moment when a relatively random collection of works, whose movement appears simply to be propelled from one creative production to the next, is at the point of becoming something more ordered and considered: an object of reflection and debate" (Hall 2001:89). Such was the case when *Antipode* was established, seemingly without reference to the radical experiments and practices that preceded the publication of the first issue (Peake and Sheppard 2014). Yet from volume 1, issue 1, and over the course of the next 50 years, *Antipode* has played an important role in both propelling and ordering radical geographical debates. From its founding, *Antipode* has sought to be an outlet for dissident voices, and statements from past editors regularly attest to the avowedly ecumenical orientation of the journal, even if at times the scope of debate was narrowed through editor discretion and decision-making, and even though at times such claims of ecumenical embrace have been a matter of vigorous dispute.[6] As Hall observes, the constitution of a journal *qua* archive is a moment that demands careful reflection. For this reason, since its founding, the journal has invited critique of its own practices (Stea 1969), and critique it has received (for example, see Castree and Wright 2005; Hague 2002; Jacoby 1987; Nast 2002; Waterstone 2002).

The forgoing discussion, like the annals of *Antipode* themselves, makes clear that radical geographical interventions have never been easy or singular or transparent; rather they have been full of contradictions, silences, and shortcomings. This is one reason careful reflection upon the archive is required. But this entails more than reflection as a form of "looking backward" across a field's oeuvre. *Antipode* should be understood as constituting "a 'living archive', whose construction must be seen as an on-going, never-completed project" (Hall 2001:89). Imperfect and unfinished, open-ended and mutable, "the journal has aspired to be a place of theoretical innovation pushing the ways in which geographers (and others) go about framing a heterodox radical politics" (Heynen et al. 2017:12). Hopefully the underlying modesty inherent in such an undertaking is clear even if it is not evinced by the boldness of the aspiration. *Antipode*, and the field of radical geography in which it is embedded, is an ongoing project. The journal is an expression of those who publish in it and of those who steward the archive at any moment in time. It is a reflection of the discipline and its cognate fields, and its specific identity is therefore indeterminate. In these senses, *Antipode* is always evolving and never fully knowable.

Antipode: **Avant la lettre**

As *Antipode* enters its sixth decade, its editorial statement has been amended. The (ever-expanding) list of radical subfields (Marxist/socialist/anarchist/antiracist/feminist/sexual liberationist) has been removed, not because these markers are unimportant or because there is a diminished commitment to any of these forms of scholarship, but because the list seems too fixed and static and incapable of reflecting radical approaches that are as of yet unnamed. Here we heed Raymond Williams' (1983:16) warning that in "periods of change" language "and concern for clarity can quickly become brittle", stifling thought and creativity. The journal's revised editorial statement continues *Antipode*'s ecumenical tradition while acknowledging the ongoing transformation of geographical analysis and the concomitant decentring of what it means to be radical, and it upholds the importance of analyses that are theoretically and empirically substantive while also providing a subtle reminder regarding the content and tenor of academic debate. It reads:

> *Antipode* publishes innovative papers that push at the boundaries of radical geographical thinking. Papers will be rigorous and substantive in theoretical and empirical terms. Authors are encouraged to critique and challenge settled orthodoxies, while engaging the context of intellectual traditions and their particular trajectories. Papers should put new research or critical analyses to work to contribute to strengthening a Left politics broadly defined.[7]

The journal thus engages radical geography by imagining it not as an unchanging sub-discipline, but rather as a site, or point of constellation, through which to engage and engender a spatially oriented Left politics. The focus emphasises intellectual and activist praxis, thus displacing the question of who is or is not "radical", and instead drawing attention to already existing and future sites of struggle, liberation, and political change (cf. Gilmore 2017). In a small way, then, *Antipode*'s commitment to publishing, presenting, and supporting the work of geographers and non-geographers imperfectly expresses this kind of vision; in the pages of the journal, through the mentorship of early-career scholars at the biannual Institute for the Geographies of Justice, in our annual lectures, and through the Antipode Foundation scholar-activist and workshop grants, we seek to honour and support radical politics and action.[8]

Lingua franca: Keywords in Radical Geographical Thought

The chapters collected in this volume commemorating *Antipode*'s 50th anniversary are short, suggestive, conversational and experimental, and they encompass a multitude of ways that radical intellectual scholarship and debate within geography and across its peripheries has unfolded, and continues to do so. They delve into the journal's intellectual history, they speculate on current conjunctures, and they reach for new, unthought horizons of critical potential. Many are playful. But play, we note, is perhaps the most creative of elans and should not be confused with political apathy, or frivolity. When play enables another way of seeing, being, or critically engaging, we should not assume it to be a- or anti-theoretical.

Given, as we have stressed above, the difficulties of knowing *Antipode*'s archive with any certainty, we have, in this volume, left the interpretation of the radical geographical project to our contributors. The 50 short essays gathered here reflect the visions, preoccupations, and not least the speculations of 50+ authors close to the journal. Their essays speak to radical geography's past-present-future in all the ways our contributors imagine them to. *Keywords in Radical Geography* is not, therefore, a dictionary of predictable or generalised words historicised and defined by each author. Instead, contributors have selected terms, concepts, or sets of ideas that resonate with them, that may be important to their research, or that simply provide for them a wormhole to a more free place or a more utopian imagination. Each author discusses the term and/or idea they have chosen in relation to radical geography. The task was simply to connect the entry to key themes and aspirations in radical geography rather than to describe and define radical geography in any sense. Eschewing any pretence of building a coherent narrative, we hope this will be a fitting testimony to the role *Antipode* has played in the generation of radical geographical engagements with the world, and the profusion of different types of radical intervention across the broader discipline of geography.

Keywords?

The title of this volume might well bring to mind Raymond Williams' classic *Keywords: A Vocabulary of Culture and Society*. First published in 1976 and revised in 1983, *Keywords* offers "an inquiry into a vocabulary: a shared body of words and meanings" used in discussions of culture and society (Williams 1983:15). This endeavour is alive and well in the 21st century. Two recent edited volumes, *New Keywords: A Revised Vocabulary of Culture and Society* (Bennett et al. 2005a) and *Keywords for Today: A 21st Century Vocabulary* (McCabe et al. 2018), similarly provide "a critical reflection on the key terms constituting our contemporary vocabulary of culture and society" (Bennett et al. 2005b:570). To be sure, some might say that these books include entries that are important to their authors, but argu-ably are not part of "our contemporary vocabulary"; their authors' interests aren't necessarily shared by others. And absent in them are words that most of us would have expected or hoped to see given their significance in current debates. What is more, one might ask, "who is this 'us'?". Sociology and anthropology, like geography, are diverse, changing fields constituted by a plurality of perspectives. So where are those perspectives from the South, subaltern perspectives, and so on? Of course, the contents of both single- and co-authored works will always be partial (and for various reasons), just as Williams himself warned that *Keywords* was "necessarily unfinished and incomplete", welcoming "amendment, correction and addition" (1983:27, 26). This notwithstanding, we can say that the tables of contents of these books are more or less predictable, non-arbitrary, more or less as they "ought" to be.

Our book is quite different. At first glance, its table of contents looks capricious, if not chaotic. Entries are organised, perhaps rather unimaginatively, via the con-vention of the alphabet. Indeed, like Williams (and Bennett et al., and McCabe et al.), we are concerned that alphabetic listing obscures the many, diverse and provocative connections between entries. But those connections are there to be made (and remade, and remade again) by readers. As Williams writes, "a book is only completed when it is read" (1983:25)—like an archive, it is not; *it becomes*. The contents of this book are surprising, to say the least, and beg questions. Glancing at it, one might wonder why "Mercury", "Enough" and "Badge", say, are present. Are they really "key terms"? (Badge? *Really*? You cannot be serious ...!). And why include these words when "Development", "Migration" "Neoliber-alism" and "War", for example, are absent? Well, our book is the product of the creative input of many rather than some version of top-down planning. As edi-tors, we afforded the book's contributors freedom to choose their own words. Like the editors of *Keywords for Today*, we see language as not only "a shared understanding", but also "a site of division" (McCabe et al. 2018:xi), of contesta-tion, of disagreement, and of struggle. Our concern is less with radical geogra-phy's "single meaning" than with its "many competing semantic elements" (ibid.). The significance of a given word is neither self-evident nor unproblematic. To make this clear, perhaps the relevance of these words within (and, indeed, beyond) our discipline ought *not to be* obvious but, rather, is there to be demon-strated by our contributors. The meaning of these words is an open question, and arguably the task of exploring meaning and usage can open new political possibilities that otherwise might go unnoticed. The case is the contributors' to make. We think they are a persuasive bunch, passionate and rigorous in their reflections on why "Love" and "Fragments", "Vulnerability" and "Monument", and all manner of other weird and wonderful words speak—*matter*—to radical geography's histories, current condition, and possible future directions. And in this sense, each entry might also be the beginning of a new conversation ... which is also what the radical project must be about.

This approach is fitting for a journal like *Antipode*, which has always welcomed the infusion of new ideas and the shaking-up of old positions through dialogue and debate. Despite the prevailing Marxist currents in parts of the journal's early life, it has never been committed to a single view of diagnosis or critique. Just look at the editorials marking its 10th, 20th, 30th and 40th anniversaries (Castree and Wright 2009; O'Keefe 1979; Peck and Wills 2000; Walker 1989): successive editorial teams have repeatedly reasserted the journal's openness and inclusivity, its commitment to broadening debate, its ecumenism and absence of dogma, its willingness to diversify and complicate. As Phil O'Keefe (1979:1) suggested in the late 1970s, despite *Antipode*'s communion with radical geography, the very equation of the journal with a specific discipline may be misleading insofar as its founders and subsequent editors were always committed to "working within and above disciplinary boundaries". And as we enter our sixth decade we would like to think we are still in a place where the new, the innovative, the creative, and the heretofore unthought radical edges of spatial theorisation and analysis can find a home. Neither unquestioningly bound to what has come before, nor wilfully distant and adrift from it, we hope to continue the tradition of striving, with passion, to know and understand the difficulties facing us without underestimating the possibilities—neither despairing about domination and oppression nor naively hopeful about resistance and alternatives.

Endnotes

[1] The Socially and Ecologically Responsible Geographers (SERGE) formed in 1971, publishing the journal *Transition* until 1986. The Union of Socialist Geographers (USG) formed in 1974; an archive of its newsletter, published until 1982, is available at https://antipodefoundation.org/2017/06/28/usg-newsletter-archive/ (see Peake and Sheppard 2014; Thatcher et al. 2017).
[2] Though *Antipode* likely was the first Geography journal to publish a feminist analysis (Walker and McDowell 1993:2), see Pat Burnett's (1973) "Social change, the status of women, and models of city form and development".
[3] The special issues were published in 1977—volume 9, numbers 1 and 3.
[4] A symposium in honor of Susan Christopherson was published in 2017, including contributions from Jennifer Clark, Amy Glasmeier, Cindi Katz, Katharine Rankin, and Rachel Weber (see: https://antipodefoundation.org/2017/10/23/on-being-outside-the-project/).
[5] Eric Sheppard and Trevor Barnes' (2019) edited book, *Spatial Histories of Radical Geography: North America and Beyond*, which is part of the *Antipode* Book Series (https://www.wiley.com/en-gb/Antipode+Book+Series-c-2222), goes some way towards this.
[6] Our archive of past editors' reflections is available online: https://onlinelibrary.wiley.com/page/journal/14678330/homepage/editor_s_past_reflections.htm
[7] This statement was written by Nik Theodore, Kiran Asher, Dave Featherstone, Tariq Jazeel, Andy Kent and Marion Werner in late 2018 and opens our guidelines for authors (https://onlinelibrary.wiley.com/page/journal/14678330/homepage/forauthors.html).
[8] For more on all these activities, see https://antipodefoundation.org

References

Amaral D J and Wisner B (1970) Participant-observation, phenomenology, and the rules for judging sciences: A comment. *Antipode* 2(1):42–51
Anderson J (1969) Moral problems of remote sensing technology. *Antipode* 1(1):54–57
Belina B, Best U and Naumann M (2009) Critical geography in Germany: From exclusion to inclusion via internationalisation. *Social Geography* 4(1):47–58
Bennett T, Grossberg L and Morris M (eds) (2005a) *New Keywords: A Revised Vocabulary of Culture and Society*. Oxford: Blackwell

Bennett T, Grossberg L and Morris M (2005b) Originals, remakes, assemblages: A retrospect on *New Keywords. Criticism* 47(4):567–571

Blaut J M (1969) Jingo geography (part I). *Antipode* 1(1):10–15

Blaut J M (1970) Geographic models of imperialism. *Antipode* 2(1):65–82

Blomley N (2006) Uncritical critical geography? *Progress in Human Geography* 30(1):87–94

Breitbart M (1972) Advocacy in planning and geography. *Antipode* 4(2):64–68

Burnett P (1973) Social change, the status of women, and models of city form and development. *Antipode* 5(3):57–62

Castree N (2000) Professionalisation, activism, and the university: Whither "critical geography"? *Environment and Planning A* 32(6):955–970

Castree N and Wright M W (2005) Home truths. *Antipode* 37(1):1–8

Castree N and Wright M W (2009) The power of numbers. *Antipode* 41(1):1–9

Chouinard V (1994) Reinventing radical geography: Is all that's Left Right? *Environment and Planning D: Society and Space* 12(1):2–6

Chouinard V and Grant A (1995) On not being anywhere near "the project": Ways of putting ourselves in the picture. *Antipode* 27(2):137–166

Christian B (1987) The race for theory. *Cultural Critique* 6:51–63

Christopherson S (1989) On being outside "the project". *Antipode* 21(2):83–89

Corey K E (1972) Advocacy in planning: A reflective analysis. *Antipode* 4(2):46–63

Donaldson F (1969) Geography and the black American: The white papers and the invisible man. *Antipode* 1(1):17–33

Earickson R (1971) Poverty and race: The bane of access to essential public services. *Antipode* 3(1):1–8

Ferretti F and Pedrosa B V (2018) Inventing critical development: A Brazilian geographer and his Northern networks. *Transactions of the Institute of British Geographers* 43(4):703–717

Finn J C and Hanson A-M (2017) Critical geographies in Latin America. *Journal of Latin American Geography* 16(1):1–15

Gilmore R W (2017) Abolition geography and the problem of innocence. In G T Johnson and A Lubin (eds) *Futures of Black Radicalism* (pp 225–240). New York: Verso

Hague E (2002) *Antipode*, Inc? *Antipode* 34(4):655–661

Hall S (2001) Constituting an archive. *Third Text* 15(54):89–92

Harvey D (1972) Revolutionary and counter revolutionary theory in geography and the problem of ghetto formation. *Antipode* 4(2):1–13

Heynen N, Kent A, McKittrick K, Gidwani V and Larner W (2017) Neil Smith's long revolutionary imperative. *Antipode* 49(s1):5–18

Jacoby R (1987) *The Last Intellectuals.* New York: Basic Books

Jarrett R and Wisner B (1969) How to build a slum, part one. *Antipode* 1(1):37–42

Johnston R (2000) Intellectual respectability and disciplinary transformation? Radical geography and the institutionalisation of geography in the USA since 1945. *Environment and Planning A* 32(6):971–990

Katz C (1996) Towards minor theory. *Environment and Planning D: Society and Space* 14(4):487–499

Mahtani M (2014) Toxic geographies: Absences in critical race thought and practice in social and cultural geography. *Social and Cultural Geography* 15(4):359–367

McCabe C and Yanacek H with the Keywords Project (eds) (2018) *Keywords for Today: A 21st Century Vocabulary.* Oxford: Oxford University Press

McDowell L (1992a) Doing gender: Feminism, feminists and research methods in human geography. *Transactions of the institute of British Geographers* 17(4):399–416

McDowell L (1992b) Multiple voices: Speaking from inside and outside "the project". *Antipode* 24(1):56–72

Mitchell K (2006) Writing from left field. *Antipode* 38(2):205–212

Morrill R (1969) Geography and the transformation of society. *Antipode* 1(1):6–9

Nast H J (2002) Prologue: Crosscurrents. *Antipode* 34(5):835–844

O'Keefe (1979) Editorial. *Antipode* 11(3):1–2

Peake L J (2015). Rethinking the politics of feminist knowledge production in Anglo-American geography. *The Canadian Geographer / Le Géographe canadien* 59(3), 257–266

Peake L and Sheppard E (2014) The emergence of radical/critical geography within North America. *ACME* 13(2):305–327

Peck J and Wills J (2000) Geography and its discontents. *Antipode* 32(1):1–3

Peet R (1969) A new left geography. *Antipode* 1(1):3–5

Peet R (1972) Editorial policy. *Antipode* 4(2):iv

Peet R (1977) The development of radical geography in the United States. *Progress in Geography* 1(2):240–263

Peet R (1985) Radical geography in the United States: A personal history. *Antipode* 17(2/3):1–7

Peet R (2000) Celebrating 30 years of radical geography. *Environment and Planning A* 32(6):951–953

Santos M (1975) Special issue on underdevelopment and domination/dependence. *Antipode* 7(1):91

Sheppard E and Barnes T (eds) (2019) *Spatial Histories of Radical Geography: North America and Beyond.* Oxford: Wiley

Sheppard E and Doherty J (1986) Editorial. *Antipode* 18(1):1–4

Stea D (1969) Positions, purposes, pragmatics: A journal of radical geography. *Antipode* 1(1):1–2

Stea D (1970) Another discourse on method. *Antipode* 2(1):52–64

Stewart R F (1969) Troubling textbooks. *Antipode* 1(1):34–36

Thatcher J, Sheppard E and Akatiff C (2017) The Union of Socialist Geographers remembered. *AntipodeFoundation.org* 1 May. https://antipodefoundation.org/supplementary-material/the-union-of-socialist-geographers-newsletter-1975-1983/ (last accessed 23 November 2018)

Walker D (1989) Guest editorial. *Antipode* 21(2):81–82

Walker R and McDowell L (1993) Editorial. *Antipode* 25(1):1–3

Waterstone M (2002) A radical journal of geography or a journal of radical geography? *Antipode* 34(4):662–666

Williams R (1983) *Keywords: A Vocabulary of Culture and Society* (2nd edn). Oxford: Oxford University Press

Wisner B (n.d.) "Notes from Underground: The Beginning of *Antipode*." https://onlinelibrary.wiley.com/page/journal/14678330/homepage/editor_s_past_reflections.htm (last accessed 16 January 2019)

A Democratic Ethos

Sophie Bond

*Te Ihowhenua / Department of Geography, Te Whare Wānanga o Ōtakou / University of
Otago, Ōtepoti / Dunedin Aotearoa / New Zealand;*
sophie.bond@otago.ac.nz

For Democracy, But Not As We Know It!

There seems to be two different ways of thinking "democracy". On one hand, there is democracy that holds the promise of a more just and fairer future, that suggests people hold the power to determine their version of a liveable life. It is a democracy that enables vibrant debate and contestation, and where different worlds coexist. On the other hand, there is actually existing democracy, which Todd May sums up nicely. He states, from his position in the US, that "we in the US were told that our God-given mission was to bring democracy to places that lack it. This was especially true of those places that have a lot of oil" (May 2010:1). Here democracy is an elitist, paternalistic set of practices and institutions that maintain a more-than-liveable life for the few. It seems clear that in terms of the first way of thinking democracy, actually existing democracy in the so-called developed world, is not, for the most part, very "democratic".

In 1991, Jean-Luc Nancy wrote of the

> world at a moment when a kind of broadly pervasive democratic consensus seems to make us forget that "democracy", more and more frequently serves only to assure a play of economic and technical forces that no politics today subject to any end other than that of its own expansion. (Nancy 1991:xxxvii)

In other words, actually existing democracy seems to only serve the expansion of economic and technical interests. More than 25 years later, similar arguments persist, framed as post-democracy (Crouch 2004), de-democratisation (Brown 2011), or depoliticisation (Bates et al. 2014; Flinders and Wood 2014). Post-politics has more recently described this closure as practices, strategies and norms that enable politics (discussion, debate, participation) but only in so far as it occurs within the narrow sphere of business-as-usual (Darling 2014; Swyngedouw 2011; Thomas 2017; Thomas and Bond 2016; Wilson and Swyngedouw 2014). Dissent outside this sphere is quickly delegitimised through a variety of tactics and modalities (Bond et al. 2018; Darling 2014) that are themselves enabled through "democratic" processes like legislation debated in a house of representatives (e.g. that increasingly criminalises protest, or encourages a culture of surveillance). In addition, a pragmatic realism constructs subjectivities that are individually oriented and disciplined into the "commonsense" of business-as-usual, consolidating TINA (there is no alternative) discourses (see Bond et al. 2018). Alternatively, political subjectivities are so precariously situated in the struggles of everyday life as to be unable to enact a politics of dissent. The result is the sedimentation of a political

Keywords in Radical Geography: Antipode at 50, First Edition. Edited by the *Antipode* Editorial Collective.
© 2019 The Authors/Antipode Foundation Ltd. Published 2019 by John Wiley & Sons Ltd.

and social order that is so cramped it makes alternatives that envision a fairer, more equitable and more just existence seem impossible (The Free Association 2010).

The failures of democracy go well beyond post-political closure and depoliticisation tactics in the public sphere. The epistemological and ontological hegemony of western systems of democracy perpetuate historical inequalities and injustices that are too often invisible to those in dominant groups. Laura Pulido (2015: 813–814) defines the components of white supremacy through an example of a polluting industry near Los Angeles with a history of non-compliance. These components are *awareness* of the harm caused by their actions; *taking*, in this case clean air and wellbeing; and demonstrating "an attitude of *racial superiority*" (emphasis added). In calling this out as white supremacy, she names the structural racism that is embedded within democratic states. Here the rule of law, and the regulatory role of the State, has enabled such racism to continue, despite knowing the inequalities and harm it results in. Of course white supremacy is nothing new, but "racial capitalism" and the space to name racism has become less visible under neoliberalism's emphasis on individualism (Pulido 2015), and I suggest, its pragmatic realism oriented toward economic efficiencies and particular racialised forms of so-called common sense.

If democracy is such a failure, servicing only a minority, being enrolled in the service of power and capital—why hold on to it as an ideal? It might be argued that democracy has never worked to produce a fairer world. Indeed, when a Māori research participant once said to a colleague and I in an interview that democracy has not been very good to us (i.e. Māori), I questioned my own adherence to a concept that is colonial, clearly doesn't work in practice, and moreover, has done and continues to do violence in perpetuating the status quo. So what does a decolonised, properly progressive ethos of democracy look like? Does it exist? Can it be facilitated such that an alternative democratic imaginary becomes possible?

Returning to the etymology of the word gives me cause for hope—demos, the people, and kratos, power—power to the people. But who are the people and to what ends do they seek and wield power? If a radical democratic ethos is the ability and right to speak out, is that sufficient to enable a democracy that addresses the inequalities that actually existing democracy perpetuates. Radical democrats like Mouffe, Rancière, Žižek, and Connolly argue in various ways for a vibrant contestatory public sphere. May (2010) highlights a key dimension of a radical democracy—the power to challenge, raise questions, and articulate alternative imaginaries. But what if these alternatives are unjust or perpetuate racist capitalism, or colonial relations of marginalisation and oppression? In many respects, this question reflects the challenge laid down by Olson and Sayer (2009) to be explicit about the normative basis of our work. We assume the orientation to justice is enough—but justice for whom? What are our normative foundations? Mouffe (2005, 2013) specifically addresses the normative question that underpins the inherent challenge to dominant practices that create or perpetuate inequalities. She argues for negotiated and situated articulations of justice, equality and liberty as the ethico-political values that underpin the situated and sometimes contested notions of a democratic ethos in particular places.

Here, I am reminded of the work of radical geographers who identify, explore and privilege the efforts of people who seek a more just, fair world, who engage and perform a politics of dissent and contestation. The contours of what is deemed just, fair and equal are evident in both the challenges made to the status quo and the alternatives imagined. For example, the work of the Community Economies Collective initially led by Gibson-Graham and now continued by Kathy Gibson, Jenny Cameron, Steve Healey, Kelly Dombroski, Gradon Diprose and many others is "committed to theorizing, representing, and enacting new visions of economy".[1] Pulido's (2015) work, mentioned above, seeks to name and make visible the ways in which racism continues to underpin capitalism and the ongoing nature of the contemporary colonial project (Bonds and Inwood 2016; Daigle and Sundberg 2017; Radcliffe 2017; Sundberg 2014; Thomas 2015). Indigenous geographers call out epistemological and ontological violence within existing hegemonies of both the academy and within other purportedly democratic institutions (Coombes et al. 2013; Daigle 2016; Hunt 2014).

These examples (and there are many more that I should acknowledge but that I haven't the space to mention here) demonstrate a democratic ethos whereby people are claiming space to live a more liveable life, and to challenge that which constrains or limits their ability to do so. These communities that radical geographers work with are typically the subjects of injustice, or seek to challenge and address injustices that harm others. Their work is underpinned by shared ethico-political principles that provide a situated but implicit normative basis around justice, equality, and a democratic space to speak. This is crucial work, highlighting the labour of groups and communities who seek to carve out spaces that once seemed impossible within the cramped contours of the current conjuncture (The Free Association 2010). Despite its significance, it seems that it highlights the work of those already enrolled in some kind of progressive democratic ethos. So how can we extend this work—or some might say scale it up (Gismondi et al. 2016)? How can we create the conditions of possibility to facilitate the spread of such a democratic ethos in those subjectivities who are so entangled within and often blind to the dominant rationalities that perpetuate harm?

I suggest that the vibrant, contestatory sphere that radical democrats argue for, while important, is insufficient to shift the epistemological and ontological hegemonies at work. While this work disrupts the order of the sensible, highlighting the contingency and cracks within these hegemonies, it seems to me that there is a need for a radical shift in political subjectivity away from individualised responsibility, blame, and liability to a more collective ethos of care and responsibility. Here I am referring to the work of Iris Marion Young in particular, but also Judith Butler and Joan Tronto, and the recent revival in radical geography of a feminist ethic of care (see, for example, Diprose 2017; Lawson 2007, 2009; Smith 2009; Williams 2017, among others). I suggest that such a care-full politics and a political responsibility for justice (Young 2011) ought to be the (always situated) normative foundation or the ethico-political basis for a radical democracy.

There are different ways of framing a feminist ethic of care, but what resonates for me is the recognition of our vulnerability to others (human and non-human) for our very existence. From the moment of birth to our everyday relationships of love and nurture (see Butler 2006), we are relational beings. Yet the tendencies for dominant discourses to individualise responsibility and privilege market rationalities, don't only reduce the space for politics and contestation. They reduce the space for thinking and acting relationally and thereby also thinking and acting care-fully. As Lawson argues, "this marginalization of care is deeply political and it bolsters relations of gender/race/ethnic inequality and restricts human flourishing" (2009:210; see also Williams 2017). Consequently, Lawson argues for a critical ethic of care and responsibility where care is not private. Rather it "begins with a social ontology of connection: foregrounding social relationships" (2007:3).

In early work on a feminist ethic of care, Tronto argues for thinking through such an ethic as the foundation for political judgments. She seeks a shift in thinking from "seeing people as rational actors pursuing their own goals and maximising their interests, ... [to] instead see[ing] people as constantly enmeshed in relationships of care" (1995:142). There are parallels here with Iris Marion Young's (2011) articulation of responsibility for justice. The model of individual rational actors that Tronto refers to invokes a liability model of justice that seeks to lay blame on individuals. While there is still a role for this in some contexts, it tends to mask the structural inequalities that create injustices. Therefore, Young argues for a conceptualisation of justice that is cognisant of the way people, typically in positions of white privilege, are entangled in societal structures that cause injustices or harm to others. This entanglement means that individuals who benefit from these structures are complicit in the resulting harm and therefore have a degree of responsibility to act to address that injustice. However, she argues this action must be a collective action. The burden is too great to bear alone. Underlying these ideas is an ethic of care and responsibility, aligned with recognition of the violence and harm of the structures of contemporary global capitalism and ongoing colonialisms. This framing of a politics of justice is spatial, relational, and situated, and is attuned to structural violence.

Radical geography has, it seems to me, always highlighted struggles for justice and voice. But as Olson and Sayer (2009) suggested, it hasn't always been explicit about the normative positioning that underpins this work. A democratic ethos is already evident in much radical geography—one that is decolonising, keenly sensitive to structural injustices, and situated in a care-full politics for a liveable life. An ethos is always negotiated and situated, but it is also guided by ethico-political principles that are shared. Let's name them.

Endnote
[1] See http://www.communityeconomies.org/about (last accessed 16 August 2018).

References

Bates S, Jenkins L and Amery F (2014) (De)politicisation and the Father's Clause parliamentary debates. *Policy and Politics* 42(2):243–258

Bond S, Diprose G and Thomas A C (2018) Contesting deep sea oil: Politicisation-depoliticisation-repoliticisation. *Environment and Planning C: Politics and Space* https://doi.org/10.1177/2399654418788675

Bonds A and Inwood J (2016) Beyond white privilege: Geographies of white supremacy and settler colonialism. *Progress in Human Geography* 40(6):715–733

Brown W (2011) We are all democrats now... In G Agamben, A Badiou, D Bensaid, W Brown, J-L Nancy, J Rancière, K Ross and S Zizek (eds) *Democracy in What State?* (pp 4–57). New York: Columbia University Press

Butler J (2006) *Precarious Life: The Powers of Mourning and Violence.* London: Verso

Coombes B, Johnson J T and Howitt R (2013) Indigenous geographies II: The aspirational spaces in postcolonial politics–reconciliation, belonging, and social provision. *Progress in Human Geography* 37(5):691–700

Crouch D (2004) *Post-Democracy.* Cambridge: Polity

Daigle M (2016) Awawanenitakik: The spatial politics of recognition and relational geographies of Indigenous self-determination. *The Canadian Geographer/Le Géographe canadien* 60(2):259–269

Daigle M and Sundberg J (2017) From where we stand: Unsettling geographical knowledges in the classroom. *Transactions of the Institute of British Geographers* 42(3):338–341

Darling J (2014) Asylum and the post-political: Domopolitics, depoliticisation, and acts of citizenship. *Antipode* 46(1):72–91

Diprose G (2017) Radical equality, care, and labour in a community economy. *Gender, Place, and Culture* 24(6):834–850

Flinders M and Wood M (2014) Depoliticisation, governance, and the state. *Policy and Politics* 42(2):135–149

Gismondi M, Connelly S, Beckie M, Markey S and Roseland M (eds) (2016) *Scaling Up: The Convergence of Social Economy and Sustainability.* Edmonton: Athabasca University Press

Hunt S (2014) Ontologies of Indigeneity: The politics of embodying a concept. *Cultural Geographies* 21(1):27–32

Lawson V (2007) Geographies of care and responsibility. *Annals of the Association of American Geographers* 97(1):1–11

Lawson V (2009) Instead of radical geography, how about caring geography? *Antipode* 41(1):210–213

May T (2010) *Contemporary Political Movements and the Thought of Jacques Rancière.* Edinburgh: Edinburgh University Press

Mouffe C (2005) *On the Political.* Abingdon: Routledge

Mouffe C (2013) *Agonistics.* London: Verso

Nancy J-L (1991) *The Inoperative Community.* Minneapolis: University of Minnesota Press

Olson E and Sayer A (2009) Radical geography and its critical standpoints: Embracing the normative. *Antipode* 41(1):180–198

Pulido L (2015) Geographies of race and ethnicity 1: White supremacy vs white privilege in environmental racism research. *Progress in Human Geography* 39(6):809–817

Radcliffe S A (2017) Decolonising geographical knowledges. *Transactions of the Institute of British Geographers* 42(3):329–333

Smith S J (2009) Everyday morality: Where radical geography meets normative theory. *Antipode* 41(1):206–209

Sundberg J (2014) Decolonizing posthumanist geographies. *Cultural Geographies* 21(1):33–47

Swyngedouw E (2011) Interrogating post-democratisation: Reclaiming egalitarian political spaces. *Political Geography* 30(7):370–380

The Free Association (2010) Antagonism, neo-liberalism, and movements: Six impossible things before breakfast. *Antipode* 42(4):1019–1033

Thomas A C (2015) Indigenous more-than-humanisms: Relational ethics with the Hurunui River in Aotearoa New Zealand. *Social and Cultural Geography* 16(8):974–990

Thomas A C (2017) Everyday experiences of post-politicising processes in rural freshwater management. *Environment and Planning A* 49(6):1413–1431

Thomas A C and Bond S (2016) Reregulating for freshwater enclosure: A state of exception in Canterbury, Aotearoa New Zealand. *Antipode* 48(3):770–789

Tronto J C (1995) Care as a basis for radical political judgments. *Hypatia* 10(2):141–149

Williams M J (2017) Care-full justice in the city. *Antipode* 49(3):821–839

Wilson J and Swyngedouw E (eds) (2014) Seed of dystopia: Post-politics and the return of the political. In J Wilson and E Swyngedouw (eds) *The Post-Political and Its Discontents: Spaces of Depoliticization, Spectres of Radical Politics* (pp 1–24). Edinburgh: Edinburgh University Press

Young I M (2011) *Responsibility for Justice*. Oxford: Oxford University Press

Agnotology

Tom Slater

School of GeoSciences, University of Edinburgh, Edinburgh, UK;
tom.slater@ed.ac.uk

> It is certain, in any case, that ignorance, allied with power, is the most ferocious enemy justice can have. (Baldwin 1998:445)

We live in an era when rigorous investigative journalism gets dismissed by the current President of the United States as "fake news"; when clear evidence of small Presidential inauguration crowd numbers is ridiculed by a senior advisor to the President as "alternative facts"; and when a successful and scandalous campaign to get Britain to leave the European Union was aided by a false statement, printed on a campaign bus, that a "Leave the EU" vote would save £350 million per week that it could instead spend on the National Health Service. Geographers, whether radical or not, have traditionally and justifiably been concerned with epistemology—the production of knowledge. But in 2018, this by itself seems insufficient. How is *ignorance* produced, by whom, for whom, and against whom?

In his swashbuckling critique of the economics profession in the build up to and aftermath of the 2008 financial crisis, Mirowski (2013:83) argues that one of the major ambitions of politicians, economists, journalists and pundits enamoured with (or seduced by) neoliberalism is to plant doubt and ignorance among the populace:

> This is not done out of sheer cussedness; it is a political tactic, a means to a larger end ... Think of the documented existence of climate-change denial, and then simply shift it over into economics.

Mirowski makes a compelling argument to shift questions away from "what people know" about the society in which they live towards questions about what people do *not* know, and why not. These questions are just as important, usually far more scandalous, and remarkably under-theorised. They require a rejection of appeals to epistemology and, instead, an analytic focus on intentional ignorance production or *agnotology*. This term was coined by historian of science Robert Proctor, to designate "the study of ignorance making, the lost and forgotten" where the "focus is on knowledge that could have been but wasn't, or should be but isn't" (Proctor and Schiebinger 2008:vii). The etymological derivation is the Greek word, agnōsis, meaning not knowing, which Gilroy (2009:np) draws upon to argue the following:

> We need a better understanding of the relationship between information and power ... a new corrective disciplinary perspective that interprets the power that arises from

Keywords in Radical Geography: Antipode at 50, First Edition. Edited by the *Antipode* Editorial Collective.
© 2019 The Authors/Antipode Foundation Ltd. Published 2019 by John Wiley & Sons Ltd.

the command of *not knowing*, from the management of forms of ignorance that have been strategically created and deployed, and institutionally amplified.

Proctor's own work provides a hugely instructive conceptual apparatus for radical geographers interested in the production of ignorance; indeed, he notes that agnotology has "a distinct and changing political geography that is often an excellent indicator of the politics of knowledge" (1995:8). It was while investigating the tobacco industry's efforts to manufacture doubt about the health hazards of smoking that he began to see the scientific and political urgency in researching how ignorance is made, maintained and manipulated by powerful institutions to suit their own ends, where the guiding research question becomes, "Why don't we know what we don't know?". As he discovered, the industry went to great lengths to give the impression that the cancer risks of cigarette smoking were still an open question even when the scientific evidence was overwhelming:

> The industry was trebly active in this sphere, feigning its own ignorance of hazards, whilst simultaneously affirming the absence of definite proof in the scientific community, while also doing all it could to manufacture ignorance on the part of the smoking public. (Proctor 2008:13–14)

Numerous tactics were deployed by the tobacco industry to divert attention from the causal link between smoking and cancer, such as the production of duplicitous press releases, the publication of "nobody knows the answers" white papers, and the generous funding of decoy or red-herring research that "would seem to be addressing tobacco and health, while really doing nothing of the sort" (Proctor 2008:14). The tobacco industry actually produced research about everything except tobacco hazards to exploit public uncertainty (researchers commissioned by the tobacco industry knew from the beginning what they were supposed to find and not find), and the very fact of research being funded allowed the industry to say it was studying the problem. In sum, there are powerful institutions that want people not to know and not to think about certain conditions and their causes, and agnotology is an approach that traces how and why this happens.

Many scholars might claim that there is no such thing as the intentional production of ignorance; all that exists are people with different worldviews, interests, and opinions, and people simply argue and defend their beliefs with passion. This claim would be very wide of the mark. To take the example of free market think-tanks, even when there is a vast body of evidence that is wildly at odds with what is being stated, and when the social realities of poverty and inequality expose the failures of deregulation at the top and punitive intervention at the bottom of the class structure, think-tank writers become noisier and even more zealous in their relentless mission to inject doubt into the conversation and ultimately make their audiences believe that government interference in the workings of the "free" market is damaging society.

An agnotological approach seeks to dissect the ignorance production methods and tactics of messengers of disinformation. There are, of course, many different ways to think about ignorance; as John Rawls (1971) did positively in his promotion of a "veil of ignorance" as an ethical method with respect to his hypothetical "original position" (whereby ignorance of how we might personally gain in a society's distribution of benefits and burdens might guarantee a kind of neutrality and balance in thinking about what a just distribution should look like). It therefore needs to be clarified that agnotology does not have a monopoly on ignorance studies, and is just one element of that nascent field (Gross and McGoey 2015), but it is something of a surprise that agnotology has hardly escaped from the disciplinary claws of science and technology studies and permeated social science, less still geography, where the relationship between evidence and policy is always contentious and sometimes tortured (Harvey 1974).

Tracking the ignorance production methods of "the outer think-tank shells of the neoliberal Russian doll", to use Mirowski's (2013:229) memorable phrasing, is a project of considerable analytic importance. For a decade I have researched the production of ignorance via the scrutiny of free market think-tanks in the UK (see Slater 2014 for a contribution I made to *Antipode*). Policy Exchange, one such think-tank based in London, is my latest analytical target. Although it purports to be "completely independent", it was founded and financed in 2002 by three Conservative Party MPs, and today it retains its very close links with that Party. It produces an astonishing amount of reports, and it sponsors many events and public statements on various policy priorities like crime and justice, immigration, education, foreign affairs, and housing, planning and urban regeneration. Its influence in all these spheres has been immense in the UK over the past decade—major political speeches and catchy slogans often originate from Policy Exchange reports—but it is housing policy in particular where it is possible to see a direct imprint of think-tank writing on what has been happening to people living at the bottom of the class structure in UK cities.

A report entitled *Making Housing Affordable*, written by neoclassical economist Alex Morton and published in 2010, is arguably the most influential document of all. Many of the proposals in this document quickly became housing policy under the coalition government (2010–2015) and subsequent Conservative governments. The report argues that social housing of any form is a terrible disaster because it makes tenants unhappy, poor, unemployed, and welfare dependent. Not only is this baseline environmental determinism, it is a reversal of causation: a very substantial literature on social housing in the UK demonstrates that the reason people gain access to social housing is *precisely because* they are poor and in need, and that it was *not social housing that created poverty and need* in the first place (e.g. Forrest and Murie 1988; Hanley 2007; Malpass 2005). Had Morton consulted the literature on housing estates across Europe, he would have discovered that nowhere is low-income housing provided adequately by the market and also that the countries with the largest social housing sectors (Sweden, France, Holland, etc.) are those with the least problematic social outcomes (e.g. Musterd and van Kempen 2007; Power 1997; van Kempen et al. 2005).

Having effectively argued that social housing is the scourge of British society, Morton then goes on to propose what he feels are solutions to the housing crisis. Most striking of all, for the purpose of this chapter, is setting those solutions against the title of the report: *Making Housing Affordable*. In addition to calling for the destruction of social housing and the removal of government support for housing associations, the report actually proposes numerous strategies to *make housing more expensive*. For example:

> What is needed are more better quality developments that both increase housing supply and raise house prices and the quality of life for existing residents in the areas that they are built. (Morton 2010:15)

> The government should scrap all density and affordable housing targets and aspirations. (Morton 2010:23)

> It is a fallacy to assume that making new homes "low-cost" will help increase affordability—it makes no difference to house prices whether you build cheap or expensive new homes. (Morton 2010:68)

> [S]ocial rents should rise to meet market rents. (Morton 2010:81)

It is difficult to imagine a more clear-cut case of the production of ignorance than a report entitled *Making Housing Affordable* recommending that affordable housing targets and aspirations should be scrapped, and social housing demolished. Viewed through an analytic lens of agnotology, we can see a complete inversion going on: the structural and political causes of the housing crisis—that is, deregulation, privatisation, and attacks on the welfare state—are put forward as desirable and necessary remedies for the crisis that will squash an intrusive state apparatus. The report won *Prospect* magazine's prestigious "Think Tank Publication of the Year" award in 2010.

Agnotology helps geographers analyse ignorance as a strategic and pernicious ploy, an active construct, "not merely the absence of knowledge but an outcome of cultural and political struggle" (Schiebinger 2004:233). A fascinating illustration, one that offers all sorts of ideas for radical geographers interested in researching ignorance production, comes from Nora Stel's (2016) *Antipode* article on the agnotology of eviction threats in south Lebanon's Palestinian refugee "gatherings" (unofficial camps). The first attempt I know of to understand the spatial dimensions of strategically imposed ignorance, she explains how the Lebanese state imposes a regime of "institutional ambiguity" on the gatherings, where responsibility and control are passed around among an array of external actors in order to maintain the temporary nature of the Palestinian presence in Lebanon (predictably sustained by the production and circulation of myths and stereotypes about refugees). In elaborate qualitative detail Stel documents how the ever-present threat of eviction is then negotiated by inhabitants of the gatherings, who generate deliberate disinformation, employ stalling tactics, and invoke both professed and real ignorance about their predicament. Stel argues that this is a reaction to their treatment by the Lebanese state, one that takes the form of ignorance as "strategic replication" of the institutional ambiguity imposed upon them. Her analysis is particularly useful as it demonstrates how marginalised spaces "both make and are made by deliberate forms of ignorance", and that ignorance production can be used as shield as well as sword.

Disinformation, fabrication, mythology and propaganda fuel the intentional circulation of ignorance, surely one of the most important political issues of our time. Ignorance takes the form of specific geographies and, when radically analysed using an agnotological approach, we can gain a closer grasp of how institutions and individuals work hard to confuse and cloud any evidence that might show us what is actually happening in particular places.

References

Baldwin J (1998 [1972]) *No Name in the Street*. In T Morrison (ed) *James Baldwin: Collected Essays* (pp 349–476). New York: Library of America
Forrest R and Murie A (1988) *Selling the Welfare State: The Privatisation of Public Housing*. London: Routledge
Gilroy P (2009) "The Crises of Multiculturalism?" Paper presented to the "Challenging the Parallel Lives Myth: Race, Sociology, Statistics and Politics" conference, London School of Economics, May. http://mcincrisis.podomatic.com/entry/index/2009-05-20T10_43_44-07_00 (last accessed 15 August 2018)
Gross M and McGoey L (eds) (2015) *The Routledge International Handbook of Ignorance Studies*. London: Routledge
Hanley L (2007) *Estates: An Intimate History*. London: Granta
Harvey D (1974) What kind of geography for what kind of public policy? *Transactions of the Institute of British Geographers* 63:18–24
Malpass P (2005) *Housing and the Welfare State: The Development of Housing Policy in Britain*. Basingstoke: Macmillan
Mirowski P (2013) *Never Let a Serious Crisis Go to Waste: How Neoliberalism Survived the Economic Meltdown*. London: Verso
Morton A (2010) *Making Housing Affordable: A New Vision for Housing Policy*. London: Policy Exchange. https://www.policyexchange.org.uk/wp-content/uploads/2016/09/making-housing-affordable-aug-10.pdf (last accessed 16 August 2018)
Musterd S and van Kempen R (2007) Trapped or on the springboard? Housing careers in large housing estates in European cities. *Journal of Urban Affairs* 29(3):311–329
Power A (1997) *Estates on the Edge: The Social Consequences of Mass Housing in Europe*. Basingstoke: Macmillan
Proctor R (1995) *The Cancer Wars: How Politics Shapes What We Know and Don't Know About Cancer*. New York: Basic Books
Proctor R (2008) Agnotology: A missing term to describe the cultural production of ignorance (and its study). In R Proctor and L Schiebinger (eds) *Agnotology: The Making and Unmaking of Ignorance* (pp 1–33). Stanford: Stanford University Press
Proctor R and Schiebinger L (eds) (2008) *Agnotology: The Making and Unmaking of Ignorance*. Stanford: Stanford University Press
Rawls J (1971) *A Theory of Justice*. Cambridge: Harvard University Press
Schiebinger L (2004) Feminist history of colonial science. *Hypatia* 19(1):233–254
Slater T (2014) The myth of "Broken Britain": Welfare reform and the production of ignorance. *Antipode* 46(4):948–969
Stel N (2016) The agnotology of eviction in south Lebanon's Palestinian gatherings: How institutional ambiguity and deliberate ignorance shape sensitive spaces. *Antipode* 48(5):1400–1419
van Kempen R, Dekker K, Hall S and Tosics I (eds) (2005) *Restructuring Large Housing Estates in Europe*. Bristol: Policy Press

Badge

Gavin Brown

School of Geography, Geology and the Environment, University of Leicester, Leicester, UK;
gpb10@le.ac.uk

"Badge" can either be a noun, referring to a device or emblem worn to signify a person's affiliation, rank or support for a cause; or it can be a verb meaning to mark something or otherwise distinguish it (OED Online 2018). In this piece, I consider how and why badges, as objects, are useful (but, largely, overlooked) devices for geographers interested in radical politics to consider. In doing so, I also consider how the social role of badges and badging has changed over the lifetime of *Antipode*.

For centuries, badges and insignia of various types have been a means of demonstrating political allegiance. In the late 20th century, cheap mass production techniques ensured that button badges became a cheap and easy way of identifying oneself with a cause. Tilly and Wood (2015) recognised the wearing of badges as one of the important ways in which the members of a social movement demonstrate their "unity" with each other. They function to "create rapid rapport and trust" within a group (Halavais 2012:357). Of course, the proliferation of cheap badges as political statements cannot be separated from the expansion of the badge as a fashion accessory in many youth subcultures from the 1960s onwards. Before going much further, and in sympathy with an international audience, it is useful to offer a word of clarification about language. The objects that I am interested in tend to be called "badges" in the UK, but "buttons" in the United States and some other countries. For Attwood (2004:1), "the two terms used on different sides of the Atlantic say something about the history and functions of these objects—and also perhaps of the two countries themselves".

In the Middle Ages, a badge was principally a way of identifying the loyalty of followers to a feudal lord. They were a physical marker of affiliation. From the 12th century onwards cheap alloy talismans, or "signs", that could be worn on the person were also produced as symbols of devotion for Christian pilgrims. Many of these medieval badges more resembled items of jewellery or medals than their contemporary counterparts. In contrast, the North American "button" has its origins in a 19th century form of novelty advertising. In the 20th century this object became ubiquitous—"a protective disc of clear plastic and a paper disc bearing a message positioned between a circular disc of metal and a metal collar, with a sprung steel pin inserted in the back" (Attwood 2004:1). Then, as now, they were most commonly one inch in diameter, as were the buttons most frequently used in clothing at the time (Lucas 2007). In their first year of production, over a million metal and celluloid badges were produced. These button badges,

Keywords in Radical Geography: Antipode at 50, First Edition. Edited by the *Antipode* Editorial Collective.
© 2019 The Authors/Antipode Foundation Ltd. Published 2019 by John Wiley & Sons Ltd.

forged from cheap aluminium, utilising new celluloid technologies, and developments in colour printing were products of the factory age of mass production. They share this origin with many of the political movements that first utilised them for propaganda purposes and as markers of mass support. They could be produced in large numbers and easily distributed for little cost.

Badges serve as a visual archive of political debates and social movements. They offer an overview of key political movements in any given period or place, but they also hint at the ways in which social movements are networked and rework slogans and tactical lessons from each other. Similarly, they can offer insights into the actions valued by movements, and the ways in which they operate and communicate in times of repression. When martial law was imposed in Poland in 1981, and it was too risky to openly wear Solidarnosc badges, some opposition supporters took to removing a common electrical component (power resistors) from domestic radios and wearing them, as a more oblique political statement and visual pun, instead—demonstrating that the wearer was in favour of resisting state power (Flood and Grindon 2014:116).

In contrast to the cheap paper and metal badges that are mass produced for many campaigns, some social movements invest in more substantial, elaborate and valuable badges for their supporters. One such example is the "Holloway Brooch" designed in 1912 by Sylvia Pankhurst and awarded to women who had been imprisoned for their campaigning for women's suffrage (Attwood 2004). These were silver and enamel badges designed with a portcullis pierced by a prison arrow in the white, green and purple colours of the Women's Social and Political Union (Sawer 2007). Experiences of incarceration that could have been viewed as "badges of dishonour" were revalued through these decorative badges.

Badge-wearers use them for a variety of purposes—to promote a cause, to identify oneself with that cause, and for decoration and adornment. Their power exceeds their communicative potential and efficiency. They hold the members of a political organisation together. They can say something about the wearer's sense of themselves and their place in the world. They perform an intimate form of scale-jumping—worn on the person, but addressing issues that affect the workplace, the community, but also national politics and (especially from the 1960s onwards, with international opposition to apartheid and the Vietnam War) matters of global concern. Button badges produced by social movements can be very effective objects for communicating subaltern geopolitical imaginations and anti-geopolitical stances (Sharp 2013).

It is not only in Europe and North America that cheap badges were mass produced. As the Cultural Revolution took hold in China, the mass production of red Mao badges reached extraordinary monthly rates (White 1994:58). In this context, badges acquired a social value well beyond their material worth as they became markers of "politically correct" revolutionary commitment. Contradictorily, they even circulated as tokens of exchange or trade within China.

Political badges (and their wearers) can appear earnest, but they are also vehicles for humour, satirising the political mainstream of the moment and other social movements. Symbols and slogans circulate, get re-appropriated, and subverted as they travel. Sometimes they get turned back on their originators.

Sometimes the simplest graphic device can be the most effective (and enduring). Gerald Holtom's peace symbol designed in 1958—combining the semaphore for CND with the symbolism of a gesture of despair—is an early example of the creative communications of the anti-nuclear movement. For 60 years it has retained global recognition as a symbol of peace (Goodnow 2006). Worn by many as a lapel badge, it became a portable statement of opposition to war.

Just as a cheap badge is easy to attach to one's clothes, they do not take up much space to store, and are light and easy to transport. Through the course of our research about anti-apartheid activists (Brown and Yaffe 2017) from the 1980s, Helen Yaffe and I noticed that many erstwhile activists continue to carry collections of badges from past campaigns with them, as personal archives of political commitment, for years afterwards. There are stories attached to each badge; memories of where they were acquired, contexts for where some were worn (but not others); thoughts on how the combination of particular badges presented an image of the activist-self.

The size, materiality and portability of badges also affects how they are (and were) worn. A solitary badge might be worn, or it might form part of a larger assemblage of political (and non-political) identifications. Badges can also be relatively easily concealed when their message is uncomfortable, hazardous or out-of-place. We heard not just about the badges that participants had worn, but how and where they remember wearing them. Some badges were considered too controversial (or too likely to expose a hidden facet of the wearer's identity) to be worn in everyday settings such as school, or the workplace, being reserved instead for the relative anonymity of a large demonstration, or the bravado of an activist meeting.

In many ways the growing popularity of badges across the 20th century (but especially from the 1960s onwards, with a distinctively anti-establishment ethos) speaks to a growing recognition that "ordinary" people had the agency to affect social and political change. This was taken even further from the 1970s onwards as a result of the do-it-yourself ethos of punk. Cheap badge-making equipment made it easy for people to produce their own badges on a small scale. This example also serves as a useful reminder that social movements seldom exist independently of wider cultural trends. On this basis, analysing political badges as a discrete category from all sorts of other subcultural ephemera from a given period might limit our understanding of their uses and significance. If badges are "disobedient objects" (Flood and Grindon 2014) in many ways, they can also challenge some of the boundaries that both activists and scholars draw around what counts as "political".

Although button badges worn on clothes and rucksacks seem to have experienced something of a resurgence of late, they have also been adapted to the social practices and material objects that are central to our contemporary lives. Political sentiments and allegiances are now as likely to be displayed in the form of cheaply produced stickers attached to the back of a laptop screen, or in digital form, as "Twibbons" temporarily adhered to our profile pictures and avatars on social media. At the same time, many websites and digital platforms now award users "badges" of various kinds to verify their identity, or as a measure of expertise, skill or reputation within a particular online community (Halavais 2012). Sometimes these digital badges communicate a belief, stance or commitment, just like button badges. However, when they reward the acquisition of a particular skill or experience, they sometimes become enrolled in game-play, as users are encouraged to work towards the next level of experience. Here there is a parallel with experiments in the use of badges in digital learning environments as a form of micro-credentialism (Fain 2014). At an organisational level, similar processes are at work in the ways in which equality and diversity policies, or best practice in the management of "human resources", become badged at an institutional level through schemes such as Investors in People or Athena SWAN charter marks (Bell et al. 2002).

In many ways the changing form, function and circulation of badges over the last five decades make them useful objects to think about alongside *Antipode*'s 50th anniversary. Their proliferation in the late 1960s was part of the explosion of youth subcultures and the New Left at that time. Today the button-badge (worn on the lapel) is less ubiquitous—perhaps a material marker of the declining social significance of the left. Instead, the symbolism of badges has increasingly become incorporated into neoliberal processes of audit, rankings and credentialism. This piece is a call to reengage with badges and use them as an archive for recalling usable pasts about social movements; but also as a warning of how easily the material culture of protest movements can be recuperated by capitalism.

References

Attwood P (2004) *Badges*. London: British Museum Press
Bell E, Taylor S and Thorpe R (2002) Organizational differentiation through badging: Investors in people and the value of the sign. *Journal of Management Studies* 39(8):1071–1085
Brown G and Yaffe H (2017) *Youth Activism and Solidarity: The Non-Stop Picket Against Apartheid*. London: Routledge
Fain P (2014) Badging from within. *Inside Higher Education* 3 January
Flood C and Grindon G (eds) (2014) *Disobedient Objects*. London: V&A Publishing
Goodnow T (2006) On black panthers, blue ribbons, and peace signs: The function of symbols in social campaigns. *Visual Communication Quarterly* 13(3):166–179
Halavais A M (2012) A genealogy of badges: Inherited meaning and monstrous moral hybrids. *Information, Communication, and Society* 15(3):354–373
Lucas G (2007) *Badge Button Pin*. London: Laurence King
OED Online (2018) "Badge, n." http://www.oed.com/view/Entry/14550 (last accessed 15 August 2018)
Sawer M (2007) Wearing your politics on your sleeve: The role of political colours in social movements. *Social Movement Studies* 6(1):39–56
Sharp J P (2013) Geopolitics at the margins? Reconsidering genealogies of critical geopolitics. *Political Geography* 37:20–29
Tilly C and Wood L J (2015) *Social Movements, 1768–2012* (3rd edition). London: Routledge
White R A (1994) Mao badges of the Cultural Revolution: Political image and social upheaval. *International Social Science Review* 69(3/4):53–70

Blues Clubs

David Wilson

Department of Geography and Geographic Information Science, University of Illinois, Urbana, IL, USA;
dwilson2@illinois.edu

Introduction: The Struggle for Blues Clubs in Urban America

Blues clubs have a long and tortuous history in capitalist urban America. Few sites for expressing the arts and the complexities of human experiences in America have been more punished and marginalised by racism and neo-colonialism. As blues performers from the Delta and South struggled with economic and political oppressiveness and moved to the industrialising Midwest and North, city clubs first appeared (Wilson 2018). Clustering in segregated, marginalised communities, clubs quickly became homes for a music-based knowledge system that served up "truths" about working class African American life (ideas that disrupted carefully constructed texts about race, class, progress, competency, citizenship, and humanity) (Woods 2006). Even as many clubs beginning in the late 1960s were partially "white popularised" to capture larger audiences, the music remained deeply connected with constructions of expressive, rights-entitled blackness. This club politicising continues today across urban America (Kobayashi 2018).

This essay discusses the state of current urban blues clubs in America as a political-economic phenomenon, focusing on what many see as America's blues capital, Chicago. Radical geographers, ever-sensitive to the elusive and changing mechanisms of race-class oppression in capitalism, need to understand the present plight of these clubs. For at these sites this oppression reveals itself in both remarkably blatant and powerfully furtive ways. Moreover, these sites also provide pointers for how adroit social movements (alive and well at these locations) may proceed. Capital and the state, moving into the city's "darkest" areas, seek to re-engineer land valorisation schemes and eradicate spaces that can cultivate counter hegemonic politics. Kinds of intervention, both elusive and transparent, reflect the turbulence, contradictions, and raw afflicting tendencies of struggling city, regional, and national economies. To glimpse what is happening at these clubs is to cast a deep gaze onto a contested race-class market rule in Chicago and beyond, replete with its hard-ass austerity and punishing stigmatisation that tumultuously pushes on.

My focus, Chicago's predominantly black and working-class South Side, is calculated. It is here that blues clubs have proliferated most in urban America, where they came to profoundly incubate innovations in blues music as an art form and epistemological carrier, and quickly interconnected with an urban and national capitalism that still struggles to manage and contain them. My comments and ideas in this essay follow from more than 10 years of closely examining and being involved with these clubs as a blues musician and regular attendee. I take up two

Keywords in Radical Geography: Antipode at 50, First Edition. Edited by the *Antipode* Editorial Collective.
© 2019 The Authors/Antipode Foundation Ltd. Published 2019 by John Wiley & Sons Ltd.

issues about these clubs: the current multi-pronged assault on their current exis-tence and how they today spawn a sly, deft politics that resists club transforma-tion. In both instances, blues clubs hook up with the flow of everyday capitalist processes, particularly with Clyde Wood's (2006) renewed plantation philosophy of urban governance, as pieces of land, social meeting grounds, and spaces of alterity.

Urban Blues Clubs and Music Under Siege

Chicago's blues clubs and their provision of blues music are currently under siege by the realities of a long-term but recently stepped-up retrenchment. Much has been written about this trend of club contraction and I will not dwell on it here. In brief, as South Side communities have become progressively poorer, a visible reflection of an alive-and-well urban neo-colonialism, their ability to financially support clubs has dwindled. Post-Fordism and the new service economy has dev-astated job bases city-wide and on the South Side (Hinz 2017). At the same time, job growth has proliferated, especially in low wage services (hotel work, day labour jobs, fast food employment, non-unionised construction jobs, dead-end retailing) (Hinz 2017). Today, seared by high joblessness and under-employment on the South Side, Chicago has the highest black unemployment rate in the nation (Strausberg 2017). Neoliberal institutions that prop up this afflicting econ-omy may be aging, tired, and out of new ideas, but they continue to shape eco-nomic conditions in South Chicago. Not surprisingly, retail-bases and one part of this, blues clubs, scramble to stay alive.

Another ominous force which has also emerged to unexpectedly besiege these clubs is the move of the city's gentrification frontier to the South Side. As the "go global Chicago drive" areally expands (Wilson 2018), Chicago's redevelopment governance strives to transform some of the South Side's poorest, most neglected blocks. To many in the city, these areas are the worst of the worst. As the South Side's physical carnage from the sub-prime mortgage fiasco persists (at least 33,000 buildings currently sit abandoned), acres of blocks (especially its blues clubs) seem ripe for a possible surge of redevelopment. Area residents, reeling from decades of public- and private-sector neglect and more recently toxic sub-prime lending, now awaken to another battle on their hands as a different kind of capital accumulation extends its tentacles. To be clear, this physical and social transformation of Chicago's South Side and its blues clubs is still highly uneven and unfolding in various stages. Clubs and their blocks north of 54th Street in the Bronzeville area (22nd to 52nd) are most significantly changing.

Not surprisingly, a growing number of blues clubs across Chicago's South Side feel the ominous winds of change. It is race-class gentrification with a vengeance. The racialised working-class and poor are to be banished from these sites while ironically being used as abstract symbols to sell the clubs. A race-class exoticisa-tion meets an edgy black carnivalising that sells the availability of "authentic", securitised culture. Poor blackness, rendered luminous and sanitised, will presum-ably be accepted by newcomers to these spaces. A proclaimed rediscovery of the black experience now increasingly centres, for bourgeois consumption, "black culture", "raw blues musical performance", and "exotic black ways". Of course, what is often called the black soul, following Frantz Fanon (1986), is a white man's artifact. Three things are celebrated in the new rhetoric: outsider's new opportunities for cultural experiences; blues clubs as a viable social model (compliant social relations) for what the South Side needs to be; and how these clubs can enrich the go global Chicago project.

In appearance, more and more clubs experience a surge of new faces, upgraded physical features, altered drink offerings, and a revised social aesthetic that changes these working-class-dominated blues joints. The new faces, often white and more affluent, trek in from other parts of Chicago, Europe, Australia, and other international locations. Swanky ales and wines once banned from South Side blues venues (associated with downtown and North Side clubs) suddenly appear on the drink docket. Newer patrons, often swelling in numbers, step up their admiration of supposed black authenticity (often imagined as transported from Africa to the Delta to Chicago). These changes, at first glance simple physical alterations, are also deep symbolic elements that communicate to many (particularly to long-term regular patrons and musicians) something disconcerting: the possibility of an "outsider" class-race takeover of a coveted social space.

Equally clear is that redevelopment has a long and tortuous post-war history of devastating working-class black communities in Chicago: this club takeover is nothing new. Federal programs, presented as crucial economic undertakings to revitalise the city, anchored city redevelopment initiatives for many decades (Vale 2013). Two post-war federal programs, Public Housing and Urban Renewal, dominated the post-war redevelopment scene through the 1970s (Vale 2013). Public housing, with its imposition of socio-spatial isolation, and urban renewal, with its devastating physical destruction, functioned as fundamentally post-war institutional fixes for deteriorating and minority swelling cities. These programs were nourished by traversing a long and complicated path of trial-and-error, dialogic feedback, and subtle modification (Wilson 2018). They never stopped being rhetorically cloaked as support of the public good and enrichment of the urban economy.

Yet, the current South Side transformation is deeply contradictory and tension-ridden (Wilson 2018). For its leaders ultimately negotiate a knife-edge path between pursuing two things, socio-racial segregation and annihilating this to pursue profitable real-estate investment. Capital accumulation through real-estate valorisation, one more time in capitalist America, struggles with incoherence and turbulence. On the one hand, a brutal and uncompromising city-wide segregation, with the South Side being a storehouse for the black poor, has enabled land values in other sections of Chicago to be cultivated and maintained. The drive to sustain this race-class segregation continues to be strong. For example, gentrification in increasingly swanky North Side and Near West Side neighbourhoods like Wicker Park, Lakeview, and Pilsen, has been enabled by cordoning off the racial poor. On the other hand, the South Side today offers emerging rent-gaps and opportunities to create property booms. Immediate capital accumulation is a real possibility for real-estate interests but would also instigate something tricky: significant residential displacement, a spectre which haunts real-estate capital.

Thus the dilemma: in perception the very profit opportunities across the city that took so long to build could quickly unravel by moving gentrification-fired redevelopment (with its residential displacing) into the South Side. So, what should be the dominant planning and policy thrust? Can the creation of fragile housing submarkets across the city successfully absorb the consequences of an active gentrifying and residential displacement on the South Side? Chicago's South Side, previously having a one-dimensional functionality as a planned space (stashing away and isolating the racialised poor), now becomes a complicated planning terrain. Behind the scenes, this ambiguity is largely concealed from public view, hidden by the decision to aggressively push South Side redevelopment (Wilson 2018). But many developers and planners are acutely aware of this tension. They love and fear South Side gentrification. It is their fondest dream and their worst nightmare as the possibility of negative fallout from redevelopment scares them. Ramped up pomp and aggression, disguising unmistakable concern, submerges a polyvocality that may yet bubble to the surface as this redevelopment further proceeds.

Reclaiming the Blues

But a struggle to thwart the commodification of these clubs, taken up especially by long-term patrons of the South Side blues clubs, is alive and well. The anger is visible if one knows how to look. Club commodification is seen by long-term patrons as disrespectful and an assault. Senses of self are intimately bound up with the feeling that these clubs are a pleasure-enabling, home space to these patrons, and are on the verge of being taken over. Cover charges, new forms of music, and altered social ways are read as race-class signifiers and visible intrusions: they communicate a race-class seizure. It is most extreme at clubs like the now temporarily closed Checkerboard: these patrons, reduced to a defensive minority, barely hold onto the threads of what they see as a once intensely local blues club. Blues flaneurs swell in numbers, and a new club ambiance of high-society consumption all but swallows up a club's South Side roots.

Without idealising or exaggerating this resistance, a distinctive kind of social formation, a "leisure as resistance" movement, has emerged. Its potential is meaningful. For the outcome, with success, can cultivate both an immediate protest consciousness and a collective capability to thwart capital's power in other domains in the future. This resistance ultimately strikes out in the short term on a specific issue and lays the groundwork to contest the structures and processes that perpetuate race-class oppression on the South Side (see Deveaux 2018). Its form is clear: Common club practices—bantering with others, bodying oneself as blues people, modes of social conduct—performatively deliver a powerful resistance-to-change politics that embeds in standard leisurely practices. As late-night revellers perform their roles as dazzling blues aficionados, grounded South Side blues fans, and drinkers and partiers, these acts multi-task as a quest to enrich and control social space. A performative, never-innocent, human intentionality ripples across these clubs as social reality is seized and manipulated to desired specifications. Resistant politics appears everywhere in the clubs: at the bars, the dance floors, the tables, the aisles.

Contradicting standard accounts of poor African-American conduct in blues clubs, regular patrons and musicians unleash a politics rooted in calculating and designing. Actions and bodying speak to something fundamental: the sense of the good club and the bad club. Realising they cannot control the decisions of club owners that will define a club's intensity of commodification, they strike out to do the next best thing: shape their perceptions by disfiguring or lampooning into oblivion the new commodification forms. Club owners, long-term patrons believe, will respond to their pleas as people who understand and relate to their needs and plight. Long-term patrons recognise an important reality: many club owners on the South Side, black and working-class, harbour strong feelings for struggling African American poor people who patronise their establishments. It is known that while these owners stand to benefit financially from a club upscaling, they also acquire immense satisfaction from servicing the desires of fellow raced-classed people and may not be willing to lose this. The struggle, it follows, has a target and a purpose.

Power here nestles in the informal. New codes and meanings are invented and circulate across the clubs that subvert and recast old symbolic regimes to eradicate key vestiges of neo-plantation traditions. A subterranean politics, innocuously cloaked from common view, grows as practices and engagements come to betray the logic that the dominant redevelopment narrative seeks to inscribe in them. Such practices and engagements—revelling in "bluesy" interactions with friends, bantering informally with musicians about club customs and traditions, humorously educating white tourists about club protocols, lampooning the white patron search for atavistic black ways, bluesifies the habitus of the club. One common practice, dancing majestically and theatrically to the music, affirms the long-term patron mastery of the club's social space and asserts a group's rights to it as a social milieu. A bodying as a to-be-seen process conveys a claim to space as a relentlessly served-up socio-physical form. These practices, flowing through the clubs, render these spaces politically charged venues of hope and possibility in the face of a possible major upscaling.

Ananya Roy (2015) references this kind of response in her work on South Side Chicago as the politics of emplacement. South Siders at "city's end" meld elegant appraisals of realities and circumstances with meaning-rich practices in the routinised shadows to stabilise lives and direct patterns of political intervention. Caring for home and curating its objects, through practices that maintain the positive symbolic content embedded in everyday domestic things, unveils a world of nurturing, creativity, hope, and entitlement to dignity. A human staging of individual rights and demands is provided a solid core. In the face of socio-spatial banishment and denied full personhood, a self-affirming politics flourishes from which all kinds of other informal and formal political movements may spring. This politics of the informal, to Roy, inaugurates a resistance that may deepen and move out to drive resistance in multiple domains of everyday life.

We should not underestimate the ambitiousness of this politics. For it is less an attempt to simply dominate a venue in the short term than a drive to produce and command space in the long term. The goal is to manufacture an enduring, nurturing space, not simply to momentarily convert a social site to a dense locale of support activity. All the while, this supportive space is to be fluid, inclusive, and impromptu as it symbolically strokes blues sensibilities, organises consciousness, and cultivates blues identities. A radical politics ultimately strives for nothing less than to socio-spatially revamp and stake a permanent claim to a social space. Yet all unfolds in a politics of elusive revealing. Outsiders see only the shadows of forceful human claims obscured by simple reverie and fun-seeking, but a politics lives in the clubs that may yet prove decisive in determining what these venues become.

A Concluding Note

As club commodification and resistance to it continues to unfold, it is clear that there is much at stake in what these clubs become. Loss of these clubs (as they are presently constituted) to either economic eclipse or major upscaling would be one more assault on the identity and life of South Chicagoans. People here, as capitalist throw-aways, continue to be assigned to stigmatised blocks and communities, the lowest rungs of local labour markets, substandard housing, and environmentally dangerous areas (Wilson 2004, 2007, 2018). City redevelopment, moreover, stands to one more time punish them: community remaking on the South Side has historically featured assaulting bulldozers, isolating high-rise constructions, blatant neglect, destructive land clearance schemes, and negligent tax increment financing districts. A repetitious product has followed: a steady diet of domination, social engineering, and human confinement. As this latest redeveloping becomes discussed and implemented across the South Side and Chicago, we must realise that this new redeveloping promises to deliver these same outcomes.

References

Deveaux M (2018) Poor-led social movements and global justice. *Political Theory* 46(5): 698–725

Fanon F (1986) *Black Skin, White Masks* (trans C L Markmann). New York: Pluto

Hinz G (2017) Chicago joblessness and crime are connected, U of I report concludes. *Crain's Chicago Business* 28 January https://www.chicagobusiness.com/article/20170128/BLO GS02/170129845/chicago-joblessness-and-crime-are-connected-u-of-i-report (last accessed 4 December 2018)

Kobayashi A (2018) "Music as Political Resistance." Unpublished manuscript

Roy A (2015) Dis/possessive collectivism: Property and personhood at city's end. *Geoforum* 80:A1–A11

Strausberg C (2017) Black unemployment in Chicago highest in nation. *Chicago Crusader* 5 June https://chicagocrusader.com/black-unemployment-in-chicago-highest-in-nation/ (last accessed 4 December 2018)

Vale L J (2013) *Purging the Poorest*. Chicago: University of Chicago Press

Wilson D (2004) Toward a contingent urban neoliberalism. *Urban Geography* 25(8):771–783

Wilson D (2007) *Cities and Race*. London: Routledge

Wilson D (2018) *Chicago's Redevelopment Machine and Blues Clubs*. New York: Palgrave Macmillan

Woods C (2006) *Development Arrested*. London: Verso

Care

Lorraine Dowler

Departments of Geography and Women's, Gender, and Sexuality Studies, Penn State University, University Park, PA, USA;
lxd17@psu.edu

Dana Cuomo

Department of Diversity and Community Studies, Western Kentucky University, Bowling Green, KY, USA

A. M. Ranjbar

Department of Women's, Gender, and Sexuality Studies, Ohio State University, Columbus, OH, USA

Nicole Laliberte

Department of Geography, University of Toronto Mississauga, Mississauga, ON, Canada

Jenna Christian

Department of Women's and Gender Studies, Bucknell University, Lewisburg, PA, USA

This essay calls for a "Manifesto of Radical Care" in Geography. The radical care that we advocate centres on non-dominant and intersectional forms of care (Lugones 2010) and challenges geographers to recognise different bodily experiences while being mindful of a commonality of vulnerability that stems from national or institutional policies and politics. This manifesto demands that geographers move beyond recognition into action, actively working to infuse radical care into our everyday interpersonal interactions and into our departmental, institutional and disciplinary policies and practices.

We assert this call with the recognition that we join similar feminist appeals for institutional change within and beyond our discipline. It is in this spirit that Melissa Wright (2001) penned a manifesto against femicide when she entangled Marxist "value" and poststructuralist "objectification" in her interrogation of institutional violence. Vicky Lawson (2009:210) applied this type of critique to the discipline of Geography when she called for the inclusion of a feminist care ethics, arguing that "[r]adical geography has much to learn from care ethics, which direct attention to the content of care, the ways in which care is marginalised and the need for ethical action". Comparably, Pat Noxolo, Parvati Raghuram and Clare Madge (2011) called for an understanding of responsibility that is informed by the exploitative nature of relationships and the need to acknowledge our complicity in the power-full relationships that sustain inequality even as we try to resist

Keywords in Radical Geography: Antipode at 50, First Edition. Edited by the *Antipode* Editorial Collective.
© 2019 The Authors/Antipode Foundation Ltd. Published 2019 by John Wiley & Sons Ltd.

it. Building on these mandates, along with a corpus of feminist geographic scholarship devoted to care theory and practice, we call for a radical care praxis that challenges the on-going devaluation of human subjectivity; recognises the persistence of institutional racism, sexism, heteronormativity, ableism, ageism, and classism within our discipline; and transforms our field by centring the needs of those most vulnerable to intersecting forms of harm (Mahtani 2006, 2014; Parizeau et al. 2016).

Our call for radical care acknowledges the Whiteness that permeates much of the feminist literature of care. Black and postcolonial theorists, such as Patricia Hill Collins (1991) and Uma Narayan (1997), have called for feminist scholars to adopt more inclusive approaches to the development of a care praxis (Dowler and Christian forthcoming). As Ahmed (2018) exposes, "[t]hose who experience the violence of a system are those who know that system most intimately". Therefore, a care praxis must employ transgressive hearing practices such as trusting those who speak about intimate vulnerabilities and institutional violences while also actively listening for, and to, silences (Ortega 2006; Schutte 1998). Consequently, when we shift the centre of analysis to those most vulnerable, a radical care practice is fluid, evolving, and responsive to experiences of different vulnerabilities as well as advocating/identifying transformative solutions. This critical engagement reveals the enormous potential for imagining Geography as a discipline of existing and changing relationalities of care (Bartos 2018a, 2018b).

The urgency of our manifesto is not inconsequential. Given the global rise of populist politics, we assert that it is critical to stand as a discipline with our colleagues who are more vulnerable to political backlash. This includes our peers who are experiencing injury for organising against the academic recasting of colonial violence, who have been legally indicted for doing humanitarian work on the US–Mexico border, who are rendered immobile because of their citizenship, and who are targets of surveillance and online trolling for researching race, ethnicity, sexuality and gender. Geography, with its disciplinary culture entrenched in fraternal codes of conduct, has left some bodies more precariously positioned than others. The radical shift in care practices that we seek must prioritise the needs of those not already benefiting from their social identities and intersecting privileges.[1]

A radical ethics of care might seem incongruous, as "radical" usually invokes far-reaching social change, while understandings of "care" often point to the private, personal and intimate. However, feminist geographers have a long history of examining the global/intimate as a single complex (Pain and Smith 2008) and with that tradition in mind, we seek to transform our discipline into one that takes the multidimensionality of the intimate seriously. Our inspiration also comes from the recent work of feminist geographers advocating for a politics of discomfort, an approach that avoids "'comfort feminism', which flattens difference and escapes a critical assessment of culpability and complicity" (Gökarıksel et al. 2017). A serious engagement with uncomfortable intimacies offers a starting point for moving beyond incomplete and superficial acknowledgments of vulnerability, and towards a more holistic and radical conceptualisation of care across difference. A politics of discomfort also creates the potential for a caring praxis in response to harms that occur closest to work and home and are perpetrated by those we know intimately, including colleagues who sexually harass or engage in micro-aggressions (Joshi et al. 2015). We see such explorations as vital if we are to avoid mainstream feminist resistance that easily folds difference into nationalist and capitalist narratives (Gökarıksel and Smith 2017). Many of us who employ a radical caring praxis are cast as "feminist killjoys", including in our disciplinary homes, a term that Sara Ahmed (2010) utilises to encapsulate detractors' silence accompanied by micro-aggressions when killjoys speak out within their institutions. These vulnerabilities place a tremendous burden on those most susceptible to institutional violence to indulge reactionary statements and superficial allies. This culture of justification and our subsequent decisions on which battles to fight —on behalf of ourselves, our colleagues, or our students—also reinforces the invisible labour that women and scholars of colour disproportionately undertake, and which can prevent tenure, delay promotions, and reduce salary increases.

Channelling our inner killjoy spirits, we call for a radical care praxis in every facet of the discipline, ranging from working in our departments and labs to conducting fieldwork, to attending annual meetings. This manifesto requires that we collectively create, employ, and fund new policies, standards, and modes of interaction that transform our discipline. As Lugones (2010) suggests, we must do more than theorise about violence and exploitation, but also form coalitions to resist interlocking oppression. In this spirit, we offer the following concrete, practical changes as a start to building a coalition of administrators, faculty, staff and students working towards a more caring discipline:

- *Developing care*: we must develop cultures of caring for and being responsible to the material conditions that shape individuals' academic praxis, teaching, and research, such as their vulnerability to their departmental and university relationships, the demands of care for themselves and others, and quality of life. This action includes the way our discipline approaches hiring practices, the job market, and the recruitment and treatment of job candidates. A culture for care includes support for disciplinary initiatives that build trauma-informed practices into our discipline, such as "mental health matters" (Mullings et al. 2016) and trauma-based services that support victims of harassment.
- *Valuing care*: we propose new guidelines for performance evaluations that extend beyond the material value that they add to an institution's capital income (e.g. research grants, number of journal articles) (Amsler 2014). Instead, evaluations should include labour that is disproportionately borne by women and faculty of colour and, far too often, rendered invisible.
- *Sharing care*: we need to institutionalise ways to celebrate community, collaboration, and participatory practices rather than emphasising individualised successes. Multiple schools of thought within the discipline currently advocate such a shift, including feminist geopolitics, geographies of care, black geographies, anti-colonial geographies, and the movement for slow scholarship.
- *Challenging systematic racism*: we must challenge systematic racism within our discipline, which requires us to advance new geographic canons that acknowledge research produced by scholars of colour. This intervention includes

university support for the creation of new syllabi and course material that prominently includes works by people of colour. Faculty should insist on broader job descriptions to create more diverse hiring pools so that our students can benefit from working with scholars of colour in their fields as role models.

- *Protecting against precarity*: we must oppose neoliberal administrative trends that replace permanent faculty positions with precarious labour. This entails protecting adjunct faculty, department staff, and graduate students from economic vulnerability by demanding fair wages and benefits for precarious workers.
- *Fighting unreasonable standards*: we must collectively fight back against the trends in academia that encourage increasing and unsustainable benchmarks. This responsibility falls most heavily upon tenured faculty who must work to challenge the expectations placed on graduate students and pre-tenure faculty.
- *Being proactive*: we must avoid reactionary, short-term and superficial fixes to the manifestations of violence. Instead, we must practice a sustained engagement with practices of radical care in our workplaces and annual meetings that proactively fight against violence and vulnerabilities in our professional communities.

The above call understands that to enact radical care in our discipline, we must practice radical and reflexive care in the complex multi-scalar contexts of where we work. Since care is an embodied experience, we must imagine how radical care in geography might "feel". It will feel uncomfortable when geographers take seriously responsibility for the discipline's colonial histories, which includes confronting the colonial present. It will feel uneasy because radical care demands that, if we actively dismantle the systemic marginalisation of people of colour, it also requires that we recognise the racism that perpetrates the discipline. This discomfort will eventually make way for feelings of hope for the kind of radical care we task Geography to realise.

Endnote

[1] The maleness (and whiteness) of Geography remains as 62% of AAG members are male to 38% female (AAG 2015). 81.3% are White, Non-Hispanic while 26.5% are Asian, 7.1% Hispanic and 4.2% African American (AAG 2016). Currently 63% of faculty in university and college geography departments are male, and 37% are female (AAG 2015). In 2016, 71% of faculty in university and college geography departments are White Non- Hispanic; of the minority groups, Asian is the most significant group represented (8%), with African American at 2.3% and Hispanic or Latino at 2.5% (AAG 2016).

References

AAG (2015) "Geographers by Gender." Disciplinary Data Dashboard, American Association of Geographers. http://www.aag.org/galleries/disciplinary-data/Geographers_by_Gender_summary_report.pdf (last accessed 20 August 2018)

AAG (2016) "Geographers by Race and Ethnicity." Disciplinary Data Dashboard, American Association of Geographers. http://www.aag.org/galleries/disciplinary-data/Geographe rs_by_Race_and_Ethnicity_summary_report_2016.pdf (last accessed 20 August 2018)

Ahmed S (2010) Feminist killjoys (and other willful subjects). *The Scholar & Feminist Online.* http://sfonline.barnard.edu/polyphonic/ahmed_01.htm (last accessed 17 October 2018)

Ahmed S (2018) Notes on feminist survival. *Feministkilljoys* 27 March. https://femi nistkilljoys.com/2018/03/27/notes-on-feminist-survival/ (last accessed 14 May 2018)

Amsler S (2014) For feminist consciousness in the academy. *Politics and Culture.* https://poli ticsandculture.org/2014/03/09/for-feminist-consciousness-in-the-academy/ (last accessed 17 October 2018)

Bartos A (2018a) The uncomfortable politics of care and conflict: Exploring nontraditional caring agencies. *Geoforum* 88:66–73

Bartos A (2018b) Relational spaces and relational care: Campus sexual violence, intimate geopolitics, and topological polis. *Area.* https://doi.org/10.1111/area.12449

Collins P H (1991) Towards an Afrocentric feminist epistemology. In P H Collins (ed) *Black Feminist Thought* (pp 269–290). London: Routledge

Dowler L and Christian J (forthcoming) Landscapes of impunity and the deaths of LaVena Johnson and Sandra Bland. *Gender, Place, and Culture*

Gökarıksel B, Hawkins M, Neubert C and Smith S (2017) "Political Geographies of Discom- fort Feminism and Uncomfortable Intimacy." Call for papers for the American Association of Geographers Annual Meeting, New Orleans, April 2018

Gökarıksel B and Smith S (2017) Intersectional feminism beyond US flag hijab and pussy hats in Trump's America. *Gender, Place, and Culture* 24(5):628–644

Joshi S, McCutcheon P and Sweet E (2015) Visceral geographies of whiteness and invisible microaggressions. *ACME* 14(1):298–323

Lawson V (2009) Instead of radical geography, how about caring geography? *Antipode* 41 (1):210–213

Lugones M (2010) Toward a decolonial feminism. *Hypatia* 25(4):742–759

Mahtani M (2006) Challenging the ivory tower: Proposing anti-racist geographies within the academy. *Gender, Place, and Culture* 13(1):21–25

Mahtani M (2014) Toxic geographies: Absences in critical race thought and practice in social and cultural geography. *Social and Cultural Geography* 15(4):359–367

Mullings B, Peake L and Parizeau K (2016) Cultivating an ethic of wellness in Geography. *The Canadian Geographer/Le Géographe canadien* 60(2):161–167

Narayan U (1997) *Dislocating Cultures: Identities, Traditions, and Third-World Feminism.* New York: Routledge

Noxolo P, Raghuram P and Madge C (2011) Unsettling responsibility: Postcolonial inter- ventions. *Transactions of the Institute of British Geographers* 37(3):418–429

Ortega M (2006) Being lovingly, knowingly ignorant: White feminism and women of col- our. *Hypatia: A Journal of Feminist Philosophy* 21(3):56–74

Pain R and Smith S (eds) (2008) *Fear: Critical Geopolitics and Everyday Life.* Aldershot: Ash- gate

Parizeau K, Shillington L, Hawkins R, Sultana F, Mountz A, Mullings B and Peake L (2016) Breaking the silence: A feminist call to action. *The Canadian Geographer / Le Géographe canadien* 60(2):192–204

Schutte O (1998) Cultural alterity: Cross-cultural communication and feminist theory in North-South contexts. *Hypatia, A Journal of Feminist Geography* 13(2):53–72

Wright M W (2001) A manifesto against femicide. *Antipode* 33(3):550–566

Children and Childhood

Cindi Katz

The Graduate Center, City University of New York, New York, NY, USA;
ckatz@gc.cuny.edu

Children and childhood are not terms that spring to mind when thinking about radical geography, but what could be more radical than imagining and making a future in which childhood was a central consideration and the creation of liveable futures was the beating heart of radical praxis? There is a pat and clichéd way that invoking "the child" calls forth futurity and insists on a healthy present, but "the child" is not children who breathe and run and think (cf. Edelman 2004). Invoking that figure is all too often bait for an aspirational future that does not reveal its narrow gauge—Eurocentric, white, middle class, heteronormative—luring people to imagine something good while enduring and creating environments that are toxic in every way. Toxic political-ecologically—think lead in the drinking water in cities like Flint, Michigan, or Newark, New Jersey; or the labour of children in the ship scrapping industries of Asia; or the polluted air of urban China; and toxic political-economically—think of the state violence in South and Central America, South Sudan, Syria or so many elsewheres that propels children on dangerous journeys to unknown places imagined as havens even as they so often are not; or the racialised state violence that kills, maims, and incarcerates young people of colour at staggeringly disproportionate rates in the US, Canada, and elsewhere; or the wars raging in many parts of the world that enlist child soldiers; or the inadequacy of social infrastructure such as basic healthcare, decent schools, and sanitation in so many places around the world, including the wealthiest (e.g. Bartlett 2018; Horton and Kraftl 2017). The list goes on and on and recognising its global sprawl and intimate effects really ought to galvanise radical geographers and spur vibrant radical geographies.

Thinking about children and childhoods connects us to social reproduction and thus to the ways knowledge and skills are produced and shared, not just to make a differentiated labour force daily and over the long haul, but to shift its grounds and find ways to create and maintain a social formation in which difference is less a means of division than a flowering of possibilities. And thinking about social reproduction, especially for geographers, calls for attention to its political ecologies—the range of settings in which it takes place, such as the household, the school, the courthouse, the public environment, and the workplace, and the sorts of material social practices through which knowledge is produced, shared, and exchanged. During childhood the practices around sharing knowledge are particularly important as children acquire and internalise the working knowledge of their communities in all of its unevenness and quirkiness—pernicious and delicious. This knowledge is not just taught directly but learned in a community

Keywords in Radical Geography: Antipode at 50, First Edition. Edited by the *Antipode* Editorial Collective.
© 2019 The Authors/Antipode Foundation Ltd. Published 2019 by John Wiley & Sons Ltd.

of practice and through the inhabited spaces of everyday life. As developed by Jean Lave and Etienne Wenger (1991), a community of practice is, as the name suggests, a group engaged in common practices wherein participation is a means of sharing knowledge. As people learn as they move from peripheral to more central participation, and in the process the community abides and regenerates. Children are part of innumerable communities of practice, and geographers have shown the multiple and overlapping scenes and practices of everyday life through which children learn the knowledge and skills of their various communities and their current and potential places in them. At the same time, as Denis Wood and Robert Beck's (1994) brilliant book, *Home Rules*, beautifully demonstrates, every object in a child's environment conveys and reinforces the customs, rules, aspirations, limits, edges, and idiosyncrasies of their inhabited worlds across scale, space, and time—making clear their "place" in society in every sense of the word.

What then of worlds saturated in possibilities, and those that convey disregard? A number of geographers have probed the variety and unevenness of children's everyday environments, marking how these can be sources of adventure, self-expression, and possibility as much as means of constriction and foreclosure both contemporaneously and for children's future life chances (e.g. Aitken 2001, 2014; Horton and Kraftl 2017; Ruddick 1996; Skelton and Valentine 1998). Radical geographers have produced critical geographies of education attending to the neoliberal practices fostering "responsibilisation" and responses to it—resistant and otherwise—while others have examined the reproduction of security regimes in educational environments (e.g. Cheng 2016; Katz 2017; Kraftl 2013; Mitchell 2018; Nguyen 2016). Percolating through this work are pressing questions of political economic presents and futures. "Living in a material world", and knowing its uneven geographies viscerally, what do young people today experience and imagine is in store for them in their futures. How can they see themselves as part of a liveable future to say nothing of one in which they might imagine thriving? The work of a group of University of Arizona geographers and their partners —public school students, teachers, and staff—in creating and sustaining school gardens in several public schools in Tucson offers a stunning example of playful work, environmental care, and affective labour that recognises the structural inequalities of contemporary life in Tucson and the US but reaches toward something else. Together they produce and reproduce vibrant communities of practice dedicated to growing vegetables, fruit, and native vegetation in ways that have altered the grounds of local schools, the diets of many students and their families, participants' subjectivities, and their collective imaginaries of and for the future, suggesting, despite the odds, that making another world is possible (Moore et al. 2015; cf. Kraftl 2013).

Radical geographers have been concerned with children and childhood since the founding of *Antipode*. James Blaut wrote about children with the same passion as he addressed such things as "the colonizer's model of the world" or the agrarian practices of peasants; indeed, for him they were of a piece in that understanding how people saw the world—and engaged their environments in practice —were not just cognitive questions as he had addressed them initially, but also and formidably political economic and cultural concerns. Blaut was curious about children's environmental thinking; how they made sense of the spatiality of their worlds, how they read the landscape and developed orthogonal thinking, and how they were mappers and map readers from earliest childhood. His and David Stea's "Place Perception Project", well-funded by the US Department of Education, supported a number of geography graduate students and others at Clark University to look at children's experiences of place (Hart 1979), children's spatial knowledge (Wood 1971), children's environmental knowledge, and children's home range (Anderson and Tindal 1972). Across the board, this work insisted upon the agency of children as subjects of knowledge production not its objects, as deep and creative knowers in their own rights, and as makers of worlds big and small.

Children were at the heart of another inspired and inspiring radical project of the late 1960s, the Detroit Geographical Expedition and Institute (DGEI). Not only was a concern for children's lived experiences in cruddy environments central to the mapping and research endeavours of the DGEI (Detroit Geographical Expedition and Institute 1971), but, depending on how you define childhood, they were among its key protagonists. Call them what you will, there were a lot of kids, young people, youth, high-school students, and school leavers who not only did a lot of the research associated with the Expedition, but also defined its contours and questions (Warren et al. forthcoming). And children were its key marker for determining the health of an environment (Bunge 1973; Detroit Geographical Expedition and Institute 1971). Among the things recorded and mapped, for instance, were the surfaces a child touches in the neighbourhood, evidence of rats and rat bites, dead trees, traffic accidents involving children, and the accumulated debris in a schoolyard. Their research provided a visceral sense of the tactile environment of a child's daily rounds along with its physical dangers, and made clear the broader social knowledge the spaces of children's everyday lives conveyed to them about their place in an unjust social formation. At the same time, doing this research and calling some of its shots incorporated the young people involved in the DGEI in an uneven but vital community of practice wherein many of them learned to do research in the interests of their community while producing and exchanging knowledge about it with outsiders, newcomers, and various interested parties, including policy-makers and future generations of radical geographers interested in urban decline and development. Their engagements also spurred similar radical projects of what we might now think of as community-based participatory action research.

Among those continuing this tradition—with perhaps fewer of the problematic power dynamics and occlusions associated with the DGEI (Warren and Katz 2014) —is Caitlin Cahill (e.g. 2000, 2007; Cahill et al. 2016) whose work with young people in places as distinct as New York City's Lower East Side and Salt Lake City, Utah, engages them in determining what they study and how they understand themselves in the uneven historical geographies they inhabit. Cahill's work with young people in New York on their spatialised survival strategies in a rapidly gentrifying neighbourhood not only cast unusual light on the uneven, taken-for-granted spaces of urban life, but also gave them tools to understand these shifts in their broad political-economic context and share it with other kids feeling the ground shift beneath them as they faced displacement and the influx of people who saw *them* as a problem. Working with young people, Cahill has turned customary analyses of contemporary processes such as gentrification and undocumented immigration inside out by focusing on young people's responses to them as they endured dispossession and learned about the ways they were scripted by those shoving them out of place or tempering the shove. Likewise, my focusing on children's deskilling in New York and rural Sudan in its generational temporality was meant to offer an alternative metric for understanding global economic restructuring than the usual ones of investment and disinvestment, financial flows, accumulations, and dispossessions. Insisting that a child's life chances and work horizons given the knowledge and skills they acquired in childhood was a way to make sense of political-economic processes too easily glossed as globalisation, and linking these slips and slides and possibilities across site and scale was intended to demonstrate that the global was always already intimate and that connecting across these intimacies might offer a different imagination of its possibilities, reworkings, and constituencies (Katz 2004; Pratt and Rosner 2012; cf. Aitken 2014; Lowe 2015).

How different might the concerns and practices of contemporary social formations be if their temporal horizon was a generation rather than the turn of a tweet, a financial cycle, or a term in office? People seem able to think generationally and act accordingly at more intimate scales, but less so at scales that feel more abstract like the nation or globe. Yet stretching the knowledge exchanges, environmental care, and affective labours associated with social reproduction to expanded scales—spatial and temporal—people collectively could begin to make habitable, safe, productive, and sustainable social ecologies reaching across difference and time. Am I rehearsing facile claims that call for action "for the children"? A little. But given that all of the work I have addressed here as projects of radical geography involve an understanding of children as social actors—it's not so much "for" young people as "with" them, and with them it's another story. With them mapping the conditions of their lives, counter-mapping gentrification and dispossession, demarcating environmental hazards, resisting racist policing, planting gardens, and playing. Play-work-protest-learning-teaching-acting that has the power to imagine and begin to make new worlds. I am haunted by the words I heard the brilliant Wole Soyinka say in a 1999 lecture about how great the loss to the individual, the community, the nation, and the world when the creativity of young people is disrupted by structural violence and other wasting practices of the state, capital, residues of colonialism, corruption, and sectarian imperatives (Soyinka 1999; cf. Katz 2011). Here's to a radical geography premised in staunching such losses—for the children … and the grown-ups too.

References

Aitken S C (2001) *Geographies of Young People: The Morally Contested Spaces of Identity.* New York: Routledge

Aitken S C (2014) *The Ethnopoetics of Space and Transformation: Young People's Engagement, Activism, and Aesthetics.* London: Ashgate

Anderson J and Tindal M (1972) The concept of home range: New data for the study of territorial behavior. In W Mitchell (ed) *Environmental Design: Research and Practice* (pp 1–7). Los Angeles: University of California Press

Bartlett S (2018) *Children and the Geography of Violence: Why Space and Place Matter.* New York: Routledge

Bunge W W (1973) The geography. *The Professional Geographer* 25(4):331–337

Cahill C (2000) Street literacy: Urban teenagers' strategies for negotiating their neighborhood. *Journal of Youth Studies* 3(3):251–277

Cahill C (2007) Negotiating grit and glamour: Young women of color and the gentrification of the lower east side. *City and Society* 19(2):202–231

Cahill C, Alvarez Gutiérrez L and Quijada Cerecer D A (2016) A dialectic of dreams and dispossession: The school-to-sweatshop pipeline. *Cultural Geographies* 23(1):121–137

Cheng Y E (2016) Critical geographies of education beyond "value": Moral sentiments, caring, and a politics for acting differently. *Antipode* 48(4):919–936

Detroit Geographical Expedition and Institute (1971) *Field Notes No. 3: The Geography of the Children of Detroit.* East Lansing: Detroit Geographical Expedition and Institute

Edelman L (2004) *No Future: Queer Theory and the Death Drive.* Durham: Duke University Press

Hart R (1979) *Children's Experience of Place.* New York: Irvington

Horton J and Kraftl P (2017) Rats, assorted shit, and "racist groundwater": Towards extra-sectional understandings of childhoods and social-material processes. *Environment and Planning D: Society and Space* https://doi.org/10.1177/0263775817747278

Katz C (2004) *Growing Up Global: Economic Restructuring and Children's Everyday Lives.* Minneapolis: University of Minnesota Press

Katz C (2011) Accumulation, excess, childhood: Toward a countertopography of risk and waste. *Documents d'Anàlisi Geogràfica* 57(1):47–60

Katz C (2017) The angel of geography: Superman, Tiger Mother, aspiration management, and the child as waste. *Progress in Human Geography* https://doi.org/10.1177/0309132517708844

Kraftl P (2013) *Geographies of Alternative Education.* Bristol: Policy Press

Lave J and Wenger E (1991) *Situated Learning: Legitimate Peripheral Participation.* Cambridge: Cambridge University Press

Lowe L (2015) *The Intimacies of Four Continents.* Durham: Duke University Press

Mitchell K (2018) *Making Workers: Radical Geographies of Education.* London: Pluto

Moore S A, Wilson J, Kelly-Richards S and Marston S A (2015) School gardens as sites for forging progressive socioecological futures. *Annals of the Association of American Geographers* 105(2):407–415

Nguyen N (2016) *A Curriculum of Fear: Homeland Security in US Public Schools.* Minneapolis: University of Minnesota Press

Pratt G and Rosner V (eds) (2012) *The Global and the Intimate: Feminism in Our Time.* New York: Columbia University Press

Ruddick S M (1996) *Young and Homeless in Hollywood: Mapping Social Identities.* New York: Routledge

Skelton T and Valentine G (eds) (1998) *Cool Places: Geographies of Youth Cultures.* London: Routledge

Soyinka W (1999) "The Open Sore of a Continent." Paper presented to Rutgers University, New Brunswick, NJ, April

Warren G and Katz C (2014) "Gwendolyn Warren and Cindi Katz in Conversation." https://vimeo.com/111159306 (last accessed 21 August 2018)

Warren G, Katz C and Heynen N (forthcoming) Myths, cults, memories, and revisions in radical geographic history: Revisiting the Detroit Geographical Expedition and Institute (DGEI). In T Barnes and E Sheppard (eds) *Spatial Histories of Radical Geography: North America and Beyond.* Oxford: Wiley-Blackwell

Wood D (1971) "Fleeting Glimpses, or Adolescent and Other Images of the Entity Called San Cristobal Casas." Unpublished MA thesis, Clark University

Wood D and Beck R J (1994) *Home Rules.* Baltimore: Johns Hopkins University Press

Classroom

John Paul Catungal

*Institute for Gender, Race, Sexuality and Social Justice, University of British Columbia,
Vancouver, BC, Canada;
catungal@mail.ubc.ca*

I write this entry on the classroom on the unceded territories of the Coast Salish peoples, for whom the geographies of teaching and learning exceed and predate formal institutional spaces of education. As a settler on these lands, I recognise traditional Indigenous territories mindful of the ethical, political and pedagogical responsibilities that arise from my status as an uninvited guest on these lands. Starting with this recognition is a forceful reminder that university classrooms as we know them are legitimised as spaces of teaching and learning in part through legal, discursive and material practices of Indigenous dispossession. Settler colonialism, underpinned as it is by dispossessive, genocidal and white supremacist logics, shapes the university classroom as an exclusionary and violent geography. In part, this has taken place through the canonisation of Western epistemological and methodological traditions, with their investments in disembodiment, objectivity and universality, along with the attendant eviction of Indigenous people and knowledge systems from classrooms and other spaces of knowledge (see Hunt 2014; Smith 1999). Masked usually as neutrality and best practice, this status quo approach treats the classroom as an apolitical ground of teaching and learning, rather than a space of power relations where bodies of knowledge are produced and negotiated.

Radical geographers and critical education scholars of various theoretical traditions have moved beyond such a treatment of the classroom as a mere stage on which teaching and learning takes place. Instead, they insist that the classroom is embedded in and shaped by institutional, governmental and interpersonal power-geometries. It constitutes one site out of many where various systems of power— e.g. white supremacy, cisheteropatriarchy, settler colonialism and capitalism— work in and through those who participate in the classroom. Radical approaches to the classroom must thus account for the socio-spatialities of power, violence and inequality that characterise its political geographies. For instance, Indigenous scholars insist that since the classroom is located in broader systems of white supremacy and settler colonialism, it is often experienced by Indigenous students and teachers as sites of anti-Indigeneity. The video *What I Learned in Class Today* powerfully documents the ways that Indigenous teachers and students feel the classroom as a space of devaluation, violence and stereotyping (https://inthecla ss.arts.ubc.ca/). Recent efforts to decolonise classrooms, under the leadership of Indigenous scholars at all levels, have thus called for the strategic centring of Indigenous bodies of knowledge in the classroom. This takes place through curricular decisions to emplace academic and cultural works on Indigeneity and settler

Keywords in Radical Geography: Antipode at 50, First Edition. Edited by the *Antipode* Editorial Collective.
© 2019 The Authors/Antipode Foundation Ltd. Published 2019 by John Wiley & Sons Ltd.

colonialism, especially by Indigenous people themselves, in the classroom (Daigle and Sundberg 2017), along with strategic hiring decisions that affirmatively bring in Indigenous scholars into university spaces. In addition, indigenising the university has also required reframing the formal classroom as a sacrosanct space of learning, recognising that land-based education constitutes an equally valid and culturally appropriate approach to Indigenous teaching and learning (Tuck et al. 2014).

Decolonial approaches to knowledge production are part of a broader effort to contest the classroom as a space invested with supposed neutrality and objectivity—key characteristics of normative forms of Western epistemologies (Smith 1999). As a response to these epistemologies and the alienation of variously minoritised subjects from canonical knowledge formations, student activists demanded the formation and institutionalisation of scholarship that we now call critical ethnic, critical race, queer and feminist studies, which also actively worked to increase representations of minoritised bodies of knowledge in classrooms. Similarly, and in tandem, feminist, queer and critical race scholars have powerfully articulated that the classroom is shaped by misogynistic, heteronormative and white supremacist logics. The feminist geographer Gillian Rose (1993) has noted that masculinist epistemologies suffuse geographical knowledge production, an insight that should be extended to classroom spaces where geographical epistemologies and methodologies circulate and reproduce. Efforts to denaturalise masculinist notions of objectivity and expertise have also translated into the increasing emplacement of positionality as a key, if complex, conceptual resource for re-shaping the classroom as a space of teaching and learning. Indeed, efforts to contest classroom geographies of power-knowledge often enact the key feminist principle that the personal is political, and thus also pedagogical. Both Browne (2005) and Bondi (2004) call for nuanced approaches that make use of teachers' and students' multiple locations and experiences as one important basis for negotiating situated knowledges in the classroom (see also Haraway 1988). In recognising the "connection between ideas learned in university settings and those learned in life practices" (hooks 1994:15), these efforts reconfigure the classroom as felt, embodied and relational spaces of power/knowledge.

One important way that power/knowledge manifests in and shapes the classrooms is through the negotiation of the relational identities of subjects—including teachers and learners—in these educational spaces. Extending Rose's insights on the gendering of knowledge production, Nash (2010) notes, for example, how gender norms affect expectations of instructors, with female professors expected to be caring and flexible and male teachers understood foremost as inscrutable content experts. In addition, classrooms are spaces of racial violence. As the Black feminist scholar bell hooks (1994:4) notes, classrooms have historically tended to be embodied sites of anti-Blackness—experienced "like a prison, a place of punishment and confinement" by Black learners—and buoyed by long circulating teaching philosophies based on "white racist assumptions that ... [Black people] were genetically inferior, never as capable as white peers, even unable to learn". Moreover, the classroom as a racialised space challenges the capacity of racialised teachers to take on the role of teacher, as "faculty of color are regularly accused of being biased when they teach about inequality, and in particular, racism ... and for that matter are more likely to be presumed incompetent to teach any subject at all" (Lerum 2012:273). Annette Henry (2015) and Minelle Mahtani (2006) have also respectively documented the ways that classrooms become sites of multiple disadvantage for Black women and women of colour, who are often variously characterised as exotic, angry, biased or otherwise incompetent (see also Muhs et al. 2012). For their part, queer geographers and geographers of sexualities have also puzzled over the politics of sexuality in the classroom, including the personal and professional consequences of including queer and sexualities content in geography courses and of "being out" in the classroom as teachers (Elder 1999; Knopp 1999; Skelton 1997). Moreover, Gill Valentine (1997) reminds us that the classroom can be a sexually violent space, in terms of being a site of both sexual harassment and homophobia and in terms of how unexamined heteronormative renderings of the world permeate course content and interactions.

Along with being a space of knowledge production, the classroom is also a workplace, characterised by uneven political geographies of (teaching) labour. As universities seek to maximise economic returns and intensify competition with others for status and revenues, the classroom becomes a site of labour exploitation of various forms. Scholars and labour activists have noted that precariously employed contract faculty perform an ever increasing portion of classroom-based work, and this pattern is gendered and racialised in uneven ways. Moreover, some have noted that the turn to peer grading in classrooms, while often framed as a more democratic and collaborative form of evaluation, maps all too nicely into neoliberal patterns that make use of students' free labour as a substitute for the paid work of teachers (Pinchin 2009). Recognising the centrality of the classroom as a teaching space to the mandate of universities, recent labour actions in UK and Canadian universities have publicised the reality of the classroom as a site of unfair work conditions, thus articulating its utility as a space for dissent and collective organising. Interrupting the everyday function of the classroom has been part of a larger strategic arsenal of academic labour movements, for whom disruptions of the institutional status quo are necessary to fight not only for more secure work conditions, but also for better learning environments. Indeed, labour activists have powerfully vocalised that the classrooms and other institutional workplaces are students' very own learning spaces, and thus work conditions are also very much learning conditions. Such assertions enable the possibilities of solidarities between workers and students. Nevertheless, these efforts battle against economistic logics that pit workers and students in the classroom, in part through the neoliberal construction of students as consumers and of teachers as service providers whose practice must be calibrated to the demands of student satisfaction (Saunders 2014). The institutional prioritisation of such neoliberal goals renders "killjoy" classroom subjects—i.e. those who name uncomfortable issues of social injustice—out of place and poses challenges to the possibility of solidarities across various classroom relations. In this context, classrooms are enrolled in the reproduction of elite learning subjects through classed inaccessibility, unaffordability and the defanging of radical classroom practice.

Radical education practitioners have sought to expand the classroom beyond formal spaces of teaching and learning in response to the multiple exclusions of educational institutions. Such efforts resist the false separation between the classroom and its supposed "outsides" (Rouhani 2012). The Detroit Geographical Expedition and Institute (DGEI) is often hailed as an exemplary, if imperfect, effort to relocate, so to speak, the classroom to Detroit's inner city, offering courses to primarily Black residents and tying educational programming and research directly to the pressing needs of the community. Rather than merely plopping down an institutional intervention into the city, DGEI participants, especially from the community itself, recognised, grounded capacities, leadership and knowledges at the community level and sought to democratise knowledge by insisting on community control and agenda setting as a key framework of the DGEI (Heyman 2007; Warren et al. forthcoming). Other efforts exist that seek to imagine the classroom within and also beyond the institutional framework of the university. Chief among them, for example, are freedom schools that centre equitable and affirming education for Black children and youth (Jackson and Howard 2014) and prison education programs that reframe exclusionary carceral spaces into affirming sites of radical representation, empowerment and knowledge production.

A key theme that threads through all the examples above is a concern with the classroom as an uneven geography of power, social differentiation and inequality. The examples discussed above name and analyse the different ways that systems and structures of power produce classrooms as sites of discursive, material and epistemological violence and exclusions. Each in their own way, they also seek to imagine the radical possibility that classrooms might be reconfigured in the spirit of decolonial, feminist, anti-racist, queer and anti-capitalist politics. They therefore evince bell hooks' (1994) desire for a transgressive pedagogy, premised as it is on an understanding of knowledge production as liberatory practice.

References

Bondi L (2004) Power dynamics in feminist classrooms: Making the most of inequalities? In Women and Geography Study Group (eds) *Geography and Gender Reconsidered* (pp 175–182). London: Women and Geography Study Group

Browne K (2005) Placing the personal in pedagogy: Engaged pedagogy in "feminist" geographical teaching. *Journal of Geography in Higher Education* 29(3):339–354

Daigle M and Sundberg J (2017) From where we stand: Unsettling geographical knowledges in the classroom. *Transactions of the Institute of British Geographers* 42(3):338–341

Elder G (1999) "Queerying" boundaries in the Geography classroom. *Journal of Geography in Higher Education* 23(1):86–93

Haraway D (1998) Situated knowledges: The science question in feminism and the privilege of partial perspective. *Feminist Studies* 14(3):575–599

Henry A (2015) "We especially welcome applications from members of visible minority groups": Reflections on race, gender, and life at three universities. *Race, Ethnicity, and Education* 18(5):589–610

Heyman R (2007) "Who's going to man the factories and be the sexual slaves if we all get PhDs?": Democratizing knowledge production, pedagogy, and the Detroit Geographical Expedition and Institute. *Antipode* 39(1):99–120

hooks b (1994) *Teaching to Transgress*. New York: Routledge

Hunt S (2014) Ontologies of Indigeneity: The politics of embodying a concept. *Cultural Geographies* 21(1):27–32

Jackson T and Howard T (2014) The continuing legacy of Freedom Schools as sites of possibility for equity and social justice for Black students. *Western Journal of Black Studies* 38 (3):155–162

Knopp L (1999) Out in academia: The queer politics of one geographer's sexualisation. *Journal of Geography in Higher Education* 23(1):116–123

Lerum K (2012) What's love got to do with it? Life teachings from multiracial feminism. In G G Muhs, Y F Niemann, C G González and A P Harris (eds) *Presumed Incompetent: The Intersections of Race and Class for Women in Academia* (pp 266–276). Boulder: University Press of Colorado

Mahtani M (2006) Challenging the ivory tower: Proposing anti-racist geographies within the academy. *Gender, Place, and Culture* 13(1):21–25

Muhs G G, Niemann Y F, González C G and Harris A P (eds) (2012) *Presumed Incompetent: The Intersections of Race and Class for Women in Academia*. Boulder: University Press of Colorado

Nash C J (2010) Gendered and sexed geographies of/in a graduate classroom. *Documents d'anàlisi geogràfica* 56(2):287–304

Pinchin K (1999) Students can't mark each other's assignments, says court. *Maclean's* 22 June. http://www.macleans.ca/education/uniandcollege/students-marking/ (last accessed 7 May 2018)

Rose G (1993) *Feminism and Geography: The Limits of Geographical Knowledge*. Minneapolis: University of Minnesota Press

Rouhani F (2012) Practice what you teach: Facilitating anarchism in and out of the classroom. *Antipode* 44(5):1726–1741

Saunders D B (2014) Exploring a customer orientation: Free-market logic and college students. *Review of Higher Education* 37(2):197–219

Skelton T (1997) Issues of sexuality in the teaching space. *Journal of Geography in Higher Education* 21(3):424–431

Smith L T (1999) *Decolonizing Methodologies: Research and Indigenous People*. London: Zed

Tuck E, Mackenzie M and McCoy K (2014) Land education: Indigenous, post-colonial, and decolonizing perspectives on place and environmental education research. *Environmental Education Research* 20(1):1–23

Valentine G (1997) Ode to a geography teacher: Sexuality and the classroom. *Journal of Geography in Higher Education* 21(3):417–424

Warren G, Katz C and Heynen N (forthcoming) Myths, cults, memories, and revisions in radical geographic history: Revisiting the Detroit Geographical Expedition and Institute (DGEI). In T Barnes and E Sheppard (eds) *Spatial Histories of Radical Geography: North America and Beyond*. Oxford: Wiley

Combination

Jamie Peck

Department of Geography, University of British Columbia, Vancouver, BC, Canada;
jamie.peck@ubc.ca

Uneven development, uneven geographical development, these have long been keywords in radical geography—sometimes to the point that their meaning *and implications* have been taken for granted, or otherwise recede into the background. In a common-sense manner, they signify the bare facts of a spatially differentiated world. In radical political economy, they point to the myriad ways in which *capitalist* worlds are relentlessly made, and made over, by transformative processes of (primitive) accumulation, dependency, exploitation, struggle, dispossession, devaluation, domination, resistance, and more. Here, uneven development signals the contradictory and countervailing tendencies, distinctive to processes of capitalist accumulation, for universalisation-cum-equalisation on the one hand and differentiation-cum-fragmentation on the other. Following this logic, Neil Smith read uneven development as "social inequality blazoned onto the geographical landscape" (2008:206). In radical geography, the formative work on uneven development took place in the early 1980s, most notably by Smith, by Doreen Massey and by David Harvey. More than three decades later, there are few, if any, uneven development deniers in this community at least. However, it has become quite commonplace for the condition to be acknowledged more in passing than as an active question, or a matter for theoretical problematisation.

David Harvey maintains that the theory of uneven development "needs further development" (2006:71), but any renovation of this ostensibly staple concept might do well to focus on two words that are often edited out in received usage, despite a clear lineage back to Leon Trotsky's idea of uneven *and combined* development (UCD). Sitting right at the heart of the concept of UCD, the pivotal but sometimes overlooked notion of combination indexes its fundamentally dialectical and relational character. "Combination", in this sense, denotes hybrid complexity, structural asymmetry, and contradictory coexistence, rather than systemic singularity, equilibrium, or purity, invoking an ontology of mutually entangled social formations fashioned from elements old and new, near and far. This is, moreover, a *landscape* ontology, the inconstant outcome of a moving and unequal matrix of articulations, (inter)relations, and unbalanced interdependencies. The superficially innocent notion of combination consequently begs a whole series of challenging questions. How, and with what effects, are heterogeneous social and political forms hybridised and articulated? How are these combinatorial forms related to one another, yielding what more-than-local patterns and interdependencies and what dynamics of domination and resistance, over time and space? What new

Keywords in Radical Geography: Antipode at 50, First Edition. Edited by the *Antipode* Editorial Collective.
© 2019 The Authors/Antipode Foundation Ltd. Published 2019 by John Wiley & Sons Ltd.

compositions and configurations might be forged, both in practice and in the political imagination, as the synthesis of multiple determinations, forces, alliances, and relations? As such, "combination" directs critical attention not only to sites of contradiction and sources of creativity, but to the always-immanent potential for radical if not revolutionary change, within and beyond capitalism. All the more curious, then, that it has become the silent C.

Making a case for not only unhiding the muted C in UCD, but recognising and problematising its constitutive and catalytic role, this entry moves first to recover and reposition this destabilising and potentially revolutionary concept before going on briefly to outline a remit for the contemporary analysis of "recombination". Combination and recombination can be seen as the connective tissue as well as the animating force of UCD, recalling as they do Doreen Massey's insistence that "[c]onnection, as well as differentiation, is what it is all about" (1995:303). As an antidote to simplistic ideas of endogenous or home-grown development, UCD stands as a caution against localism and methodological internalism. It also issues a challenge to the presumptions of teleological stage models, where "lagging" regions are seen to emulate and follow the pre-existing pathways carved by their supposedly more "advanced" peers. Instead, the axiom of combination problematises questions of interdependence, interconnectivity, situatedness, positionality, and the immanent potential for finding and shaping *new* configurations and pathways. Conceived as an active and politicised process, UCD can furthermore be seen as a corrective to those readings of uneven development that in their abbreviated form foster the impression of an inert landscape positioned only in the background, or which summon only abstract or metaphysical conceptions of capitalist dynamics. Recombination consequently deserves a special place in relational methodologies, and those that recognise and actively work with sociospatial difference and ontological multiplicity. And it has an important (if neglected) role to play in the interrogation of those variegated, polymorphic, and contradictory modes (and conditions) of development that define the restructuring present ... and which open horizons to alternative futures.

Revolution

A germinal concept in Trotsky's work on the political economy of "permanent revolution", UCD was originally posited as "the most general law of the historic process" (Trotsky 2008:5). Critical of (some) "stagist" arguments about transitions to socialism by way of advanced capitalism, although himself hardly immune to teleological thinking, Trotsky recognised that the mercurial geographies of bourgeois reformism and revolutionary potential were shaped in conjunction with uneven and combined development, such that the (pre)conditions for radical transformation might well be found in those economies, like Russia, with certain "advantages of backwardness". Predicated on an understanding of uneven processes of integration into the emergent world system, this implied a fundamentally non-linear understanding of political, economic, and social development, and a repudiation of those sequentialist models premised assumptions of the "repetition of the forms of development". Trotsky's phrasing may have been antique, but his vision was a revelatory one:

> Although compelled to follow after the advanced countries, a backward country does not take things in the same order. The privilege of historic backwardness—and such a privilege exists—permits, or rather compels, the adoption of whatever is ready in advance ... skipping a whole series of intermediate stages. Savages throw away their bows and arrows for rifles all at once, without traveling the road that lay between those two weapons in the past ... The development of historically backward nations leads necessarily to a peculiar combination of different stages in the historic process. Their development as a whole acquires a planless, complex, combined character. (Trotsky 2008:4)

Trotsky had a distinctive take on the "advantages" of late or what he called slow-tempo development, suggesting that ostensibly lagging regions possessed the advantage of being in a position to select from, and respond to, "the latest conquests of capitalist technique and structure", contesting and combining these in such a way to forge "amalgam[s] of archaic with more contemporary forms" (Trotsky 1976:583), and raising the possibility of "skipping" what elsewhere had been intermediate steps or stages. There is some degree of latent developmentalism in this idea of jumping stages, but the (re)formulation of UCD is an attempt to escape that, to recognise structural constraints and forces along with a political economy of emergence. Viewed from the perspective of the evolving geography of the world system, rather than from a single region, the development process is duly revealed as multipolar, interdependent, "planless, complex, [and] combined".

This, to be sure, is no receipt for parsimonious theorising or for the formulation of crisply stylised categories of analysis—one reason, perhaps, why UCD was destined to languish, for much of the 20th century, among that "class of Marxist notions who suggestiveness is equaled only by their elusiveness" (Elster 1986:56). Rejuvenated and repurposed by radical geographers in the 1970s and 1980s, albeit for the most part through the truncated nomenclature of uneven development, the concept informed a creative surge of theoretical development from a range of (neo)Marxist standpoints, not least due to its place in (new) explanations of the real-time gestalt shift in the spatial ordering of capitalist development and crisis—accelerating deindustrialisation across the global North, the selective industrialisation of parts of the global South, new international (and spatial) divisions of labour, and such. Moving into the 1990s, when these reordered circumstances began to assume the shape of a new normal of sorts, explicit recourse to the concept of uneven development once again began to wane, albeit while retaining mostly uncontested status in radical geography circles. Economic geography though, for its part, was shifting its focus of attention to growth nodes and global networks, and by the same token to endogenous causality and productive articulations, which at best amounted to a lop-sided treatment of "uneven development", but also to the less benign neglect of crisis, inequality, dislocation, and (inter)dependency. UCD became the back stage. Twenty years after Elster's dismissive remark, Neil Smith (2006:180) reflected that—notwithstanding his own efforts, alongside those of Massey, Harvey, and others—the concept had been "subject to a remarkable lack of serious analysis".

Recombination

Perhaps it was the disorienting fallout of the global financial crisis beginning in 2008, or proliferating sociospatial inequalities, or the concurrent delegitimation of neoliberal truth claims, or the continuing ascendancy of China as a quasi-capitalist superpower, coupled with the divergent political and economic development of the remaining BRICS, or populist revolts against open borders, free-trade, and the orthodox globalisation project, or some combination of all these things, but the pervasive sense that the capitalist world may be in the throes of another gestalt shift, comparable in scope and significance to that of the 1970s and 1980s, may account for what has been a recent return to the active theorisation of UCD. But this time it has been initiated by theorists in international political economy, critical development studies, anthropology, and historical sociology (rather than by radical geographers per se), and this time the middle C is intact (see, for example, Anievas and Nisancioglu 2015; Kasmir and Gill 2018; Makki 2015; Rosenberg 2016). Notwithstanding its quite profound contemporary resonances, a great deal of this more recent conversation has however been conducted on the terrain of the *longue durée*. Rather less attention, to date, has been paid to the restructuring present and the problematic of real-time recombination. This raises the possibility of a different kind of interdisciplinary conversation around the problematics of capitalist and more-than-capitalist crisis and transformation, mobilising new readings of UCD in relation to the phenomena of real-time recombination, polycentric restructuring, and heterogeneous development, in all their variegated, conjunctural, *and relationally interdependent* forms. In this context, the recovery and repurposing of that neglected C may come to assume a new significance. Allinson and Anievas (2009:49), for example, see in the uniquely protean idea of combination scope for "real theoretical innovation", while Gavin Smith (2016:224) maintains that "*combination* as a current condition of reality is *the* challenge presented to us by the organic crisis".

One way to think about the Rubik's Cube-like problematic that is (re)combination in territorial terms, at the regional or national scale, is to go back to a particularly arresting metaphor once used by David Stark and László Bruszt in their work on the postsocialist transformations of the 1990s. A friend of theirs in Budapest had once recounted a tale about the communist party-approved version of the *Monopoly* game, called *Gazdálkodj Okosan!* ["Economize Wisely"], which had been played with the quotidian objective of securing a job and apartment, and then saving for items of furniture. (Ernő Rubik himself, who was born in Budapest in 1944, surely played it as a child.) Apparently, some of those old enough to remember the arch-capitalist version of the game, *Capitaly*, which had been played in Hungary before the Second World War, had passed down stories about the alternative configuration, with its rules (and value system) of ruthless accumulation. These folk memories would later enable the dissident practice of flipping over the authorised edition of the board, safe from the prying eyes of the party state, so as to pencil in an approximation of the capitalist playing field on the other side, after which bricolaged variants of the game were developed with repurposed pieces. "The notion of playing capitalism with noncapitalist pieces strikes us as an apt metaphor for the postsocialist condition", Stark and Bruszt (2001:1129–1130) reflected: "the ruins of communism were not a tabula rasa, and so the new hybrid game was played with institutions cobbled together partly from remnants of the past that, by limiting some moves and facilitating other strategies".

Needless to say, the recombinant capitalisms of postsocialist Eastern Europe do not come close to exhausting the practically infinite list of actual (not to say imaginable) configurations—witness the apparently *sui generis* pathways that are being constructed in China, for example. The wide (and moving) spectrum of actually existing variegation is a reminder that what Trotsky memorably called the "whip of external necessity"—no matter how strong that whip hand, be it in the "invisible" form of coercive competition or in much more visible guises like structural adjustment—is hardly a unilateral or automatic determinant of political-economic outcomes at the "local" scale. It is not to underestimate the force-fields of dependency and discipline, with their binding constraints and unequal terms of trade, or for that matter the geopolitical entrenchment of dominant models of development and distribution, to make the case that none of this nullifies the potential for unruly divergence and recombinant discovery—albeit with the potential for all manner of reformist, reactionary, and revolutionary outcomes. Social innovation, political struggle, not to mention the work of countless bricoleurs, all matter in this context; there is no mechanical predetermination. It is a reminder that the geographies of the restructuring present are (being) made; they are not pregiven.

Yet the implications of UCD go beyond this. That middle C is more than merely a placeholder for the recognition of geographical difference and interdependence. As Justin Rosenberg (2006:318), has explained, it derives from an "ontological premise of 'more-than-one' ... [which] stretches the referent of the term 'development' across the conceptual space of multiple instances", not to say multiple sites and multiple pathways. Unevenness may be a stubborn fact of socioeconomic life, and precondition and prerequisite for combination, but the middle C presents as a potentially boundless source of "multiplier" effects in both socioeconomic and explanatory terms. In a static or cross-sectional sense, the idea of combination suggests that national or regional economic formations should be seen as "amalgams", admixtures of inherently "local" or *sui generis* features and those born of the multiplicity of connections into (and relations with) wider social fields; from trading blocs and neoimperialist incursions to diasporic networks, from techno-organisational infrastructures and transnational social movements to corporate supply chains. In dynamic and historical terms, however, this idea begs yet more searching questions about interdependence, interactivity, multiplicity, mutuality, responsibility, and complex coexistence.

"Combination", in this latter sense, is a signifier of condition of deep (inter)relationality and social multiplier effects, the restless outcomes of which will include the remorseless reproduction of familiar terrains, to be sure, but will also open up some previously uncharted territories. If UCD is to mean more than seesawing inequality and the relentless remaking of core–periphery relations—the business of uneven capitalist development "as usual"—then the disruptive potential of the middle C must be not only recognised but actively grasped. If the actually existing present of globalising capitalism is to be understood not as some unitary system or tendential monoculture, as a competitive ecosystem pulled along in the wake of "advanced" forms and forces, but as a contradictory and mutating matrix of unruly articulations and recombinant developments (each with the potential for all manner of Rubik-style next steps, in turn yielding their own interaction effects), then the ultimate promise of recombination is the open remit that it provides for making and thinking new configurations. Pathways towards this open horizon are not freely or voluntaristically chosen, of course, and some will be more realisable and sustainable than others, but recovering the middle C should at least be a safeguard against pragmatist necessitarianism and nihilistic foreclosure. "The scientific task, as well as the political", Trotsky (2004:193) once wrote, "is not to give a finished definition to an unfinished process, but to follow all its stages, separate its progressive from its reactionary tendencies, expose their mutual relations, foresee possible variants of development, and find in this foresight a basis for action".

References

Allinson J C and Anievas A (2009) The uses and misuses of uneven and combined development: An anatomy of a concept. *Cambridge Review of International Affairs* 22(1):47–67

Anievas A and Nisancioglu K (2015) *How the West Came to Rule*. London: Pluto

Elster J (1986) The theory of combined and uneven development: A critique. In J Roemer (ed) *Analytical Marxism* (pp 54–63). Cambridge: Cambridge University Press

Harvey D (2006) *Spaces of Global Capitalism*. London: Verso

Kasmir S and Gill L (2018) No smooth surfaces. *Current Anthropology* 59(4):355–377

Makki F (2015) Reframing development theory: The significance of the idea of uneven and combined development. *Theory and Society* 44(5):471–497

Massey D (1995) *Spatial Divisions of Labour* (2nd edn). Basingstoke: Macmillan

Rosenberg J (2006) Why is there no international historical sociology? *European Journal of International Relations* 12(3):307–340

Rosenberg J (2016) Uneven and combined development: "The international" in theory and history. In A Anievas and K Matin (eds) *Historical Sociology and World History* (pp 17–30). New York: Rowman and Littlefield

Smith G (2016) Against social democratic angst about revolution: From failed citizens to critical praxis. *Dialectical Anthropology* 40(3):221–239

Smith N (2006) The geography of uneven development. In B Dunn and H Radice (eds) *100 Years of Permanent Revolution* (pp 180–195). London: Pluto

Smith N (2008) *Uneven Development* (3rd edn). Athens: University of Georgia Press

Stark D and Bruszt L (2001) One way or multiple paths: For a comparative sociology of East European capitalism. *American Journal of Sociology* 106(4):1129–1137

Trotsky L (1976 [1938]) Revolution and war in China. In L Evans and R Block (eds) *Leon Trotsky on China* (pp 578–591). New York: Pathfinder

Trotsky L (2004 [1937]) *The Revolution Betrayed*. New York: Dover

Trotsky L (2008 [1932]) *History of the Russian Revolution*. Chicago: Haymarket Books

Community Economy

Community Economies Collective[1]

http://www.communityeconomies.org

Community economy unites two terms that have long been understood as mutually exclusive. Throughout the 20th century, theorists and folk on the ground alike have understood the modern capitalist economy as an expanding and unitary system that, among other things, has been a force displacing and undermining community. Traditional and localised economies were seen and experienced as entangled with and embedded in social relations that, often, resonated with the positive aspects of community such as mutual care, interdependence, recognition, collective wellbeing, and a sense of place; while the ever-expanding capitalist economy appeared as insisting upon individual utility maximisation, self-preservation, anonymity in market exchange, and the homogenisation of culture and place. This relationship of opposition relies upon an understanding of economy as a singular and expanding capitalist system dominating and shaping the social, a system to which community is, at best, subordinate and, at worst, an obstacle. Furthermore, it reduces community to an archaic and pre-modern form of social organisation distinct from economy yet always beholden to its needs, shaped by its force, and penetrated by its dynamic.

Using community to modify economy, however, signals a decidedly different understanding of economy as something modifiable, differentiated, and perhaps beholden to the needs and desires of community. It suggests that community may itself be a mode and form of economy distinct from other modes and forms (e.g. capitalist economy, slave economy, or household economy). To differentiate it from other economies, we define a community economy as a set of economic practices that explicitly foregrounds community and environmental wellbeing. Indeed, from a community economy perspective, such wellbeing is the purpose rather than the hoped for and, at best, secondary outcome of economy.

Building upon the progressive potential of community, community economies are sites of economic decision making, negotiation, and experimentation; they are also sites and starting points for research and activism that not only inventory and assess the specific dynamics of such economies but also foster and amplify their potential to be more durable practices and transposable models. As a dynamic and emergent form of economy rather than simply an economy that is "community-based", community economy signals a novel research trajectory within economic and radical geography and, increasingly, an activist agenda to "take back the economy" for communities and the environments upon which they depend.[2]

Activating community economy as an object of analysis and economic practice requires a rethinking of economy where the economy loses its power to structure

Keywords in Radical Geography: Antipode at 50, First Edition. Edited by the *Antipode* Editorial Collective.
© 2019 The Authors/Antipode Foundation Ltd. Published 2019 by John Wiley & Sons Ltd.

and figure all other processes (e.g. community) as well as a rethinking of community as other than a static and bounded entity based upon principles of exclusion.

Rethinking Economy

The work of "rethinking economy" by Gibson-Graham (2014), Mitchell (2008), Callon (2007) and others is transforming how we conceptualise the economic and its relationship with other processes, practices, and actors. Rethinking economy sees "the" economy as an outcome or effect of economic discourse, metrics, calculations, and socio-technical devices rather than an overarching system, entity, or force which operates via a set of universal laws, progressing and moving independent of other processes (e.g. culture or community). The economy, therefore, is a decidedly more contingent assemblage of processes, practices, and actors (human and non-human) that make possible the production and distribution of goods and services.

Activists are also productively rethinking the economy as an indeterminate site open to intervention, local action, and possibility rather than inevitability (de Sousa Santos 2006a, 2006b; Escobar 2009, 2018). Around the world and networked via movements such as the World Social Forum there exist myriad enactments of economic difference and diversity that build upon the successes of cooperative production, fair trade, democratic budgeting, peasant and indigenous peoples' livelihoods, alternative food and craft networks, and, generally, production and consumption practices that foreground community and environmental wellbeing (Roelvink 2016). These alternative economic practices require an alternative imaginary of economy as a site of possibility and ethical concerns rather than a global and totalising system beyond intervention.

So, where economic geography sees relationality and embeddedness, perhaps even explaining local variations of a global capitalism (Boggs and Rantisi 2003), the economy as rethought suggests that the multiplicity of economic practices all around us, as well as those we might imagine and enact in the future, do not add up to any unified form or system, capitalist or otherwise. Informal and non-monetised forms of exchange, independent or cooperative production, household or community-based labour, state sectors and nationalised industries, even alternative corporate practices are all elements of the *diverse economy* (Gibson-Graham 2006).[3]

Rethinking Community

To think of community as an adjective for economy, such that a "community economy" might be possible, requires a rethinking of community as well as economy. Indeed, rather than a sign of tradition, homogeneity, or exclusion (Young 1990), we must recognise and magnify community's "progressive and ethical force", its potentials for economic innovation and experimentation, and its ability to shape and transform local livelihoods and wellbeing (Gibson-Graham 2006).

While the use of the term community in economic geography is diverse, it most often suggests a set of relations embedded in a particular place that either shape or are shaped by economic (read capitalist) practices (cf. Amin and Roberts 2008). In this sense community is essentially local, and it suggests boundaries, however blurred, to both its location and those who "belong" to it. This aligns with the common conception of community as a place or homogenous group of people, and it tends toward an imaginary of community as conservative, exclusionary, and essentially based upon some commonality or "common being" (cf. Joseph 2002).

Yet community can be other than commonality; it can also be an acknowledgement and practice of "being in common"—that understanding of the individual not as a singularity but as always being with others. Nancy (1991a, 1991b) suggests that this understanding of being gives us a foundation to rethink community as an always emergent process and a practice of co-dependence and mutuality. Beginning from this conception of being and community, we might start to see the potentials of community desires, ethics, and dispositions (rather than just individual drives) to be central to an economic dynamic and to guide economic decision-making (Popke 2010). We might begin to imagine a host of economic practices beholden to community rather than communities displaced or beholden to "the economy".

Activating Community Economies
Emerging from these core insights, the community economies project today constitutes an ongoing effort to make other worlds possible. These efforts resonate with and advance a number of themes in economic and radical geography.

Economic Difference "Here and Now"
While economic and radical geography have long documented the unevenness and variability of the economic landscape, the heterogeneity revealed invariably exists within or relative to a singular and now global capitalist economy (Gibson-Graham 1996). Similarly, despite challenging the systemic coherence and unity of "the" economy by foregrounding how economic worlds are made by and through a range of actors, associations, and processes, "rethinking economy" scholars most often document and thereby come to perform essentially capitalist practices, markets, and economies (cf. Butler 2010). From a community economies orientation, however, the goal of revealing economic difference is to produce a rich reservoir of examples that work to foster an "imagining and enacting [of] noncaptialist futures". Indeed, the "capitalocentrism" of much economic and radical geography, from a community economies perspective, works to stunt our imaginations of economic difference and stifle our desires to enact other economies "here and now" (Community Economies Collective 2001).

By contrast, building upon feminism and queer theory, community economies scholars and activists approach economic difference as a resource for thinking and doing non-capitalism rather than as a remnant of some pre-capitalist economic form or an oversight of an otherwise comprehensive and ever expanding capitalism. Doing so allows them to engage in a politics of possibility, revealing the political potency of a proximate and always emerging economic diversity. As feminists and queer theorists have done across scales and to great effect, community economies scholars make visible and account for that which had been invisible and unaccounted in order to, in this case, perform new worlds beyond the confines of capitalism where alternative economic subjectivities and practices might thrive.

For the community economies project, the inventory of economic difference is a profoundly political act that reminds us of existing and possible economic difference not in spite of capitalism's power, but as an outgrowth of the performative and proliferating nature of economic practice. A whole range of such economic practices are successful, many of which foreground community livelihoods (Safri 2015; Safri and Graham 2010; Sweet 2016), environmental sustainability (Emery and Pierce 2005; Gabriel 2011; Hurley and Emery 2018; Poe et al. 2013), and struggles for class, racial and gender justice (Borowiak et al. 2018; Heras and Burin 2014; Huron 2015). Furthermore, this ontological starting point, of economic diversity, beckons us to see and assess not only the occurrence of economic difference, but also any particular economy's range of economic dynamics, conditions of durability and transposition, and alignments with ethical concerns for human and environmental justice (Roelvink et al. 2015; Sarmiento 2017).

The Possibilities of Economic Justice

By and large, radical geography's core interest in economic justice has traditionally been pursued through the important documentation of the injustices which emerge from capitalism's essential dynamic, much of which has been traced in the pages of *Antipode*. While a community economy perspective acknowledges the many injustices that result from capitalist practices, it seeks to avoid the melancholia and paralysis that often result by focusing on such injustices alone (Gibson-Graham 2006). Indeed, performing only capitalism's injustices will only reproduce, to borrow a phrase, the "brutal energy" of capitalism, reifying its totalising power (Roy 2011; see also Roelvink 2016). Locating and amplifying economic difference, as noted above, counters such powerful narratives and works to perform an economic "otherwise". Yet, making visible the diverse economy does not in itself foster economic, social, or environmental justice. What it does do, insofar as the economy is differentiated and therefore open to possibility, is to make clear that economies can be (and may be already) sites of ethical negotiation and decision-making whose dynamics can be assessed and reworked by our research practices, activist initiatives, and communities themselves.

But assessment and reworking of economic practices, specifically on behalf of community and environmental wellbeing, will require a set of entry points for inquiry, entry points that open economic decision-making to concerns for human and environmental justice. Those coordinates which we might use to recognise, evaluate, and/or foster community economies include the following:[4]

- *Survival:* What do we really need to survive well? How do we balance our own survival needs and well-being with the well-being of others and the planet?
- *Surplus:* What's left after our survival needs have been met? How do we distribute this surplus to enrich social and environmental health?

- *Transactions:* What are the ranges of ways we secure the things we cannot produce ourselves? How do we conduct ethical encounters with human and non-human others in these transactions?
- *Consumption:* What do we really need to consume for our well-being? How do we consume sustainably and justly?
- *Commons:* What do we share with human and non-human others? How do we maintain, replenish, and grow this natural and cultural commons?
- *Investment:* What do we do with stored wealth? How do we invest this wealth so that future generations may live well?

While these coordinates cannot act as a blueprint for "the" just economy, they nevertheless prompt us (communities and researchers alike) to interrogate, reimagine, and struggle to bring into being more just economies (Dombroski 2016; Dombroski et al. 2016; Morrow and Dombroski 2015).

Amplification and Action

Finally, we wish to highlight the contribution of community economies research to radical geography's long-standing commitment to political praxis. Imagining and desiring something other than capitalism (e.g. an economy informed by ethical commitments to community and environment) has required scholars and activists to loosen their affective investments in economic explanations that presume the ever-present dominance of neoliberal capitalism, and instead pursue an open and reparative stance towards economic life that refuses to know too much, and yet also takes responsibility for the economic worlds that are made and remade through practices of research, activism, writing, and representation.

This attentiveness to how research can be deliberately structured to not only document and inventory economic difference but bring into being other economic worlds (and community economies in particular) has led to the development of a rich set of methods beyond the conceptual tools of the community economies coordinates. For example, community economies scholars have rethought participatory action research in terms of its capacity to initiate shifts in economic subjectivity and imaginaries of economic possibility (e.g. Cameron and Gibson 2005; Gibson-Graham 2008). They have explored where and how metrics and maps can, along with others, perform other economies (e.g. Borowiak et al. 2018; Safri and Graham 2010; Snyder and St Martin 2015). And their work has intersected with that of artists and other activists creatively engaging with communities to rethink and rework their economies (e.g. Hwang 2003).[5] Across this variety of research and more activist-oriented projects, there is a common commitment to not only making community economies visible, but also to actively extend their reach and influence, multiply the sites where they exist, and amplify the work they do on behalf of community and environment.

Conclusion

Community economy, since the mid-1990s, has signalled an expanding and evolving project within radical geography that resonates with a host of initiatives taking place around the world. Indeed, the call to "take back the economy" for community and environment is being heard and alternatives are emerging, and it is imperative that academics and activists alike harness and extend the traditions of radical geography to align with and foster such action. The community economies project does so by foregrounding the politics of making visible economic difference and its attendant subjectivities, creating openings in the economy for ethical negotiation, taking responsibility for those economies we wish to see thrive, and working with others (human and more-than-human) across a range of sites and scales to bring community economies into being.

Endnotes

[1] Oona Morrow, Department of Social Sciences, Wageningen University, Wageningen, The Netherlands, oonamorrow@gmail.com; Kevin St Martin, Department of Geography, Rutgers University, Piscataway, NJ, USA, kstmarti@rutgers.edu; Nate Gabriel, Department of Geography, Rutgers University, Piscataway, NJ, USA, nategab@geography.rutgers.edu; Ana Inés Heras Monner Sans, CEDESI Humanidades UNSAM, Instituto para la Inclusíon Social y el Desarrollo Humano, and CONICET, Argentina, herasmonnersans2@gmail.com

[2] The term "community economies" as understood here emerges from the influential body of work by the feminist economic geographer J.K. Gibson-Graham and that of the Community Economies Collective (see http://www.communityeconomies.org). Key texts by Gibson-Graham include *Class and Its Others* (Gibson-Graham et al. 2000), *The End of Capitalism* (Gibson-Graham 1996), *A Postcapitalist Politics* (Gibson-Graham 2006), *Take Back the Economy* (Gibson-Graham et al. 2013), and *Making Other Worlds Possible* (Roelvink et al. 2015).

[3] The diverse economies project clearly overlaps with and complements the burgeoning literature on "alternative economies" in economic geography (e.g. Fuller et al. 2010; Leyshon et al. 2003). However, the former emphasises the non-capitalist practices and potentials resident in various sites, whereas the latter more often discovers such sites to be home to a hidden or nascent capitalism (for discussion, see Healy 2009).

[4] For an introduction to community economies coordinates, see Gibson-Graham (2006: Chapter 4). For a guide on how to activate these coordinates with others, see Gibson-Graham et al. (2013). For a recent update and discussion, see Gibson-Graham and Community Economies Collective (2017).

[5] See, for example, the projects "Arts and Community Economies" (https://artsandcommunityeconomies.wordpress.com) and "Redrawing the Economy" (https://redrawingtheconomy.info).

References

Amin A and Roberts J (2008) *Community, Economic Creativity, and Organization.* Oxford: Oxford University Press

Boggs J S and Rantisi N M (2003) The "relational turn" in economic geography. *Journal of Economic Geography* 3(2):109–116

Borowiak C, Safri M, Healy S and Pavlovskaya M (2018) Navigating the fault lines: Race and class in Philadelphia's solidarity economy. *Antipode* 50(3):577–603

Butler J (2010) Performative agency. *Journal of Cultural Economy* 3(2):147–161

Callon M (2007) What does it mean to say that economics is performative? In D Mackenzie, F Muniesa and L Siu (eds) *Do Economists Make Markets? On the Performativity of Economics* (pp 311–357). Princeton: Princeton University Press

Cameron J and Gibson K (2005) Participatory action research in a poststructuralist vein. *Geoforum* 36(3):315–331

Community Economies Collective (2001) Imagining and enacting noncapitalist futures. *Socialist Review* 28(3/4):93–135

de Sousa Santos B (2006a) *The Rise of the Global Left: The World Social Forum and Beyond.* London: Zed

de Sousa Santos B (ed) (2006b) *Another Production is Possible: Beyond the Capitalist Canon.* London: Verso

Dombroski K (2016) Hybrid activist collectives: Reframing mothers' environmental and caring labour. *International Journal of Sociology and Social Policy* 36(9/10):629–646

Dombroski K, McKinnon K and Healy S (2016) Beyond the birth wars: Diverse assemblages of care. *New Zealand Geographer* 72(3):230–239

Emery M R and Pierce A R (2005) Interrupting the telos: Locating subsistence in contemporary US forests. *Environment and Planning A* 37(6):981–993

Escobar A (2009) Other worlds are (already) possible. In J Sen and P Waterman (eds) *World Social Forum: Challenging Empires* (pp 393–404). London: Black Rose Books

Escobar A (2018) *Designs for the Pluriverse: Radical Interdependence, Autonomy, and the Making of Worlds.* Durham: Duke University Press

Fuller D, Jonas A E G and Lee R (eds) (2010) *Interrogating Alterity: Alternative Economic and Political Spaces.* Farnham: Ashgate

Gabriel N (2011) The work that parks do: Towards an urban environmentality. *Social and Cultural Geography* 12(2):123–141

Gibson-Graham J K (1996) *The End of Capitalism (As We Knew It): A Feminist Critique of Political Economy.* Minneapolis: University of Minnesota Press

Gibson-Graham J K, Resnick S A and Wolff R D (2000) *Class and Its Others.* Minneapolis: University of Minnesota Press

Gibson-Graham J K (2006) *A Postcapitalist Politics.* Minneapolis: University of Minnesota Press

Gibson-Graham J K (2008) Diverse economies: Performative practices for "other worlds". *Progress in Human Geography* 32(5):613–632

Gibson-Graham J K (2014) Rethinking the economy with thick description and weak theory. *Current Anthropology* 55(s9):s147–s153

Gibson-Graham J K, Cameron J and Healy S (2013) *Take Back the Economy: An Ethical Guide for Transforming Our Communities.* Minneapolis: University of Minnesota Press

Gibson-Graham J K and Community Economies Collective (2017) Cultivating community economies. *The Next System Project.* https://thenextsystem.org/cultivating-community-ec onomies (last accessed 9 August 2018)

Healy S (2009) Alternative economies. In N Thrift and R Kitchin (eds) *The International Encyclopedia of Human Geography* (https://doi.org/10.1016/b978-008044910-4.00132-2). Oxford: Elsevier

Heras A I and Burin D (2014) Para que las diferencias no se transformen en desigualdad. *Revista Idelcoop* 213:72–112

Hurley P and Emery M (2018) Locating provisioning ecosystem services in urban forests: Forageable woody species in New York City, USA. *Landscape and Urban Planning* 170:266–275

Huron A (2015) Working with strangers in saturated space: Reclaiming and maintaining the urban commons. *Antipode* 47(4):963–979

Hwang L (2013) Rethinking the creative economy: Utilizing participatory action research to develop the community economy of artists and artisans. *Rethinking Marxism* 25(4):501–517

Joseph M (2002) *Against the Romance of Community.* Minneapolis: University of Minnesota Press

Leyshon A, Lee R and Williams C C (eds) (2003) *Alternative Economic Spaces.* London: Sage

Mitchell T (2008) Rethinking economy. *Geoforum* 39(3):1116–1121

Morrow O and Dombroski K (2015) Enacting a postcapitalist politics through the sites and practices of life's work. In K Strauss and K Meehan (eds) *Precarious Worlds:*

Contested Geographies of Social Reproduction (pp 82–98). Athens: University of Georgia Press

Nancy J-L (1991a) *The Inoperative Community*. Minneapolis: University of Minnesota Press

Nancy J-L (1991b) Of being-in-common. In Miami Theory Collective (eds) *Community at Loose Ends* (pp 1–12). Minneapolis: University of Minnesota Press

Poe M R, McLain R J, Emery M and Hurley P T (2013) Urban forest justice and the rights to wild foods, medicines, and materials in the city. *Human Ecology* 41(3):409–422

Popke J (2010) Ethical spaces of being-in-common. In S J Smith, R Pain, S A Marston and J P Jones (eds) *The Handbook of Social Geography* (pp 435–454). London: Sage

Roelvink G (2016) *Building Dignified Worlds: Geographies of Collective Action*. Minneapolis: University of Minnesota Press

Roelvink G, St. Martin K and Gibson-Graham J K (eds) (2015) *Making Other Worlds Possible: Performing Diverse Economies*. Minneapolis: University of Minnesota Press

Roy A (2011) Slumdog cities: Rethinking subaltern urbanism. *International Journal of Urban and Regional Research* 35(2):223–238

Safri M (2015) Mapping noncapitalist supply chains: Toward an alternate conception of value creation and distribution. *Organization* 22(6):924–941

Safri M and Graham J (2010) The global household: Toward a feminist postcapitalist international political economy. *Signs: Journal of Women in Culture and Society* 36(1):99–125

Sarmiento E R (2017) Synergies in alternative food network research: Embodiment, diverse economies, and more-than-human food geographies. *Agriculture and Human Values* 34 (2):485–497

Snyder R and St Martin K (2015) A fishery for the future: The midcoast fishermen's association and the world of economic being-in-common. In G Roelvink, K St Martin and J K Gibson-Graham (eds) *Making Other Worlds Possible: Performing Diverse Economies* (pp 26–52). Minneapolis, MN: University of Minnesota Press

Sweet E L (2016) Locating migrant Latinas in a diverse economies framework: Evidence from Chicago. *Gender, Place, and Culture* 23(1):55–71

Young I M (1990) The ideal of community and the politics of difference. In L Nicholson (ed) *Feminism/Postmodernism* (pp 300–324). New York: Routledge

Contract

Kendra Strauss

The Labour Studies Program, Simon Fraser University, Burnaby, BC, Canada;
kstrauss@sfu.ca

Contract shapes everyday life in myriad ways. Legal contracts structure economic and social relations, including between workers and bosses, spouses, landlords and tenants, and producers or sellers and consumers of commodities. This rather banal observation signals the extent of the "legal constitution of social reality" (Blomley et al. 2001:xv). But contract is more than a legal process and technology. It is also fundamental to liberal conceptualisations of state and polity, to citizenship—and hence the regulation of membership and mobility rights—and to the normative (political) and organisational logics of institutions. Across all of these domains, contract is thus essential to the space-economy and spatial relations, the social construction and political meanings of scale, and relational place-making.

Contract is also constituent of liberal Western ontology going back to Rousseau, Hobbes, and Locke. In other words, it is fundamental not only to forms and practices of sovereignty, citizenship and economic organisation (and thus territorialisation), but also to understandings of the nature of their existence—the way that the liberal subject and economic rationality are naturalised (Strauss 2008). This is all the more true when viewed through the lens of contract deficit: the way that informal relations and non-Western legal orders are elided and pathologised for not treating promises, commitments and claims according to the same norms of recognition (in particular, of private property) and exchange.

Contract thus features in work by economic and labour geographers, especially on the changing organisation and relations of production, and regimes of labour (Cockayne 2016; Fudge and Strauss 2014; Gidwani 2015; Herod 1997; Potts 2016; Terry 2009). It is also increasingly present in work in legal geographies on commodification and the production of value (Fannin 2011; Quastel 2017; Robinson and Graham 2018). And yet, contract is often in the background of geographical research and theorising in these sub-fields, rather than foregrounded as a subject of analysis in its own right. There is no geographical theory of contract.

This may, of course, be excusable. Theories of contract relate to, and shape the very idea of, different and separable domains of life—the economic, the political, the social—making a unified geographical theory of contract undesirable in a pluralistic discipline. It is in any case a chaotic concept. In political theory, the *social* contract "has been used in radically different ways—the contract as literal, metaphorical, historical, hypothetical, descriptive, prescriptive, prudential, moral, constitutional civil, regulative ideal, device of representation" (Pateman and Mills 2007:81). Thus political geographies of the state, particularly in an era of climate change and disaster politics, grapple with the concept of the social contract (Castree 2016; Harris 2017; see also Pelling and Dill 2010; Wainwright and Mann 2018) while feminist geographers interrogate the sexual contract (McDowell 2014). However, in part because theoretical explorations of contract are rare relative to empirical and descriptive accounts in economic and labour geography, the imbrication of social, political and economic domains—including in classical political economy—is often unexamined. This means inter alia that the spatial implications of separable domains are not epistemologically connected to foundational notions of contract in much geographical research.

Carol Pateman makes these connections very clear in her feminist analyses of contract, *The Sexual Contract* (1988) and, with Charles Mills, *Contract and Domination* (2007). Pateman uses the concept of the original contract to explore the three interrelated dimensions of the social, sexual and racial contracts (Thompson et al. 2018)—and her work on the sexual contract explicitly links conjugal and labour contracts, patriarchy and the development of capitalism, private and public spheres. As Pateman (1988:1) writes about the foundational role of contract theory in modern political and economic thought:

> The story, or conjunctural history, tells how a new civil society and a new form of political right is created through an original contract ... The attraction of the idea of an original contract and of contract theory in a more general sense, a theory that claims that free social relations take a contractual form, is probably greater now than at any time since the 17th and 18th centuries.

This is as true today as it was in 1988, when *The Sexual Contract* was published. The degradation of the standard employment relation (SER), the expansion of the marriage contract, and the reshaping of the sexual contract (see Caretta and Borjeson 2015; Forsberg and Stenbacka 2017; Fudge and Vosko 2001; Grimsrud 2011; McDowell 2017; Sola 2016) do not indicate a decline in the centrality of contract to social and economic relations. The weakening or recasting of divisions between public and public spheres, the emergence of new time-spaces and places of work, the rapid evolution of online spaces, and the recent unravelling of the post-war geopolitical order and related trade liberalisation agenda all *strengthen and reinscribe* the centrality of contract (Potts 2016; Radin 2017; Sandford 2017). They also signal social struggles that rearticulate relations like subordination (in labour law), for example through the lens of precarity (Strauss 2018).

Critiques of, and struggles against, forms of domination embedded in contractual relations are not new. Marx, in *Capital Volume 1*, described the sphere of circulation or commodity exchange, with irony, as:

> the exclusive realm of Freedom, Equality, Property and Bentham. Freedom, because both buyer and seller of a commodity, let us say of labour power, are determined only by their free will. They contract as free persons, who are equal before the law. (Marx 1990:280)

But this ironic statement of the presumption of equality and economic rationality in Marx fails to unpack *who* is included in the category of the possessor of labour power. Pateman and Mills explicitly and meticulously argue, through spotlighting the "domination contract" (Mills 1997), that the exclusion and subordination of women, of the slave and racialised workers more broadly, and of Indigenous peoples under relations of colonialism (the "settler contract"), are fundamental to the original contract.

There are some parallels, despite important epistemological and theoretical differences, in the radical, path-breaking work in black feminist geographies on the relationship between subjectivity, the production of space and the category of labour (King 2016; McKittrick 2006, 2011). McKittrick explicitly links political domination to *geographies* of domination. King builds on (among other things) the work of Jackson (2012) to question the category of labour for unpacking "geography's attendant project of human-making". What Jackson highlights is the way that the creation of the post-colonial state and settler-social contract is spatialised:

> Following Mills and Pateman, I suggest that in Guyana the settler contract applies to the coastal area, while the interior exists at the limits of that contract, more broadly falling within a racial contract ... that justifies differential treatment of its inhabitants. (Jackson 2012:11–12)

Her arguments build on Mills' argument that raced space marks the geographic boundary of the state's obligations, and his insight that a consequence of the racial contract is that the political space of the polity is not co-extensive with its geographical space (Mills 1997:42).

Contract and its relationship to the state, to the process of the commodification of labour, to international law, colonialism, and to the spatial struggles against domination and subordination, play out in profoundly spatial ways—or in McKittrick's (2006:xiv) words, are revealed by "the ties and tensions between material and ideological dominations and oppositional spatial practices". Oppositional practices include those that seek to expose processes by which some are excluded from, and stigmatised by, mainstream relations of contract. These are just a few examples of how an explicit engagement with contract can enrich critical geographic scholarship, if geographers take up the challenge.

References

Blomley N, Delaney D and Ford R T (eds) (2001) *The Legal Geographies Reader: Law, Power, and Space.* Oxford: Blackwell

Caretta M A and Borjeson L (2015) Local gender contract and adaptive capacity in smallholder irrigation farming: A case study from the Kenyan drylands. *Gender, Place, and Culture* 22(5):644–661

Castree N (2016) Geography and the new social contract for global change research. *Transactions of the Institute of British Geographers* 41(3):328–347

Cockayne D G (2016) Sharing and neoliberal discourse: The economic function of sharing in the digital on-demand economy. *Geoforum* 77:73–82

Fannin M (2011) Personal stem cell banking and the problem with property. *Social and Cultural Geography* 12(4):339–356

Forsberg G and Stenbacka S (2017) Creating and challenging gendered spatialities: How space affects gender contracts. *Geografiska Annaler: Series B, Human Geography* 99(3):223–237

Fudge J and Strauss K (2014) *Temporary Work, Agencies, and Unfree Labour: Insecurity in the New World of Work*. New York: Routledge

Fudge J and Vosko L (2001) By whose standards? Reregulating the Canadian labour market. *Economic and Industrial Democracy* 22(3):327–356

Gidwani V (2015) The work of waste: Inside India's infra-economy. *Transactions of the Institute of British Geographers* 40(4):575–595

Grimsrud G M (2011) Gendered spaces on trial: The influence of regional gender contracts on in-migration of women to rural Norway. *Geografiska Annaler: Series B, Human Geography* 93(1):3–20

Harris L (2017) Political ecologies of the state: Recent interventions and questions going forward. *Political Geography* 58:90–92

Herod A (1997) Labor's spatial praxis and the geography of contract bargaining in the US east coast longshore industry, 1953–1989. *Political Geography* 16(2):145–169

Jackson S N (2012) *Creole Indigeneity: Between Myth and Nation in the Caribbean*. Minneapolis: University of Minnesota Press

King T L (2016) The labor of (re)reading plantation landscapes fungible(ly). *Antipode* 48 (4):1022–1039

Marx K (1990 [1867]) *Capital Volume 1*. London: Penguin

McDowell L (2014) The sexual contract, youth, masculinity, and the uncertain promise of waged work in austerity Britain. *Australian Feminist Studies* 29(79):31–49

McDowell L (2017) Youth, children, and families in austere times: Change, politics, and a new gender contract. *Area* 49(3):311–316

McKittrick K (2006) *Demonic Grounds: Black Women and the Cartographies of Struggle*. Minneapolis: University of Minnesota Press

McKittrick K (2011) On plantations, prisons, and a black sense of place. *Social and Cultural Geography* 12(8):947–963

Mills C W (1997) *The Racial Contract*. Ithaca: Cornell University Press

Pateman C (1988) *The Sexual Contract*. Cambridge: Polity

Pateman C and Mills C W (2007) *Contract and Domination*. Cambridge: Polity

Pelling M and Dill K (2010) Disaster politics: Tipping points for change in the adaptation of sociopolitical regimes. *Progress in Human Geography* 34(1):21–37

Potts S (2016) Reterritorializing economic governance: Contracts, space, and law in transborder economic geographies. *Environment and Planning A* 48(3):523–539

Quastel N (2017) Pashukanis at Mount Polley: Law, eco-social relations, and commodity forms. *Geoforum* 81:45–54

Radin M J (2017) The deformation of contract in the information society. *Oxford Journal of Legal Studies* 37(3):505–533

Robinson D F and Graham N (2018) Legal pluralisms, justice, and spatial conflicts: New directions in legal geography. *Geographical Journal* 184(1):3–7

Sandford M (2017) Signing up to devolution: The prevalence of contract over governance in English devolution policy. *Regional and Federal Studies* 27(1):63–82

Sola A G (2016) Constructing work travel inequalities: The role of household gender contracts. *Journal of Transport Geography* 53:32–40

Strauss K (2008) Re-engaging with rationality in economic geography: Behavioural approaches and the importance of context in decision-making. *Journal of Economic Geography* 8(2):137–156

Strauss K (2018) Labour geography 1: Towards a geography of precarity? *Progress in Human Geography* 42(4):622–630

Terry W C (2009) Working on the water: On legal space and seafarer protection in the cruise industry. *Economic Geography* 85(4):463–482

Thompson S, Hayes L, Newman D and Pateman C (2018) *The Sexual Contract* 30 years on: A conversation with Carole Pateman. *Feminist Legal Studies* 26(1):93–104

Wainwright J and Mann G (2018) *Climate Leviathan: A Political Theory of Our Planetary Future*. London: Verso

Corruption

Sapana Doshi

School of Geography and Development, University of Arizona, Tucson, AZ, USA;
sdoshi@email.arizona.edu

Malini Ranganathan

School of International Service, American University, Washington, DC, USA;
malini@american.edu

Corruption as a Keyword in Late Capitalism

Corruption appears to be erupting everywhere. Newspaper headlines are saturated with high level scandals such as the Panama Papers, which indicted top business leaders for tax evasion and fraud. Meanwhile anti-corruption has mobilised millions seeking political accountability, an end to elite abuses of power, and substantive democracy through movements from the "Arab Spring" to the "Indian Summer" in 2011. Corruption allegations have toppled heads of state from South Korea to Brazil to Italy and taken centre stage in electoral contests across the world, including both the 2016 Trump and Sanders campaigns in the US. Though not new, what is significant about the contemporary conjuncture is the frequency and scale of outrage over corruption across the world. Today's concern with corruption is not limited to the so-called "Third World" where it has historically been posited as the norm that besets venal states and where anti-corruption efforts have served as a disciplinary tool to scale back government. Rather, corruption is being spotlighted in the heart of ostensibly clean and transparent western market economies. This shift seems to signal an instantiation of what Jean and John Comaroff (2011) refer to as Euro-America "evolving" towards the ostensibly degenerate societal dynamics of Africa and other parts of the South.

Why should radical geographers care about corruption politics? Today's corruption politics reveal the limits of economistic approaches of international development institutions and rational choice theorists who define corruption as the "abuse" of public or entrusted power for private gain—something that is considered aberrant to healthy capitalism. While dominant scholarly discourse posits a clean separation between the state and market and between public and private realms as necessary for the elimination of corruption, a critical geographical lens would see these divides as inherently blurred in actually existing "normal" capitalism, even—or especially in—advanced economies. What is dismissed as the normal workings of capitalism in one instance, then, is called corruption in the next. We thus posit that corruption is not a fixed set of practices, but rather an interpretive rubric that serves to make sense of and distinguish what is ethical or not,

Keywords in Radical Geography: Antipode at 50, First Edition. Edited by the *Antipode* Editorial Collective.
© 2019 The Authors/Antipode Foundation Ltd. Published 2019 by John Wiley & Sons Ltd.

what is harmful or not, and what matters or not (and to whom) in the ordinary processes of wealth accumulation and dispossession that define capitalism as we know it. We suggest that corruption is a capacious and slippery language put to a variety of opportunistic uses. Ironically, talk of corruption may be wielded by those who are most guilty of it. In some instances, corruption may also serve as a tool for progressive actors to critique capitalist forces, as we have shown through empirical work in Mumbai and Bangalore (Doshi and Ranganathan 2017). We illustrate our argument with the story of a transnational real estate deal involving US president Donald Trump who has earned a reputation for both corruption and anti-corruption.

Corruption, Capitalism, and the Mumbai Trump Towers

A recent journalistic exposé by Anjali Kamat (2018) zeroes in on Mumbai, one of the most expensive real estate markets in the world, where skyscrapers donning the Trump brand name have become mired in scandal. The story of the Mumbai Trump towers is both exceptional and ordinary for Indian cities. The towers are among dozens of luxury high rises and malls that have transformed the face of Mumbai's former industrial district of Lower Parel where thousands of acres of publicly owned textile mill properties have been siphoned off for redeveloped through closed-door negotiations since the 1990s. Yet the involvement of the US president has allowed for the story to gain traction in international media circuits and shed light on common but ethically questionable practices involving public officials, developers, and transnational capital. Kamat painstakingly reviews the myriad irregularities and dubious dealings of local developers, investors, politicians, and the Trump family. The project was launched with fanfare in 2010 through a licensing agreement between the Trump group and Rohan Developers Private Limited in Mumbai, which enabled the US president to claim to be free of conflicts of interest and the developer the right to use the Trump brand in exchange for millions in royalties. The veneer of the luxury tower began to be tarnished as residents of the older building set to be demolished to make way for the tower began protesting the project, demanding on-site compensation and reporting threats by the developer. The recalcitrant residents were soon removed by a mysterious fire. Such practices of intimidation are not unique to this project; one of the key developers involved in the Rohan company, Harresh Mehta, is well known in Mumbai for amassing fortunes on slum redevelopment schemes in which land has been secured through bribing officials, fraudulent consent documentation, and intimidation of residents refusing to consent to projects. Mehta also has close family-like connections to the state's most powerful elected officials, architects, and financiers, many of whom affectionately call him *Harreshbhai* (Brother Harresh). Next a number of dangerous and illegal infractions of permissible construction area and fire safety regulations revealed fraudulent permits were cleared by paying off officials. These blatant illegalities created a liability for politicians facing an avalanche of allegations during a period of intense anti-corruption mobilisation in India in 2011 (the aforementioned "Indian Summer"). As a result, the first tower has since been abandoned. A second golden-hued Trump tower also steeped in corruption was launched in 2013 and is now set to be released to investors by the end of 2018.

The Mumbai towers are not the only instance of murky real estate dealings involving conflicts of interest with the US president. Since 2007, the Trump organisation has made more deals in India than anywhere else outside of the US as Kamat notes. Several other projects in and around Mumbai and Delhi involving the Trump group, public officials, and developers are also embroiled in corruption investigations and lawsuits for illegal land acquisition, bribery, fraud, money laundering, tax evasion, and intimidation. Donald Trump and Trump Jr have played an active role in bolstering deals that have been threatened by such violations. They have found willing allies among members of India's right-wing Hindu Nationalist Bharatiya Janata Party who promise to "make Mumbai great again" through vigorous corporate-driven development and hardline actions against Muslims and other ostensibly anti-national groups.

Though on the surface the case of the Mumbai Trump towers appears to be another sensationalist story of Trumpian exploits, it illustrates key aspects of corruption discourse and power under contemporary global capitalism that we seek to highlight in our definition of corruption. First the towers exemplify commonplace practices in urban land use and development throughout the world that blur the boundaries between state and society, legality and illegality, and public and private. Where does developer power end and state autonomy start? Where do the duties of the president of the US to his citizens give way to his wealth amassing empire? The meshing of private and public interests and identities in the advancement of plunder, after all, are not supposed to be the logics by which liberal democratic and capitalist polities work. And yet, they are. Such dynamics are nothing new to scholars of colonialism and post-colonial development in Asia, Africa, and Latin America where indirect rule through private and so-called customary agents and "conflicts of interest" on the part of colonial agents were always and continue to be central to wealth extraction and accumulation (Mamdani 1996; Scott 1969; Watts 2003). Urban theorists (Roy 2005) have conceptualised similar dynamics focusing on the notion of "informality"—practices of governing territory, claiming and using land, providing services, and organising labour that exceed the boundaries of law, regulation, and public–private spheres. Informality is commonly associated with the urban poor who endure highly precarious and often criminalised housing, livelihoods, and access to services because they lack property titles or documentation. Yet it is clear from the Trump story and countless others, not just in India but in Lagos, Jakarta, and Ho Chi Minh City, that informality is also a key logic in the domain of elite projects and transnational business more broadly. In general, high-priority projects like the towers that are riddled with regulatory fraud, bribery, money laundering, and criminal violence often enjoy impunity and tacit acceptance through *ex post facto* legalisation. Once it is built, it is untouchable.

By highlighting hand-in-glove collusions between the state and favoured capitalist actors and elite informality as part and parcel of actually existing capitalist development, the story of Mumbai's Trump towers complicates the ubiquitous assumption that corruption and its logics of informality are endemic to the Third World where capitalism has "not yet" arrived in its proper form. Such notions of corruption have a history. While 17th and 18th century European thinkers expressed concern over the fundamental corrupting force of speculative finance, this perspective shifted to normalise such practices (De Goede 2005; Pocock 2016). Specific strands of mid-20th century modernisation theory came to see corruption as a feature of "traditional societies" that would disappear with capitalist modernisation, all the while eliding the effects of European colonialism on regimes of plunder and influence peddling. Corruption later emerged as a problem to be solved by international development agencies in the late 1980s and early 1990s. Thus followed a decade of aggressive efforts in the global South to "fix" corruption-ravaged economies through neoliberal, market-oriented paradigms of privatisation and deregulation (Wedel 2012). The results were often disastrous, particularly in Latin America, where Cold War geopolitics combined with aggressive neoliberal policies and market-oriented "anti-corruption" efforts that served to further entrench criminal and patronage networks (Brown and Cloke 2004).

Crucially, most mainstream anti-corruption reformers posited the US and European countries as free of corruption because they had successfully instituted a clean break between market and state. Yet new marriages of public–private interests flourished in the West, such as through campaign financing, corporate lobbying, tax breaks, and shell companies in real estate, to name just a few modes of what has been deemed as "legal corruption" (Kaufmann 2001). Still the term "corruption" rarely appeared in Western media, electoral campaigns, or public discourse outside of exceptional cases. Meanwhile, in countries like India experimenting with market reform, state kick-backs to the private sector, massive money in elections, and blurred roles between the private and government sectors ballooned in the post-reform era. Nonetheless market reformers and elites decried corruption as a means of disciplining lower rungs of the state and undermining poor and lower caste groups who had begun to claim power through changing electoral regimes (Teltumbde 2012).

Returning to the Trump towers case, a second notable aspect that demonstrates our larger argument is that the practices that would otherwise be identified as "corruption" (but are seldom) and the blurring of state and market roles also serve to advance capitalist accumulation by dispossession, in this case through real estate pilfering by a nexus of state and capital power brokers. Indeed, Marxist scholars have long highlighted the rapacious and often fraudulent land and resource plundering that undergirds the creation and ongoing maintenance of market economies (Harvey 2003). The Trump tower case thus highlights the class struggles inherent in urban development when the informal land uses of the poor are criminalised, through slum evictions, for instance, while the state sanctions and condones the legal and illegal land grabs of the elite. As we have demonstrated elsewhere, anti-corruption movements among the poor and lower middle classes have called out such hypocrisy in their struggles against dispossession (Doshi and Ranganathan 2017).

Relatedly, our final point exemplified by this case is that discourses of corruption are ultimately about making meanings and ethical assessments about the contemporary political and economic order. There comes a tipping point, perhaps, when normalised processes of actually existing capitalism are finally called out as corruption. In other words, talk of corruption sometimes provides a moral, albeit subtle, indictment of capitalism. Corruption stories are thus tricky. On the one hand, Kamat's exposé of the Mumbai Trump towers reveals the corrupt underpinnings of everyday capitalist development where state agents not only serve the interests of elites, but they often occupy the role of investors and developers who enjoy impunity and regulatory laxity. Thus, for some, capitalism itself may be seen as a corrupt system. On the other hand, corruption is also deployed as a normative discourse about the abuse of power. In these instances, actions such as the blurring of public and private boundaries or the infraction of accepted norms of legality are always implicitly positioned relative to a perceived normal or previously "uncorrupted" state of affairs. Thus, even as corruption talk serves to critique the business-as-usual of capitalism, it also may serve to bolster the notion that a better and purer capitalism may be achieved.

Finally, corruption discourse has recently also been deployed in the service of ethno-nationalist, populist and revanchist capitalist projects steeped in racialised oppression. That officials themselves mired in corruption have managed to capture public support through anti-corruption promises (think Trump's "drain the swamp" campaign) attests to the potency and malleability of corruption discourse.

In sum, corruption talk plays a critical political role in distinguishing which agents and boundary-blurring transactions are designated immoral and which are seen as acceptable. Thus, idioms of corruption provide a revealing lens into the spatial struggles of contemporary global capitalism and their attendant cultural politics of racial-ethnic supremacy, patriarchy, and nationalism. Diverse and interconnected symbolic, material, and territorial formations of power—forged through race, class, gender, and other hierarchical relations of difference—are especially influential in determining which actions, places, and bodies are deemed corrupt and which publics are imagined to be harmed by corruption at different historical and political-economic junctures. Critical to this understanding is how and when certain practices are thrown into question and become deviant or "corrupt". Put simply, corruption is a narrative that does work in the struggle over power, territory, and resources. Ultimately, corruption discourse is a terrain of conjunctural struggle in the Gramscian sense. We must pay attention not simply to the "fact" of corruption, but rather *how* corruption talk is being used to suit different agendas. Radical scholars would do well to pay attention to the political life of this keyword.

References

Brown E and Cloke J (2004) Neoliberal reform, governance, and corruption in the South: Assessing the international anti-corruption crusade. *Antipode* 36(2):272–294
Comaroff J and Comaroff J (2011) *Theory From the South, or, How Euro-America is Evolving Toward Africa*. Boulder: Paradigm
De Goede M (2005) *Virtue, Fortune, and Faith: A Genealogy of Finance*. Minneapolis: University of Minnesota Press

Doshi S and Ranganathan M (2017) Contesting the unethical city: Land dispossession and corruption narratives in urban India. *Annals of the American Association of Geographers* 107(1):183–199

Harvey D (2003) *The New Imperialism*. Oxford: Oxford University Press

Kamat A (2018) Political corruption and the art of the deal. *The New Republic* 21 March. https://newrepublic.com/article/147351/political-corruption-art-deal (last accessed 4 June 2018)

Kaufmann D (2001) Rethinking the fight against corruption. *Brookings* 29 November. https://www.brookings.edu/opinions/rethinking-the-fight-against-corruption/ (last accessed 6 June 2018)

Mamdani M (1996) *Citizen and Subject: Contemporary Africa and the Legacy of Late Colonialism*. Princeton: Princeton University Press

Pocock J G A (2016) *The Machiavellian Moment: Florentine Political Thought and the Atlantic Republican Tradition*. Princeton: Princeton University Press

Roy A (2005) Urban informality: Toward an epistemology of planning. *Journal of the American Planning Association* 71(2):147–158

Scott J C (1969) The analysis of corruption in developing nations. *Comparative Studies in Society and History* 11(03):315–341

Teltumbde A (2012) Caste in the play of corruption. *Economic and Political Weekly* 47(47/48):1011

Watts M (2003) Development and governmentality. *Singapore Journal of Tropical Geography* 24(1):6–34

Wedel J R (2012) Rethinking corruption in an age of ambiguity. *Annual Review of Law and Social Science* 8:453–498

Counterhegemony

Andrea Gibbons

School of Health and Society, University of Salford, Salford, UK;
a.r.gibbons1@salford.ac.uk

Counterhegemony—a growing and diverse alliance of forces not just in resistance to the existing, oppressive alliance ranged against them, but exploding beyond it in a great scattering of sparks. New visions, new ways of being, burning through the limits of what can be thought. New and better worlds collectively imagined and created through sedimentations of small victories or with the rubble remaining after a victorious revolution. Whether the old world must first be taken over, smashed, or simply shed like a skin may be a point of contention. But the old world has had its day.

In this, counterhegemony somewhat belies its name. To be in true resistance it must think, imagine and dream beyond hegemony, breaking through its false limits. Yet it is a word belonging to our long war of position. For those like Pablo González Casanova (1984), writing optimistically in the midst of Central American revolutions of the 1980s, there is little value in the idea of "counterhegemony". In the wars of manoeuvre in Nicaragua and El Salvador, freedom fighters equipped with AK-47s and machetes required a theorising of the ways in which the left could build a shattered hegemony anew. Roque Dalton sang of love, denounced dictators and first dreamed then fought for revolution and the creation of a new world (as did Agostinho Neto and so many other poets in the face of colonial, neo-colonial and imperial forces). The "counter" existed in El Salvador both in open armed resistance in the turmoil of an old hegemony failing, and in the guerrilla schools teaching literacy and *conscientisizacion* beneath the mountain trees, developing better tools than guns for the new world to be created. Even in battle, the counter could not be defined simply by what it stood against, but its violence made stark the sides and the stakes. The limitations of our language set utopian struggle for a better world in definition against "hegemony" as it stands in the present. Yet although counterhegemony is founded in resistance to an oppressive present, it cannot remain limited to it.

The richness of hegemony as a term in theorising revolution comes initially from Antonio Gramsci, although Perry Anderson (1976) shows its previous use by intellectuals of the Russian Revolution and the Third International. A trade-union organiser, journalist and party activist, Gramsci lived through a time when it seemed the revolutionary moment opened up by the Bolsheviks might spread through the militant industrial councils of 1920s Turin. Unlike González Casanova, he only found time and space for theorising hegemony in a moment of that movement's most profound defeat. He wrote his notebooks in a fascist prison, struggling to understand not just the reknitting of a new, increasingly violent

Keywords in Radical Geography: Antipode at 50, First Edition. Edited by the *Antipode* Editorial Collective.
© 2019 The Authors/Antipode Foundation Ltd. Published 2019 by John Wiley & Sons Ltd.

hegemony under Mussolini and Hitler, but also the politics of the rural, poverty-stricken Mezzogiorno, where he grew up, in relation to the wealthy, industrialising North of Italy.

In reaching to understand these events through a Marxist framework already 80 years old, Gramsci opened up a new way to think theoretically about how power was built and maintained. The revolutionary struggle is not as simple as one class exerting force against another, an exploited majority of workers and producers in constant opposition to a wealthy minority owning the means of production. Instead the forces of domination are a forged conglomeration, an alliance of groups and forces that come together in their own interest to support those in power. Where power is maintained purely through force it is fragile. Hegemony only exists where there is a balance of force and consent enforcing rule over the majority, expressed and buttressed through media, associations, and popular culture. This balance between domination and acceptance brings about a construction of common sense that works to limit imagination itself to "a particular way of seeing the world and human nature and relationships". Raymond Williams (1983:145) explains what this means for how we understand revolution, which requires "the overthrow of a specific hegemony ... an integral form of class rule which exists not only in political and economic institutions and relationships, but also in active forms of experience and consciousness". A new, counter-, alternative hegemony must be created to replace the old, both internally and externally—the war of position must be carried out on economic, political and cultural fronts.

This expansion of revolutionary activity and agency along axes of culture, race, gender, nationality etc.—characteristics that cannot be reduced to class—allows the many complexities of oppression and strategies of control to be understood as they articulate with economic and political structures. It removes a reductionist blockage and spatialises revolution. Counterhegemony has a location—the subaltern. The most well known development of this aspect of Gramsci's theories undoubtedly emerged from the Subaltern Studies Group in South Asia, founded in 1982. As Veena Das (1989:312) writes, part of their importance lay in establishing the "centrality of the historical moment of rebellion in understanding the subalterns as subjects of their own histories". They are the historical agents of change.

Also wrestling with these issues was the subaltern studies group in Latin America, translating texts and undertaking their own reclamation of voice and history through the work of, among others, Silvia Rivera Cusicanqui and Rossanna Barragan. In 1997 they published translations of key texts of the South Asian Subaltern Studies Group, but their work had long drawn on the same concepts, theorising subaltern subjects in historical moments of counterhegemonic rebellion (see also Mignolo 2012). Stuart Hall in the UK and Laura Pulido in the US worked to understand the subaltern positionality internal to what Paul Gilroy (1999) calls the "overdeveloped" nations, as well as external to it within the global South. These insights brought together with those emerging from Black and Indigenous feminisms—Angela Davis, bell hooks, Patricia Hill Collins, Andrea Smith, for example—signal a possible definition of subaltern as a category in which any one individual's place (or positionality) must be understood as a relational constellation rather than a fixed identity, shaped by the intersectionalities of class, race, gender, sexuality, ability, and immigration status, language and all the multitude of differences used as bases for oppression. In the balance of force and consent, these are the positionalities upon which the full force of domination is often expended. Cast outside of the community of consent, the demonisation of the poor, the working class, the immigrant, the young people of colour serves to bind the alliance of privileged groups ever tighter, especially for those whose relative privilege is precarious. For this very reason, it is not simply from precarity but from subaltern locations that the clearest view of hegemony can be found.

Ernesto Laclau and Chantal Mouffe (1985) describe the possibilities of counter-hegemonic alliance working across difference, multiple identities coming together in chains of equivalence to recreate the world. Theorised through Derrida, difference becomes rooted wholly in discourse. Nothing is fixed, any alliance is possible. For those drawing on the work of subaltern studies, however, Gramsci's own method demands a deep rooting in the concrete histories of place and specific hierarchies of oppression. In allowing for the cultural constructions of race and gender, there is still no denying their terrible, death-dealing materiality (Gilmore 2002). Theorising counterhegemony must escape abstraction, stretch to understand the world as it is (and long has been)—global, dominated first by Europe and now by the United States. Counterhegemony has a location—in the camps and segregated communities of colour in the North, and across the Global South. The overdeveloped nations still appropriate a majority of global resources, still dominate along political, economic and cultural fronts through a combination of force and consent.

Charles Mills (1997:18) writes that the global system of white supremacy contains "an epistemology of ignorance, a particular pattern of localized and global cognitive dysfunctions (which are psychologically and socially functional)". Such ignorance produces the "ironic outcome that whites will in general be unable to understand the world they themselves have made" (Mills 1997:18). Racial and patriarchal logics articulating with the logics of capitalism have worked to place the subaltern outside of concepts of community, citizenship, and at times of humanity itself. This process has taken place as a project of (neo)colonial and imperial domination through the work of the World Bank, IMF, WTO alongside black ops, pre-emptive strikes and illegal processes of extradition for imprisonment and torture. But it has also worked within national boundaries. The continuing murderous assault on black bodies in the United States displays graphically the deadly force applied with impunity against those considered outside the (white) communities of consent. Likewise, environmental racism has long inflicted prosperity's costs of pollution and waste upon subaltern communities least able to bear them. As all of us hurtle towards global environmental catastrophe, these longer-term costs of global transformation are now falling due. Again, it is the subaltern who is already paying without ever having enjoyed prosperity.

It is not from the racialised and gendered positions of privilege that such hegemonic formations can be best understood, but from the subaltern positionality with its daily experience of hegemonic violence and oppression. To be of the community of consent is to be subject to a kind of blindness that must be actively undone. This is not to argue that blindness belongs to the community of consent alone, but only that it is maintained through a weight of both conscious and unconscious privilege that must be painfully acknowledged and picked apart. This is a process which can be undertaken in choosing a new positionality in relation to the interlocking systems of oppression, but it must be undertaken in all humility as a collective project, with a willingness to commit to, and to learn side by side with, others in struggle. Paulo Freire (1993) describes just such a process in *Pedagogy of the Oppressed*, by which the subaltern working collectively and in dialogue names reality in order to transform it. This requires of them also a conscious choice, but to them belongs the ability to see the world most truly.

We cannot also forget that in many colonial spaces where domination has ruled above consent, strong connections to culture, as well as political and economic structures never fully incorporated within the hegemonic, colonial project of European modernity, continue to develop and grow. Out of these reservoirs are emerging movements of vibrancy and hope distinct from Europe's old left (Mbembe 2015; Rivera Cusicanqui 2010; Santos 2016). We must welcome the new organic intellectuals, the ferment of theory and praxis emerging from subaltern communities and countries in a multitude of languages and idioms. This is where counterhegemony will be built and the new world ushered in.

References

Anderson P (1976) *The Antinomies of Antonio Gramsci*. New York: Verso

Das V (1989) Subaltern as perspective. In R Guha (ed) *Subaltern Studies VI* (pp 310–324). New Delhi: Oxford University Press

Freire P (1993) *Pedagogy of the Oppressed*. New York: Continuum

Gilmore R W (2002) Fatal couplings of power and difference: Notes on racism and geography. *The Professional Geographer* 54(1):15–24

Gilroy P (1999) *The Black Atlantic: Modernity and Double Consciousness*. London: Verso

González Casanova P (1984) *La Hegemonia del Pueblo y la Lucha Centroamericana*. Buenos Aires: Editorial Contrapunto

Laclau E and Mouffe C (1985) *Hegemony and Socialist Strategy*. New York: Verso

Mbembe A (2015) *On the Postcolony*. Johannesburg: University of Wits Press

Mignolo W D (2012) *Local Histories/Global Designs: Coloniality, Subaltern Knowledges, and Border Thinking*. Princeton: Princeton University Press

Mills C (1997) *The Racial Contract*. Ithaca: Cornell University Press

Rivera Cusicanqui S (2010) *Ch'ixinkax utxiwa: Una reflexión sobre prácticas y discursos descolonizadores*. Buenos Aires: Tinta Limon

Santos B d S (2016) *Epistemologies of the South: Justice against Epistemicide*. Abingdon: Routledge

Williams R (1983) *Keywords: A Vocabulary of Culture and Society*. London: Fontana

Decolonial Geographies

Michelle Daigle

Department of Geography, University of British Columbia, Vancouver, BC, Canada;
michelle.daigle@geog.ubc.ca

Margaret Marietta Ramírez

Stanford Arts Institute, Stanford University, Stanford, CA, USA;
mmrez@stanford.edu

If I am against colonialism in particular, then I must also be against colonialism in general. (Maracle 1996:123)

I am struggling to find the language for this work, find the form for this work. Language and form fracture more everyday. (Sharpe 2016:19)

Throughout our five years of collaboration, we have been building a language for decolonial geographies. Yet, as reflected through Christina Sharpe's quote above, there is an impossibility to defining this work, for there is no clearly defined structure that neatly traces and binds decolonial geographies. Harkening the words of Gloria Anzaldúa (Moraga and Anzaldúa 1981), we are building this bridge as we walk; these are constellations in formation.

Part of this impossibility lies in the incompatibility of decolonial geographies with colonial knowledge projects. It is counter-intuitive to attempt to classify or systematise Indigenous, Black and other cultural knowledge systems into neat synopses. We also acknowledge that we, a cis Mushkegowuk (Cree) woman and a cis Xicana settler woman, do not have the authority to proclaim a definition of decolonial geographies that will steadily travel through time and space. We both operate within the North American context, and our theorising is rooted and routed in the places and genealogies we inhabit: Michelle from her home territory of the Mushkegowuk nation located in what is now commonly known as northern Ontario, Canada, and who has been living as an uninvited visitor on Coast Salish territories for the past decade; and Magie from the Ohlone territories now called Oakland, California, a place defined by its Blackness for the past 70 years where Black residents are currently undergoing rampant dispossession. We understand decolonial geographies to be a diverse and interconnected landscape grounded in the particularities of each place, starting with the Indigenous lands/waters/peoples from which a geography emerges, and the ways these places are simultaneously sculpted by radical traditions of resistance and liberation embodied by Black, Latinx, Asian and other racialised communities. The decolonial shapeshifts depending on the land you stand upon, including the differential decolonize desires layered into a place (Tuck and Yang 2012); this is therein the central anchor and tension of decolonial geographies that informs our thinking and praxis.

Keywords in Radical Geography: Antipode at 50, First Edition. Edited by the *Antipode* Editorial Collective.
© 2019 The Authors/Antipode Foundation Ltd. Published 2019 by John Wiley & Sons Ltd.

In the following, we theorise decolonial geographies as a constellation in formation. We trace this language from Nishnaabeg scholar, activist and artist Leanne Betasamosake Simpson's (2017) work on constellations of co-resistance, while our spatial framing stems from bringing Indigenous geographies into dialogue with the geneologies of Black geographies envisioned by Katherine McKittrick and Clyde Woods (2007). We situate decolonial geographies within embodied theories and praxes of liberation to elucidate the connective fabric of various decolonial struggles. We conclude by elaborating on constellations in formation, as embodied in the present, to envision radical spatial visions of the future.

Spatial Weavings

Constellations exist only in the context of relationships; otherwise they are just individual stars. (L.B. Simpson 2017:215)

Drawing from Leanne Betasamosake Simpson, we conceptualise decolonial geographies as constellations of co-resistance and liberation. Simpson anchors her thinking on constellations within Nishinaabeg cosmologies, and draws on Cree media maker and writer Jarrett Martineau's (2015) work on affirmative refusal, as well as Stefano Harney and Fred Moten's (2013) work on fugitivity, to reflect on how constellations of co-resistance provide a flight "out of settler colonial realities into Indigeneity" (L.B. Simpson 2017:217). Centring relationship-building across Indigenous, Black and other racialised communities, she asks Indigenous peoples who they/we should be in constellation with. In doing so, she cautions that Indigenous resurgence risks "replicating anti-Blackness without solid, reciprocal relationships with Black visionaries who are also co-creating alternatives under the lens of abolition, decolonization, and anti-capitalism" (L.B. Simpson 2017:228–229).

We understand constellations in formation as the embodied knowledge of Indigenous peoples coming into dialogue and relationship with those of Black and other dispossessed peoples. For Indigenous peoples, this includes a knowledge of the place, the lands and waters, that they have roots in and have been the caretakers of for generations (Coulthard and Simpson 2016; L.B. Simpson 2014). This includes urban space, reaffirming that these sites are part of Indigenous geographies despite relentless reframings by white proprietary logics and practices. As Simpson states, decolonial geographies must form "as place-based constellations in theory and practice", to foreground Indigenous intelligence that is generated from the ground up (L.B. Simpson 2017:231). These grounded practices remain a crucial tenant to Indigenous geographies as they are pedagogical pathways into relationalities with the human and non-human world. Embodied intelligence systems, Simpson says, make up the radical resurgence that has always existed across diverse and interconnected Indigenous landscapes.

Indigenous geographies are spatially woven in relation to those of other dispossessed peoples, and Black geographies in particular inform our theorising of decolonial geographies. McKittrick and Woods rendered the language of Black geographies, building a rich body of spatial theory that both locates the plantation economy as a central node of racial capitalism in the Americas, and demonstrates how "the geographic knowledges that black subjects impart ... inform black lives" (2007:6–7). As McKittrick articulates, "the knotted diasporic tenets of coloniality, dehumanization and resistance" structure a Black sense of place, "wherein the violence of displacement and bondage, produced within a plantation economy, extends and is given a geographic future" (2011:949). We situate the theorising of decolonial geographies, drawing from McKittrick and Woods' spatial method, from this notion that space is perceived and produced differentially. A constellation in formation thus takes the shape of multiple stars, multiple roots and routes—embodied conceptualisations of space coming into formation in pursuit of a decolonial vision.[1]

The decolonial, we contend, is an affirmative refusal of white supremacy, anti-blackness, the settler colonial state, and a racialised political economy of containment, displacement and violence (Coulthard 2014; A. Simpson 2014a; L.B. Simpson 2017). As Mohawk scholar Audra Simpson (2014a) elucidates, a politic of refusal fundamentally repudiates colonial dispossession and violence on Indigenous lands and bodies and makes up the very foundation of Indigenous nationhood. As reflected through Indigenous and Black movements for liberation throughout time and space, this refusal requires the dismantling of systems of oppression such as the racial capitalist carceral system, state militaries, white proprietary landholdings and colonial capitalist resource exploitation (Fabris 2017; Gilmore 2007; Hunt 2015; Maynard 2017; Robinson 1983). Refusal is also a comprehensive politics of resistance that is vigilant to the ways interconnected violences against racialised lands, spaces and bodies get reproduced in the most mundane and practical ways (Cowen 2017; Million 2013; A. Simpson 2016).

Refusal *is* liberation from the violent fractures of settler colonialism and white supremacist structures. Yet, liberation also builds on refusal through a resounding affirmation and embodiment of alternative relationalities. We contend that abolitionism, a liberatory praxis rooted in the Black radical tradition, needs to be understood in relation with decolonisation when weaving decolonial geographies in the Americas (Anderson and Samudzi 2018; Johnson and Lubin 2017; Kelley 2002). Fuelled by Black geographies, abolitionism is a struggle for a society free of prisons and policing; as a movement it seeks to uproot the anti-blackness inherent in the white supremacist power structure and envisions inter-relational alternatives that are premised on the affirmation of Black humanness (Davis 2003; Harney and Moten 2013). As Anderson and Samudzi state, "Black liberation poses an existential threat to white supremacy because the existence of free Black people necessitates a complete transformation and destruction of the settler state" (2018:8). Meanwhile, Indigenous movements for radical resurgence call for a comprehensive transformation of the settler colonial present guided by Indigenous political and legal orders that have been renewed throughout time and space in spite of colonial dispossession and violence. Together these movements envision a society beyond criminalisation, extraction, militarised borders and violence, by demanding and embodying structures and relationships centred on principles of restorative and transformative justice and relational accountability (Berger 2014; Kaba 2012; Loyd et al. 2012).

Drawing on the brilliance of Indigenous and Black feminist, queer and Two-Spirit theorists, we understand liberation as a refusal of heteropatriarchal violence and the surveillance and criminalisation of gender variance and intimacies in Indigenous and Black communities (Crenshaw and Ritchie 2015; Hunt 2015; Maynard 2017; Women's Earth Alliance and Native Youth Sexual Health Network 2016; L.B. Simpson 2017). Although in many ways distinct, Indigenous and Black women, youth, queer, Two-Spirit and trans individuals continue to be subjected to interconnected forms of gendered colonial and anti-black violence in settler colonial contexts, thus bodily sovereignty is essential to liberation, particularly as it is a crucial site of knowledge and self-determination (Simpson and Maynard 2918). As Audra Simpson (2014b) powerfully asserts, Indigenous women's bodies *are* political orders and by their very existence, pose a threat to the settler colonial order. Along similar lines, Latinx feminism theorises how Latinx peoples imagine and create decolonial spaces from the same colonial fissures that dismember their lands and bodies (Anzaldúa 1987; Pérez 1999; Sandoval 2000). Colonialism created the fissured mestizx body, and Latinx migration across colonial borders further complicate their relationship to land and place. Latinx queer and feminist writing emphasises how the Latinx body refuses the violence of the colonial border through its embodied in-betweenness, and how decolonial possibilities emerge from the in-between or borderland space that Latinx peoples inhabit. As Alan Pelaez Lopez powerfully articulates, the X in Latinx is a visible wound that forces the Latinx diaspora to contend with the legacies of settler colonialism, anti-blackness, the femicide of cisgender and trans women, and a collective inarticulation due to colonial histories of erasure. As Pelaez Lopez (2018) explains, "the X is attempting to speak to the violences of colonization, slavery, against women and femmes, and the fact that many of us experience such an intense displacement and silence that we have no language in which to articulate who we are", encouraging Latinx peoples to envision liberatory futures that also centre Black, Indigenous and Trans peoples.

In weaving a fabric of decolonial geographies in the North American context, it is necessary to consider these multiple geographies of Black, Brown and Indigenous peoples in relation to one another to illuminate the interconnected struggles for land and space, and how these become foundational sites for self-determination and freedom. In the following we consider how the interconnected terrain of decolonial geographies is embodied as constellations in formation.

Constellations in Formation

In the wake of the Trump administration's anti-Muslim travel ban, Indigenous activists such as Melanie Yazzie (Dine) and Nick Estes (Kul Wicasa, Lower Brule Sioux Tribe) embodied powerful assertions of Indigenous jurisdiction at the LAX airport by demanding "no bans on stolen lands" (Monkman 2017). Led by the Tongva people, the original and lawful caretakers of this area, they proceeded to carry out welcoming ceremonies for a number of refugee and Muslim families. In doing so, they made legible the interconnected struggles of xenophobic racism, settler colonial policing of borders, and the ongoing dispossession of Indigenous lands and bodies. Indigenous presence was crucial as anti-Muslim protests across the country risked erasing Indigenous authority. As Estes stated:

> [i]t means that the United States, as a settler nation, does not have the final say on who or what comes into the country because it's not theirs to own ... When we do that as Indigenous people, it's reclaiming our sovereignty, our citizenship, and more importantly our kinship. (Quoted in Monkman 2017)

Holding welcoming ceremonies at LAX airport, Indigenous activists forged constellations of co-resistance through an ethic of relational accountability to their lands and peoples. This ethic is a mutual one in which Indigenous peoples are urged to practice what some scholars such as Glen Coulthard (Dene) (2017) have called "radical hospitality". In this way, the neoliberal economic logics of the settler colonial border and what Iyko Day (2016:172) calls the "contradictory promise of settler colonial hospitality" are refused. Radical hospitality compels Indigenous peoples to welcome other dispossessed peoples into their/our homelands, according to their/our own laws, as they become displaced through the violence of racial capitalism. However, as South Asian activist and writer Harsha Walia (2012) argues, racially displaced migrants' struggle for justice must also be in solidarity with Indigenous resistance against settler colonialism, and embody a thorough and comprehensive accountability to place and its people.

As Coulthard (2017) reminds us, these constellations, like the ancestral knowledges and political traditions that they are based on, are nothing new. Through tracing the global anti-imperialist dialogue that emerged between Black and Red political thought in the 1960s and 1970s, he elucidates the spatio-temporality of radical critiques of colonialism and visions of liberation that have co-emerged across Indigenous and Black geographies. Recent dialogues on Cedric Robinson's theory of racial capitalism build on these understandings by not only tracing the global and interconnected terrain of radical thought and resistance across transatlantic Black geographies (Kelley 2017), but by, crucially, reasserting the African continent as Indigenous space.

Constellations are in formation all around us, re-envisioning and re-embodying a politics of place by interweaving spatial practices of resistance, refusal and liberation. These historical and always emerging relationships across decolonial struggles transcend colonial boundaries by disclosing the interconnected terrain of racial capitalism, colonialism and white supremacy from one space to the next. More than this, through the spatial concept of constellations, differentially situated peoples are renewing and creating futures that have always been present in their/our own communities. These spatial formations of resistance and creation draw from the histories and geographies of Black, Brown and Indigenous peoples to re-root and re-route toward more accountable relations. And this what we see as the heart of decolonial geographies: that these stars form constellations to guide us toward decolonial futures.

Endnote
[1] Our intention is not to foreclose the expansiveness, mobility and complexity of Indigenous geographies, nor the embodied land-based knowledge of Black and other racialised

peoples. Rather, our aim is to centre the spatial weavings of decolonial geographies as they take form on stolen and occupied Indigenous lands and waters.

References

Anderson W and Samudzi Z (2018) *As Black as Resistance: Finding the Conditions for Liberation.* Chico: AK Press

Anzaldúa G (1987) *Borderlands: La Frontera.* San Francisco: Aunt Lute

Berger D (2014) *Captive Nation: Black Prison Organizing in the Civil Rights Era.* Chapel Hill: University of North Carolina Press

Coulthard G S (2014) *Red Skin, White Masks: Rejecting the Colonial Politics of Recognition.* Minneapolis: University of Minnesota Press

Coulthard G (2017) "Fanonian Antinomies." Paper presented to Simon Fraser University, Vancouver, BC, Canada

Coulthard G and Simpson L B (2016) Grounded normativity / place-based solidarity. *American Quarterly* 68(2):249–255

Cowen D (2017) Infrastructures of empire and resistance. *Verso Books Blog* 25 January. https://www.versobooks.com/blogs/3067-infrastructures-of-empire-and-resistance (last accessed 15 August 2018)

Crenshaw K W and Ritchie A J (2015) *Say Her Name: Resisting Police Brutality against Black Women.* New York: African American Policy Forum

Davis A Y (2003) *Are Prisons Obsolete?* New York: Seven Stories

Day I (2016) *Alien Capital: Asian Racialization and the Logic of Settler Colonial Capitalism.* Durham: Duke University Press

Fabris M (2017) Decolonizing neoliberalism? First Nations reserves, private property rights, and the legislation of Indigenous dispossession in Canada. In M H Bruun, P J Cockburn, B S Risager and M Thorup (eds) *Contested Property Claims: What Disagreement Tells Us About Ownership* (pp 185–204). New York: Routledge

Gilmore R W (2007) *Golden Gulag: Prisons, Surplus, Crisis, and Opposition in Globalizing California.* Berkeley: University of California Press

Harney S and Moten F (2013) *The Undercommons: Fugitive Planning and Black Study.* New York: Minor Compositions

Hunt S (2015) "Violence, Law, and the Everyday Politics of Recognition." Paper presented to the Native American and Indigenous Studies Association (NAISA), Washington, DC, USA

Johnson G T and Lubin A (2017) *Futures of Black Radicalism.* London: Verso

Kaba M (2012) Transformative justice. *Prison Culture Blog* 12 March. http://www.usprisonculture.com/blog/transformative-justice/ (last accessed 15 August 2018)

Kelley R D G (2002) *Freedom Dreams: The Black Radical Imagination.* Boston: Beacon

Kelley R D G (2017) What did Cedric Robinson mean by racial capitalism? *Boston Review* 12 January. http://bostonreview.net/race/robin-d-g-kelley-what-did-cedric-robinson-mean-racial-capitalism (last accessed 15 August 2018)

Loyd J M, Mitchelson M and Burridge A (eds) (2012) *Beyond Walls and Cages: Prisons, Borders, and Global Crisis.* Athens: University of Georgia Press

Maracle L (1996) *I Am Woman: A Native Perspective on Sociology and Feminism.* Vancouver: Press Gang

Martineau J (2015) "Creative Combat: Indigenous Art, Resurgence, and Decolonization." Unpublished PhD thesis, University of Victoria

Maynard R (2017) *Policing Black Lives: State Violence in Canada from Slavery to the Present.* Winnipeg: Fernwood

McKittrick K (2011) On plantations, prisons, and a black sense of place. *Social and Cultural Geography* 12(8):947–963

McKittrick K and Woods C (eds) (2007) *Black Geographies and the Politics of Place.* Toronto: Between the Lines

Million D (2013) *Therapeutic Nations: Healing in an Age of Indigenous Human Rights.* Tucson: University of Arizona Press

Monkman L (2017) "No ban on stolen land", say Indigenous activists in US. *CBC News 2 February.* http://www.cbc.ca/news/indigenous/indigenous-activists-immigration-ban-1.3960814 (last accessed 15 August 2018)

Moraga C and Anzaldúa G (1981) *This Bridge Called My Back: Writings by Radical Women of Color.* Watertown: Persephone

Pelaez Lopez A (2018) The X In Latinx is a wound, not a trend. *EFNIKS* 13 September. http://efniks.com/the-deep-dive-pages/2018/9/11/the-x-in-latinx-is-a-wound-not-a-trend (last accessed 9 October 2018)

Pérez E (1999) *The Decolonial Imaginary: Writing Chicanas into History.* Bloomington: Indiana University Press

Robinson C (1983) *Black Marxism: The Making of the Black Radical Tradition.* London: Zed

Sandoval C (2000) *Methodology of the Oppressed.* Minneapolis: Minnesota University Press

Sharpe C E (2016) *In the Wake: On Blackness and Being.* Durham: Duke University Press

Simpson A (2014a) *Mohawk Interruptus: Political Life across the Borders of Settler States.* Durham: Duke University Press

Simpson A (2014b) "The Chief 's Two Bodies." Paper presented to the Annual Critical Race and Anticolonial Studies Conference, Edmonton, AB, Canada. https://vimeo.com/110948627 (last accessed 15 August 2018)

Simpson A (2016) "Reconciliation and its Discontents: Settler Governance in an Age of Sorrow." Paper presented to the University of Saskatchewan, Saskatoon, SK, Canada. https://www.youtube.com/watch?v=vGl9HkzQsGg (last accessed 15 August 2018)

Simpson L B (2014) Land as pedagogy: Nishnaabeg intelligence and rebellious transformation. *Decolonization: Indigeneity, Education, and Society* 3(3):1–25

Simpson L B (2017) *As We Have Always Done: Indigenous Freedom through Radical Resistance.* Minneapolis: University of Minnesota Press

Simpson L B and Maynard R (2018) Leanne Betasamosake Simpson in conversation with Robyn Maynard. *CKUT 90.3FM* 23 April. https://soundcloud.com/radiockut/leanne-betasamosake-simpson-in-conversation-with-robyn-maynard (last accessed 15 August 2018)

Tuck E and Yang K W (2012) Decolonization is not a metaphor. *Decolonization: Indigeneity, Education, and Society* 1(1):1–40

Walia H (2012) Decolonizing together. *Briarpatch* 1 January. https://briarpatchmagazine.com/articles/view/decolonizing-together (last accessed 15 August 2018)

Women's Earth Alliance and Native Youth Sexual Health Network (2016) *Violence on the Land, Violence on Our Bodies: Building an Indigenous Response to Environmental Violence.* Berkeley: Women's Earth Alliance / Toronto: Native Youth Sexual Health Network. http://landbodydefense.org/uploads/files/VLVBReportToolkit2016.pdf (last accessed 15 August 2018)

Digital

Jen Jack Gieseking

Department of Geography, University of Kentucky, Lexington, KY, USA;
jgieseking@uky.edu

I think I've lost my phone.

In the issuance of that sentence, did you just feel sorry for me? Frustrated? A little jealous? Delighted in my liberation? Regardless of the specific reaction, an affective and emotional pull likely just rippled through your body at the absence of the most mundane and totalising extension of self: the device and its digital affordances. Donna Haraway (1990) long ago dubbed the imbrication of the digital and the human as the "cyborg". Today, "the digital" is everywhere as the cyborg has become everybody/every body. Watches, phones, refrigerators, houses, cities, supply chain networks, and even periods of history have been dubbed "smart". Far from a resultant liberation, data and the algorithms used to collect, sort, organise, and analyse data have become more invasive and with profound ramifications upon democracy. Digital geographies have become more totalising to the point of being an essential component to the study and understanding of geography as both field and experience. Geography as a discipline bears special responsibility in critically examining and intervening in the spread of digital devices, platforms, and code as they infiltrate, define, and shape the spaces and experiences of all human and animal life.

The term "digital" embraces the data and algorithms, software and hardware, and the affective, political, economic, social, and physical effects on human bodies, objects, and spaces, as well as the structural oppressions and systems of power that are encoded into these elements. Sarah Elwood and Agnieszka Leszczynski (2018:629) define the "digital" as a term that covers "digital systems that encode, store and manipulate data; the forms of the material objects that mediate environments and human engagements with digitality; the structurings of everyday life through digital praxes; and the knowledges that secure and reproduce digitality" (see also Ash et al. 2016). These two (of many) definitions depict how "digital" is an encompassing term, framed across multiple scales, lenses, and approaches even in the field of geography alone (see also Wilson 2018). As "algorithmic citizenship" increasingly relies on codification (Bridle 2016), geography has responded with feminist, queer, critical race, postcolonial, and other critical perspectives (cf. Browne 2017; Cockayne and Richardson 2017; Datta 2018; Elwood and Leszczynski 2011, 2018; Kwan 2002). This scholarship has attended to projects related to data studies, algorithm studies, the internet of things, cybergeographies, spatial humanities and digital humanities, critical GIS and geoweb studies, geographies of the internet, communications, and telecommunications (ICT), social justice, digital ethics, and critical geographic thought, theory, and method.

Keywords in Radical Geography: Antipode at 50, First Edition. Edited by the *Antipode* Editorial Collective.
© 2019 The Authors/Antipode Foundation Ltd. Published 2019 by John Wiley & Sons Ltd.

Even before the "digital turn" came to geography (Ash et al. 2016), considerations of the digital and space were both a digital geography of everywhere, and a geography of digital everywhere. Rob Kitchin and Martin Dodge (2011) contend that "code/space" surrounds us in the mutual constitution of software and the spatiality of everyday life. The ubiquitous geographies of code/space reach everyone regardless of gender, race, class, or age through state satellite surveillance, using computers for homework (or the inability to use them or access the internet), cheap digital burner phones shared between poor families, online forms and background checks required for minimum wage jobs, digital transmission of funds from urban workers back to rural families, and Facebook state reconnaissance (with a quarter of the world as members) among other digital platforms. In fact, code, social media, and data now play a central role in shaping gender and racial identities, reinforcing sexist and racist stereotypes into all aspects of spaces ranging from the courtrooms and jails to local pharmacies and job applications (Browne 2017; Cheney-Lippold 2017; Noble 2018; Thatcher 2013).

In the early 2010s, the academy, politicians, and markets became fixated on data, specifically the promises of big data alongside conversations about open data and data justice (cf. Burns and Thatcher 2014; Dalton and Thatcher 2014; Johnson 2014). By the end of the decade, the public's obsession with algorithms and their effects (including artificial intelligence and machine learning) reached a point of "algorithmic fetishism" (Monahan 2018). For example, for those with access to smart phones and other GPS devices, spatial cognition has vastly decreased in the last decade due to the rise of and dependence on geoweb software and geolocation apps (Grabar 2014). In the midst of this transformation, the most-prized form of data for algorithmic output emerged as geodata. In other words, the data captured to represent our identity, transportation behaviours, health, purchasing preferences, and even our sexual preferences are deemed most valuable to researchers, government agencies, and corporations when they can be geolocated in real time. Critical GIS interventions encourage attention to the ethics, processes, and practices that afford the mapping of these and other data (Thatcher 2013, 2017; Wilson 2017).

Along with the geographical role in data analysis came the creation of "smart cities", whereby data and predictive algorithms promised and still promise to predict flows of traffic, speed up emergency responder pathways, and lessen crime. Revealing the depth and breadth in smart city interests, Ayona Datta (2018:1) addresses how India claims to be embarking upon a "smart urban age" in using the "'future' as a blueprint for social power relations in postcolonial urbanism". Geographers have also critically addressed issues of surveillance and privacy, securitisation and privatisation at stake in smart cities, as well as the heavy financial burden placed upon citizens to employ such technology (cf. Cowley and Caprotti 2018; Gabrys 2014; Kitchin 2014; Leszczynski 2016; Shelton et al. 2015).

GIS has also proved a core concern of the growing interdisciplinary field of the digital humanities. Historians, literary scholars, language scholars, and art historians eagerly adopted and adapted GIS technologies to craft cutting-edge mapping projects: walkthroughs of ancient cities and the first towns of freed slaves in South Carolina, and the personal histories of urban renewal and more recent urban evictions (Byrd 2015; Digital Scholarship Lab 2017; Anti-Eviction Mapping Project 2018). With the use of qualitative and quantitative methods in dialogue, projects in digital humanities promote continued interdisciplinary conversation, which can support significant headway for future digital geographic research (Gieseking 2018). Like the project of feminist digital geographies that seek to "unsettle the masculinist epistemologies undergirding much of digital work", digital humanists, like geographers, are at the front lines of colonial, white supremacist, ableist, heteronormative epistemologies evident in many digital projects as well (Elwood and Leszczynski 2018:630; see also Datta 2018; Gieseking 2017; Hamraie 2017; Hicks 2017; Noble 2018).

Most urgently, the "digital" requires attention to the space and time of everyday life in order to attend to the ways that the co-production of space, people, and the digital coalesce. Stephen Graham (2005) described many of these phenomena as "software-sorted geographies" to record how behaviour patterns are shaped through computer code intended to sort "good" consumers and citizens from those that are not compliant with hegemonic ideals. The real-world effects of geo-based algorithms are further widespread than most people imagine, which includes shaping if not forcing the outcome of elections, amplifying surveillance as a solution, redefining currencies and governance, laying out poorly paid Uber routes, and pushing the virality of misinformation/actual fake news (Bakir and McStay 2018; Browne 2017; Persily 2017; Rosenblat 2018; Zook and Blankenship 2018). While computing reads as "objective" science that results in truth, algorithms are not neutral. They reproduce structural oppressions of racism, sexism, homophobia, transphobia, ableism, and colonialism in the ways they target individuals for policing, hiring or firing in the workplace, and even lead to suicide and violence (Noble 2018; Zook and Blankenship 2018).

The scholarship of geographic thought, theory, and method is in a prime position to attend to issues of digital ethics and offer intervention for digital activisms. Geographic studies are prime to argue against "digital dualism", the idea we behave differently online and IRL (or "in real life", as the youth say); the online world actively shapes our everyday world, and vice versa. While much attention has been devoted to online/IRL identities, geographers have much yet to attend to in regard to the *where* of the internet and the co-production on online/IRL spaces.

One heavily theorised internet geography is that of the body. Beyond the actual technology in our bodies (teeth, hips, eyes, knees, organs), there is the cyborg we produce on Instagram, in PlayStation 4's "Call of Duty", and even in the pattern of communications in texts and emails. The spread of search engines and social media have prompted the growth of different networks, and also shifted the spatiotemporality of knowledge, news, and even social movement resistance as immediate. These platforms and devices also alter the mind itself; teenagers who heavily use devices are prone to increased anxiety, depression, and loneliness (RSPH and Young Health Movement 2017).

Many scholars still see the digital as beyond their research focus, yet the digital has permeated most surfaces, bodies, and imaginaries: from the stock exchange to recipe exchanges, from Fitbits on wrists to phone screens in pockets (should it ever actually be designed to fit in women's pockets, should women actually be gifted pockets) (Farokhmanesh 2018). The internet was once imagined as a "technotopia" or platform for radical democracy, but has fallen wildly short of those promises. Instead it has more often become a tool to support and promote racist, sexist, ableist, heteronormative, and colonial neoliberal capitalism. Geography's attention and intervention are required.

As sociologist Karen Gregory (2018) wrote: "We are in need of new visions, new imaginaries of what a digital world can or should be. The nightmare versions are exhausting (literally, exhaustion is the theme of our time)." Only in embracing the digital as part of the larger production of space can we begin to reckon with the rich, complex geographies of the world. As simple as it would be to sort out our feelings and ideas about finding and losing our phones, such an accomplishment lies ahead in the study of (digital) geographies.

References

Anti-Eviction Mapping Project (2018) "List of Maps." https://www.antievictionmap.com/maps/ (last accessed 12 December 2018)

Ash J, Kitchin R and Leszczynski A (2016) Digital turn, digital geographies? *Progress in Human Geography* 42(1):25–43

Bakir V and McStay A (2018) Fake news and the economy of emotions. *Digital Journalism* 6(2):154–175

Bridle J (2016) Algorithmic citizenship, digital statelessness. *GeoHumanities* 2(2):377–381

Browne S (2017) Race, communities, and informers. *Surveillance and Society* 15(1):1–4

Burns R and Thatcher J (2014) What's so big about big data? Finding the spaces and perils of big data. *GeoJournal* 80(4):445–448

Byrd D (2015) Tracing transformations: Hilton Head Island's journey to freedom, 1860–1865. *Nineteenth-Century Art Worldwide* 14(3) http://www.19thc-artworldwide.org/autumn15/byrd-hilton-head-island-journey-to-freedom-1860-1865 (last accessed 15 November 2018)

Cheney-Lippold J (2017) *We Are Data: Algorithms and the Making of Our Digital Selves.* New York: NYU Press

Cockayne D G and Richardson L (2017) Queering code/space: The co-production of socio-sexual codes and digital technologies. *Gender, Place, and Culture* 24(11):1642–1658

Cowley R and Caprotti F (2018) Smart city as anti-planning in the UK. *Environment and Planning D: Society and Space* https://doi.org/10.1177/0263775818787506

Dalton C M and Thatcher J (2014) What does a critical data studies look like, and why do we care? *SocietyandSpace.org* 12 May. http://societyandspace.org/2014/05/12/what-does-a-critical-data-studies-look-like-and-why-do-we-care-craig-dalton-and-jim-thatcher/ (last accessed 15 November 2018)

Datta A (2018) Postcolonial urban futures: Imagining and governing India's smart urban age. *Environment and Planning D: Society and Space* https://doi.org/10.1177/0263775818800721

Digital Scholarship Lab (2017) "Renewing Inequality: Urban Renewal, Family Displacement, and Race, 1955-1966." American Panorama: An Atlas of United States History, Digital Scholarship Lab, University of Richmond. http://dsl.richmond.edu/panorama/renewal/ (last accessed 15 November 2018)

Elwood S and Leszczynski A (2011) Privacy, reconsidered: New representations, data practices, and the geoweb. *Geoforum* 42(1):6–15

Elwood S and Leszczynski A (2018) Feminist digital geographies. *Gender, Place, and Culture* 25(5):629–644

Farokhmanesh M (2018) Women's pockets are the only reason to buy an iPhone XS. *The Verge* 13 September. https://www.theverge.com/2018/9/13/17854750/iphone-xs-womens-pockets-small-apple (last accessed 15 November 2018)

Gabrys J (2014) Programming environments: Environmentality and citizen sensing in the smart city. *Environment and Planning D: Society and Space* 32(1):30–48

Gieseking J J (2017) Size matters to lesbians too: Queer feminist interventions into the scale of Big Data. *Professional Geographer* 70(1):150–156

Gieseking J J (2018) Where are we? The method of mapping with GIS in digital humanities. *American Quarterly* 70(3):641–648

Grabar H (2014) Smartphones and the uncertain future of "spatial thinking". *CityLab* 9 September. http://www.citylab.com/tech/2014/09/smartphones-and-the-uncertain-future-of-spatial-thinking/379796/ (last accessed 15 November 2018)

Graham S D N (2005) Software-sorted geographies. *Progress in Human Geography* 29 (5):562–580

Gregory K (2018) "We are in need of new visions..." Twitter post by @claudiakincaid, 3 September. https://twitter.com/claudiakincaid/status/1036534481572831233 (last accessed 15 November 2018)

Hamraie A (2017) *Building Access: Universal Design and the Politics of Disability.* Minneapolis: University of Minnesota Press

Haraway D J (1990) *Simians, Cyborgs, and Women: The Reinvention of Nature.* New York: Routledge

Hicks M (2017) *Programmed Inequality: How Britain Discarded Women Technologists and Lost Its Edge in Computing.* Cambridge: MIT Press

Johnson J A (2014) From open data to information justice. *Ethics and Information Technology* 16(4):263–274

Kitchin R (2014) The real-time city? Big Data and smart urbanism. *GeoJournal* 79(1):1–14

Kitchin R and Dodge M (2011) *Code/Space: Software and Everyday Life.* Cambridge: MIT Press

Kwan M P (2002) Feminist visualization: Re-envisioning GIS as a method in feminist geographic research. *Annals of the Association of American Geographers* 92(4):645–661

Leszczynski A (2016) Speculative futures: Cities, data, and governance beyond smart urbanism. *Environment and Planning A* 48(9):1691–1708

Monahan T (2018) Algorithmic fetishism. *Surveillance and Society* 16(1):1–5

Noble S U (2018) *Algorithms of Oppression: How Search Engines Reinforce Racism.* New York: NYU Press

Persily N (2017) The 2016 US election: Can democracy survive the internet? *Journal of Democracy* 28(2):63–76

Rosenblat A (2018) *Uberland: How Algorithms are Rewriting the Rules of Work.* Oakland: University of California Press

RSPH and Young Health Movement (2017) "#StatusofMind: Social Media and Young People's Mental Health and Wellbeing." Royal Society for Public Health, London. https://www.rsph.org.uk/our-work/campaigns/status-of-mind.html (last accessed 15 November 2018)

Shelton T, Zook M A and Wiig A (2015) The "actually existing smart city". *Cambridge Journal of Regions, Economy, and Society* 8(1):13–25

Thatcher J (2013) Avoiding the ghetto through hope and fear: An analysis of immanent technology using ideal types. *GeoJournal* 78(6):967–980

Thatcher J (2017) You are where you go, the commodification of daily life through "location". *Environment and Planning A* 49(12):2702–2717

Wilson M W (2017) *New Lines: Critical GIS and the Trouble of the Map.* Minneapolis: University of Minnesota Press

Wilson M W (2018) On being technopositional in digital geographies. *Cultural Geographies* 25(1):7–21

Zook M A and Blankenship J (2018) New spaces of disruption? The failures of Bitcoin and the rhetorical power of algorithmic governance. *Geoforum* 96:248–255

Doom

Geoff Mann

*Department of Geography & Centre for Global Political Economy, Simon Fraser University,
Burnaby, BC, Canada;
geoffm@sfu.ca*

Distraction and denial might occasionally obscure it, but the straits look pretty
dire these days. It seems increasingly sensible to be not merely "concerned" that
climate change, inequality, racist nationalism and the collapse of anything resem-
bling democracy might be driving the world over a precipice, but rather to feel
fairly certain of it. At the very least, it would appear the most likely destination of
the current trajectory. An "optimistic" insistence that the growing chorus of
doom-sayers might be wrong (or, even more Pollyannaish, that they have a "responsibility" to hope because otherwise there is no hope) does little to brighten
the gloom. It is true that there is a long history of Chicken Littles running pan-
icked beneath skies that have not fallen. But frantic desperation is not the prevailing mode of contemporary doom-saying, at least for now. It is much closer to
resignation, or disappointment, a doom almost indistinguishable from tragic fate.

This judgment of our moment in history and its trajectory is tragic not only
in the sense that we seem to be heading toward catastrophic ends. It is also
tragic in a dramatic sense. In *Antigone* or *Macbeth* or *Death of a Salesman*, the
tragedy is always witnessed and recognised by a detached spectator who can
do nothing to prevent it. We watch the tragically flawed character march
toward their doom, sometimes when he or she is aware of its approach long
before the moment arrives. Something similar can be said for our era, in which
many of the most privileged (which would include many self-identified radical
geographers) observe ourselves, often literally from afar—via satellite images,
newsfeeds, atmospheric carbon graphs, etc.—recognising what is wrong and
what must be done to begin to address it, but doing nothing, or the look-
busy equivalent of nothing.

Often, too, the tragic figure is like Antigone herself, a "good" person, or at least
someone who understands themselves as doing the right thing. There is, therefore, an understandable temptation to attribute tragedy to evil actors or forces
driving toward the abyss, and they are not hard to find if we go looking—the fossil fuel cartel and Donald Trump, or, more broadly, capitalism, colonialism, patriarchy and the hubris of the Euro-American modernity they underwrite. But the
tragic sensibility does not require a Big Bad Wolf. On the contrary, it tends to
adopt something closer to a "structuralist" perspective: fate's ultimate causal
mechanisms are so un-entanglable that by the time we recognise the tragic for
what it is, it is effectively inevitable. From a distance, we see ourselves wandering
distractedly toward apocalypse, as unable to intervene as an audience watching
Oedipus murder his father.

Keywords in Radical Geography: Antipode at 50, First Edition. Edited by the *Antipode* Editorial Collective.
© 2019 The Authors/Antipode Foundation Ltd. Published 2019 by John Wiley & Sons Ltd.

None of which is to affirm these judgments. I emphatically reject the claim that we are done for, whoever "we" are. The point, rather, is that the development of radical strategies, the implementation of political tactics, the cultivation of hope—all of this is much less effective if it does not also confront the very real feeling among many that all might already be lost. This idea of doom is, however tacitly, inescapable for those who take up the task of writing, teaching and doing radical geography today, because it presents fundamental challenges to the progressivist faith that animates radical critique. It forces us to ask what the collapse (or imminent collapse) of that narrative means for the project of radical geography as it has emerged and proliferated over the last half century. The challenge is not simply that we have reached the end of the emancipatory teleology of which Marxism is so often (wrongly) accused—partly because even the rather "orthodox" Marxisms of dominant strands of radical geography have never been that naïve, and partly because Marxism, while certainly crucial, is only one strand in an increasingly expansive web of radical geography.

In other words, the problem is neither that the field is "Marxist" nor that radical geographers are utopian in the sense Marx and Engels attacked in the *Manifesto*. The problem is that the entire field is founded on the potential for massive social transformation in the relations among humans and between humans and the non-human worlds in which we are inextricably embedded. As David Stea put it 50 years ago, on the very first page of the very first issue of *Antipode*, "We believe that revolutionary change in the social climate and physical environment are necessary and possible" (1969:1).

Transformation of that kind does not have to be guaranteed or even likely to provide radical geography with its foundation. It just has to be possible, and possibility is not determined *ex ante*. To say something is "possible" is an act of (reasoned) expectation; possibility is by definition a state that is *deemed* possible. If better futures—which in this case must mean more just, secure, free and joyful futures—are no longer deemed possible or are understood to be increasingly beyond our reach (regardless of the objective accuracy of this assessment), then the possibility of radical geography itself is in question. The sense of an impending doom loosens the stitching of historical expectation that provides much of radical analysis, pedagogy and politics with its safety net. This expectational foundation is most visible in Marxism, based as it is on a *salto mortale* (mortal leap) justified by the wager that somehow, someway, the struggle to produce our history will eventually be rewarded (e.g. Harvey 2006:68). But a similar fidelity to the future is central to feminist, decolonial, antiracist and other radical geographies, whether they are tied to its mostly-Marxist origins or not (e.g. Gibson-Graham 2006). As Martin Luther King (1965) famously said to the people of Montgomery in Spring 1965, "the arc of the moral universe is long, but it bends toward justice".

But what if it doesn't? The (perceived) unravelling of the spatiotemporal safety net disrupts the hopefulness of possible future times and spaces, while simultaneously suggesting that what is to come, since it represents some "decline" or "reversal", cannot contain a radically progressive realisation of history (hence widespread fears, on the Left as much as the Right, of "descent" into a Hobbesian state of nature). Or alternatively, the progressivism so central to modernity seems to have succeeded in completely detaching the present from the past, and the future looks more and more like inevitable (if paradoxical) global civil war (Agamben 2015; Koselleck 2004:264).

One does not need to share this sense of doom to be forced to take it seriously. First, because it is very geographically specific: just as colonialism is not a past event, but an ongoing process, doom is not confined to the future, but has a long history among the planet's poorest and most exploited peoples, both within and beyond the societies that call themselves "modern". Some doom-sayers have already been proven right. Second, the fact that it is somehow possible that calamity might not be on the horizon is much less meaningful than the critic of "defeatism" would like to believe. It is certainly fair to suggest, with Hannah Arendt (1951:144), that the "defeatists" have merely "replaced the superstition of progress with the equally vulgar superstition of doom". But, just as the joy in life is not undone if an injustice ultimately ends it, the experience of a present that appears to have no future to redeem it cannot be retroactively erased, even if that experience turns out to be founded in mistaken anticipation.

In other words, contemporary radical geography cannot avoid the question of doom. It is already a part of every conversation in which it is engaged. Almost 10 years of research on climate change politics tells me this eminently rational dread —the acknowledgment of the likelihood of the futility of efforts to avoid catastrophe—is fundamental to contemporary social life. I find it particularly common among students, and in my experience it is now inescapable in pedagogical contexts. If we do not name it, we ignore an elephant in the room. What kinds of teaching and research, let alone social and political responses, might we attempt when the tunnel appears to have little or no light at the end? These are certainly not the only questions we should be asking, but they must be among them.

Part of the problem we face lies in the shifting possibility of redemption, but of an historical rather than religious sort. Suffering, calamity, crisis—these are, to a significant extent, the empirical "stuff" of radical geography—find in their narrative virtually all their meaning in the commitment to a future that promises to redeem them—that makes them worthwhile, compensates their injustice, or even just demonstrates they were unnecessary. Without it, we have only absurdity. This might even be said of our individual lives as well. The celebrate-able finitude of a "long and full life" disappears in the tragic mode, because it offers no future to which that life can bequeath its fullness. It is just an end. This has catastrophic effects on people's experience of the meaningfulness of their lives; not because we want to live forever in paradise, but because life's meaningfulness as a whole is only possible in the context of other meaningful lives, past, present and future, striving for the things we (hope they) do: love, joy, dignity, security, justice, etc. If these don't seem possible in the future, their status as measures of life's worth erodes. The wager that living is redeemed in commitment becomes unjustifiable, and with it the normative basis of radical geographies' ethical claims.

This confrontation with a tragic collective doom undermines much of what many citizens of capitalist liberal democracy understand as the elemental building blocks of politics (Shotwell 2016). It denies individual or collective autonomy, it makes social institutions palliative care mechanisms at best, and it refuses the very basis of solidaristic organisation: positive change. Which means that times of tragedy are—or can easily become—profoundly anti-democratic. What kinds of politics make sense in those moments? Are there any that do not imply violence? I do not think radical geography, at least as I currently understand it, has much to tell us about what happens to politics when this sensibility seeps into individual and collective life, but perhaps it is better prepared to do that than many other fields (e.g. Gibson-Graham 1996; Gidwani 2008; Ishiyama 2003; Pasternak 2016).

One place I think the answer does not lie is in the concept of crisis, around which much research has understandably revolved since "radical geography" became a term that made sense to people. We are living through what feels like a staccato of crises, in which the re-citation of Benjamin's (2006) supposedly permanent "state of emergency" has become a commonplace. It seems worth asking what happens when crisis becomes tragedy. What is the relationship between the two? At first glance, it might seem that tragedy is simply crisis past some point-of-no-return threshold. But surely emergency and crisis lose their meaning in a state of permanence. Naming the crisis and declaring the emergency is an early-warning system for tragedy, but it also renders tragedy difficult to understand. Indeed, the whole point of the idea of crisis is to deny tragedy: crisis is the word for a moment that is by definition supposed to be transitional. The condition of permanent or ubiquitous crisis is not the unfortunate outcome of history or human nature, but a performative declaration of victory on the part of those with the power to name it. This, I think, is Benjamin's real point regarding the state of emergency: the "tradition of the oppressed teaches us that the 'state of emergency' in which we live is not the exception but the rule. We must attain to a conception of history that accords with this insight" (2006:392). The inauguration of a scare-quoted "emergency" or crisis merely obscures the truth of what our histories can be.

In other words, the novelty of the particular problems we face at present (global warming in particular) can sometimes lead us to forget that this is far from the first time the world is going to end, or the first time that tragic dread clouds the future. Calamitous developments and doom have loomed over many communities in the past. The stories of those who managed to live through what they never imagined they would live through have much to tell us. Those moments sometimes drove those communities to radically different ways of organising and making sense of their collective commitments, and sometimes they helped them discover new resources in existing ways of life.

How did these communities organise a collective world in these conditions, knowing or expecting that world would not last? One of the great challenges we currently face is how to give meaning to teaching, research and politics *even if* the world is going to fall apart. By "we", I do not mean to suggest we are all one big happy family, but only that no community can escape this challenge. And despite what seems to me a common assumption that in moments of unfolding calamity we will all turn into selfish violent survivalists, it ain't necessarily so. As long as we live together with others in some degree of solidarity, we will still care about things like distribution, justice, equality, and citizenship, however differently we define them. We must avoid the temptation of nostalgia or cliché ("they fought till the end"; "they never gave up")—stories that are too easily, even instantly, redeemable in exactly the way tragedy refuses—but rather to think about how people tried to maintain a sense of collective responsibility in the shadow of doom, even if (as I do of course hope) we can undo the calamities of the present.

References

Agamben G (2015) *Stasis: Civil War as a Political Paradigm*. Palo Alto: Stanford University Press

Arendt H (1951) *The Origins of Totalitarianism*. New York: Harcourt

Benjamin W (2006 [1940]) On the concept of history. In H Eiland and M W Jennings (eds) *Selected Writings, Vol. 4: 1938–1940* (pp 389–400). Cambridge: Belknap Press

Gibson-Graham J K (1996) *The End of Capitalism (As We Knew It)*. Minneapolis: University of Minnesota Press

Gibson-Graham J K (2006) *A Postcapitalist Politics*. Minneapolis: University of Minnesota Press

Gidwani V (2008) *Capital, Interrupted: Agrarian Development and the Politics of Work in India*. Minneapolis: University of Minnesota Press

Harvey D (2006) *Spaces of Global Capitalism: Towards a Theory of Uneven Geographical Development*. London: Verso

Ishiyama N (2003) Environmental justice and American Indian tribal sovereignty: Case study of a land-use conflict in Skull Valley, Utah. *Antipode* 35(1):119–139

King M L (1965) "Our God Is Marching On!" Speech at Montgomery, AL, 25 March. https://kinginstitute.stanford.edu/our-god-marching (last accessed 2 October 2018)

Koselleck R (2004) *Futures Past: On the Semantics of Historical Time* (trans K Tribe). New York: Columbia University Press

Pasternak S (2016) *Grounded Authority: The Algonquins of Barriere Lake Against the State*. Minneapolis: University of Minnesota Press

Shotwell A (2016) *Against Purity: Living Ethically in Compromised Times*. Minneapolis: University of Minnesota Press

Stea D (1969) Positions, purposes, pragmatics: A journal of radical geography. *Antipode* 1(1):1–2

Earth-Writing (Spaciousness)

Sharad Chari

Department of Geography, University of California at Berkeley, Berkeley, CA, USA
and

Wits Institute for Social and Economic Research, University of the Witwatersrand,
Johannesburg, South Africa;
chari@berkeley.edu

I

I begin with the question concerning "geography" that takes apart the "geo" and the "graphia" to ask "what is this form of writing?" The question contains a yearning for Geography as many forms of Earth-writing, some buried or drowned and yet all-pervasive, if only we have a spacious enough imagination. If this sounds cryptic, perhaps that is necessary for our moment of apparent transparency and immediacy in a planet of walls and rising seas in which geographical matters are, more than ever, political, futurist, embodied cries for a different relation to our oceanic planet. We certainly know this already from Édouard Glissant (2010) and Katherine McKittrick (2006); and yet "we" as collective, collectively imperilled, have scarcely begun to ask the question, let alone face its consequences.

I ask this question while concluding a book on South Africa called *Apartheid Remains*, on temporal and material remains of racial capitalism and struggle in our politically beleaguered age. The book concludes with the insights of Black documentary photographers Peter McKenzie and Cedric Nunn who elaborate on an audio-visual Black radical tradition on Indian Ocean shores; I think with Woods' (1998) powerful geographic critique of the blues as subaltern critique of the plantation complex in its changing forms, and with Moten's (2003) reading of what he calls the sound of a photograph. I argue that McKenzie and Nunn play the Blues through their camera lenses, darkrooms and digital manipulations in bittersweet engagements with a past too slow to recede, and a future too slow to be born. In their film and photography, these critics articulate a particular kind of blues aesthetic at the confluence of the Black Atlantic and the Indian Ocean, an active practice of imagining other relations to our planetary future (Chari 2017a, 2017b). This is what I mean by "spaciousness". And *Antipode* has sought to be a vehicle for spaciousness in changing ways: *viva*!

I cannot think this without Ruthie Gilmore's refrain that the precise sense in which Cedric Robinson (2000) deploys the term racial capitalism is as a reminder that capitalism always renovates prior forms of racial and, following Françoise Vergès (1999), gender/sexual authority. Robinson offers as an alternative a Black Marxism committed to thinking of capitalism (as long as we need reminders, "racial/gendered/sexual/otherwise-differentiated capitalism") always in dialectical relation to differentiated and differenced forms of opposition. Gilmore (2008: 42) frames her praxis and the praxis of her interlocutors as of "stretching" theory and struggle. This geo-graphical imperative prompts my second key question concerning Earth-writing as imagining spaciousness: what does it mean to stretch Marxism, not just in a subjective or ideological register. Cedric Robinson's answer is that Marxism is stretched when we "read" it (in theory and struggle) in interaction with the Black radical tradition, just as Gilmore (2007) stretches and reformulates David Harvey's (2007) notion of capital's spatial fix through Black struggle to explain and contest what she calls the "prison fix".

Putting these questions together: what kind of geography as Earth-writing is adequate to the task of an anti-racist critique in which the reproduction of spatial fragmentation and embodied difference are key mechanisms not just for the *hope* of capital's reproduction but also for opposition with the aim of abolition, at which point we might imagine an actualisation of spaciousness in a universal sense (Gilmore 2017).

▌▌

The seeming intractability of racism in habits of mind and heart, in second natures and built environments, in political dreams and embodied struggles requires keeping space for a slower form of attention alongside the real-time reactivity we have grown accustomed to: an earthly writing that connects a diagnosis of multi-layered social domination with many embers of critique that point beyond the rotten state of politics, including Left politics, across our planet.

For one way out of our hide bound, terrestrial understandings of the rebirth of authoritarian populism in our time, I suggest retracing our global imaginations. My own practice of reading against the present takes me from the Antilles to the Southern African Indian Ocean, from the Haitian Revolution and its calls for a new human, to its effects in the reconstitution of plantation power in the consolidation of US capitalism as well as in the ricocheting struggles in the aftermath of slavery in the New World, and to the effects of the abolition of slavery and the slave trade on racial capitalisms of the Pacific and Indian Oceans, the making and unmaking of settler democracies, and to various strands of radical critique that have engaged with African and Asian decolonisation and what comes next. This is epigraphic, I grant, and I beg the reader's indulgence.

I turn for this task to Frantz Fanon, reading him as theorist of the human to come, as Fanon scholars like Sekyi-Otu and Sylvia Wynter have argued, but also as the writer of an Earth to come. I suggest that *Damned of the Earth* can be read as a work of Earth-writing in three senses, as I interpret references to "*la terre*" in Fanon (1961, 1967) as, first, a defence of the land in peasant revolutionary struggle for land, bread and dignity; second, at other points, the text seethes in outrage against "*la terre brulee*" or colonial scorched earth warfare (I write this just after the California fires have been extinguished, this time); and third, this is my claim, beyond the referential sign, the text points beyond this burning planet to an emergent Earth prophesied by *les damnés*, the one we might all inherit when the rotten state of postcolonial politics in former colonies and metropoles has been laid bare, as it must, again and again. Mirroring Robinson (2000) and

Gilmore (2007), Fanon (1967:31) argues quite early in the text that "Marxist analysis should always be slightly stretched every time we have to do with the colonial problem". In fact, all these thinkers stretch Marxist analysis by reading in dialectical relation to the praxis of the oppressed. What they point to is an Earth-writing forged through the labour of earthly reading, indebted to the arguments of "the damned". We might read in this an argument that "spaciousness" is a product of translation across multiple radical intellectual currents, including those subalternised or treated as merely cultural.

There is more to the section preceding Fanon's injunction, in which he argues in seemingly rigid terms that the "colonial world is a world cut in two", two zones, two towns, two peoples (Fanon 1967:29). There are different ways to read this passage and others like it in Fanon's oeuvre. I find persuasive Ato Sekyi-Otu's (1996:4–8, 76) argument that when Fanon writes about embodied experience, space or political economy in terms that *seem* ontologically dichotomous (and that are read at face value by strands of US Afropessimism; cf. Chari 2017b: note 4), he means to *dramatise* an unfolding dialectical method. One clue about the politics prefigured here are in Fanon's (1967:30) characterisation of "the native town, the Negro village, the medina" as "a world without spaciousness". But if spaciousness is the fundamental and universal demand of *the damned*, it is a mysterious one. Fanon does not so much as offer a definition; he leaves it as a clue for our beleaguered age.

In our time of planetary enclosure, it is not immediately obvious what an actually spacious world might be: there are no blueprints. We know it must be "post-occidentalist" in the sense that Fernando Coronil (1996) argues, as a refusal of all forms of geographical fetishism and partition. And we have a world of differentiated yearnings, expressed in music, dance, poetry, food, mundane forms of *marronage*, and occasional insights from people we call "thinkers". And so, I turn to some "thinkers", earth-writers by other names, for their intertwined explorations of spaciousness, and their attempts at stretching and connecting traditions of the oppressed.

III

The first was the great Indian thinker, politician, social reformer, jurist and economist, Bhimrao Ramji Ambedkar, whose life from the late 19th century to the middle of the 20th spanned the end of colonialism and the remaking of Indian national, ethno-racial capitalism, and the problem of "untouchable" or Dalit freedom. Ambekdar's life is of dramatic transformation from poverty and oppression as a person labelled "untouchable", through a series of exceptional turns to Bombay, on for graduate studies to Columbia University and then to the London School of Economics, to his return to India as a leading anticolonial figure and rival to Gandhi, and later architect of the Indian Constitution. We don't know what he collected from the bookshops of New York that he scoured in the 1910s, because the ship carrying his beloved library home was torpedoed, but he wrote to W.E.B. DuBois in the 1940s that he was familiar with the Negro question and was interested in its lessons for the fight against untouchability in India.

In an age obsessed with industrialisation, Ambedkar was a defender of the agrarian question and food security. There is no question of spaciousness without a full stomach. He defended birth control and full equality for women. And around his 60th birthday, in the last five years of his life, the architect of the Indian Constitution saw the fundamental limits to a jurisprudential route to Dalit freedom, and he turned to an existential critique through Buddhism. While the man called the Buddha, or the awakened, had been dead for over 550 years, and Buddhism largely absent in mid-century India, Ambedkar effectively turned the Dalit question into one of ontological refusal of Hindu-dominated racial capitalism, and a renovation of Buddhist thought as "Navayana" or "new vehicle" (as opposed to extant Mahayana or Theravada) Buddhism.

In his last essay, "The Buddha or Karl Marx", Ambedkar reads "the creed of the Buddha" against "the Marxian creed" to argue that while both aim at liberation, they differ fundamentally in their means (Ambedkar 1956). Quite simply, Ambedkar could not condone the ongoing violence implied by the notion of "the dictatorship of the proletariat". In a thoughtful elaboration, Ajay Skaria (2015:452) returns to the last line of this text, which poses that both traditions "work … towards the promise of a world organised by equality, liberty, and fraternity, with equality as the key term", and that for Ambedkar, Marxism was perhaps the most intense "moment of the striving to keep that promise". Yet, in the face of Soviet violence, Ambedkar clarifies his own view, and "[t]o that renunciation of violence, to that other universality, Ambedkar gives in the essay the names 'religion' and Buddhism" (Skaria 2015:463). Our time of ethnoracial violence by putatively Buddhist states clarifies Ambedkar's Navayana Buddhist renunciation as requiring what Skaria (2015:465) aptly calls "participation without a part, without sovereignty". Skaria's concern is with the figure of the minor in this imagination of politics after liberation. What I offer is that just days before his death, the architect of the Indian Constitution proposed a practice of dialectical reading of Marx and Buddhism that requires of both the renunciation of sovereignty. What is the praxis of spaciousness that this renunciation might yet enable?

IV

When he decided that Dalit conversion to Buddhism was key to Dalit liberation, Ambedkar turned to the most senior modern Buddhist monk in India, a young white man in his 20s. Dennis Linwood was born in 1920s working-class South London, he travelled to India as a soldier in World War II, where he decided to emulate the life of the Buddha by tearing up his passport and taking to the streets. He wandered around South India, spending time with several mystics, including Ramana Maharishi; he learnt Pali, the language of the historical Buddha and immersed himself in Buddhist thought; renamed Sangharakshita upon ordination as a Theravada monk, he became a Buddhist scholar at a major crossroads in the Himalayas, Kalimpong, on the route that all the exiled Tibetan Buddhists took to India; he was there to receive the Dalai Lama when he went into exile, and he learnt from high monks across Theravada, Tibetan Mahayana and Cha'an schools of Buddhism before co-writing an important book about Buddhism in

English. This text brought him into contact with Ambedkar. He declined to convert Ambedkar, referring him to his teacher; though wide apart in age and experience, they became "guru-bhais", kin through a common teacher.

Then, in the 1960s, after spending 20 years in India, Sangharakshita returned to London and was shocked to find that Buddhism in London was doctrinal, divided, and totally disconnected from the experiments in spaciousness that permeated the countercultural revolt of the 60s. He dove into the world of social and sexual liberation in 1960s London; it was soon clear that he was queer and drew in a gay following. But he also, inadvertently, was part of a process of reviving a late-Victorian white male homosocial world with its combination of cross-class, intergenerational homosexual experimentation and misogyny.

I tried to interview the ailing 90-something in late 2016 in northern England. In a gender-segregated community, I was mistakenly housed in the women's dormitory, and met some remarkable Buddhists. Waiting for an interview with the nonagenarian, I was also struck by Sangharakshita's library, perhaps the only collection with signed material to a person from both Ambedkar and Alan Ginsberg. In a beautiful building facing a meadow, one side of the library houses books on Buddhist Dharma, including legendary conversations between the young Sangharakshita and his countercultural interlocutors; the other side houses a "non-Buddhist" collection filled with books Sangharakshita read, including the works of modern Indian spiritualists like J. Krishnamurti and also the gamut of white gay English modernists like Isherwood, Forster, Auden, and so on. The arrangement of Buddhist and non-Buddhist texts offers a kind of dialectical imagination with its own conceptions of liberation, conditioned by specific limits, but not necessarily in a dogmatic sense. What was going on around the ailing old man was what one woman called a post-charismatic moment, in which women, including Black women, as well as trans youth and queer Dalits have been struggling to transform the homonormative misogyny and racism in contemporary Buddhist praxis.

Sangharakshita died in November 2018, and yet his flawed, remarkable life unearths a set of interconnected, ongoing experiments in connecting social and existential conceptions of spaciousness. In purely social terms, several figures in these networks continue to make space to refuse Dalit subjection in old-new India, while others find ways to critique whiteness, masculinism and neoliberal individualism in contemporary Buddhism. As a dialectical tradition linking the social and ontological—like the blues in Woods' (1998) rendition—Buddhist praxis refuses to ontologise social struggle, its relation to the past haunted by a deeper yearning for a new kind of collectively embodied and earthly future, a spaciousness to come.

V

The third figure I turn to is an iconic Black Marxist-feminist, veteran of the US Black Freedom movement, whose praxis stretches from the embers of Jim Crow racial violence to the militant activism of the Black Power years, mediated by Frankfurt School critical theory, and turning to global struggles for prison abolition and against global apartheid. What is fascinating about Angela Davis' praxis is what I see as her multi-spatial political imagination that is at once biographical and always about forging solidarities that scale walls. Davis' (1998) book on Black women Blues singers shows us that forms of Earth-writing that have imagined a more spacious world are pervasive, in linked, far-flung, variously sensory philosophies of life that surround us (McKittrick and Woods 2007; Woods 1998). What is fascinating is not just the ways in which Davis has journeyed out from her own experience of incarceration, but also how she stretches an abolitionist imagination to include a commitment to veganism as part of a critique of industrial agriculture. There is more to ask about Davis' vision of planetary ecology, and on the real and imagined global connections it mobilises as an extension and elaboration of the Black radical tradition as it confronts global subaltern struggles in a time of ecocide. What I suggest is that Davis picks up the question of spaciousness alongside, and not just after, the deadly articulation of sovereignty and capital.

I conclude this short piece with an event that highlighted the ways in which stretching the Black Radical Tradition forces us to confront trans-spatial, subaltern traditions that surface linked calls for spaciousness through earthly and oceanic connections that already exist. To be attentive, we, Geographers, must learn new ways of reading and writing.

When Angela Davis came to South Africa in September 2016 to deliver the Steve Biko Lecture, she ended her talk by calling to the stage a 13-year-old girl who had made the news by defying the racist rulebook at Pretoria Girls School which did not allow Black girls to wear their hair natural; Mark Hunter (2019) explains the politics of education at work at this moment. In her powerful talk, Davis pointed to the Haitian Revolution as the first truly universal claim that Black lives matter, through an inclusionary and political notion of Blackness. She reminded the audience that the Haitian Constitution made the powerful argument that anyone who wanted to stay to rebuild post-revolutionary society could do so, but they would have to subscribe to a notion of Blackness that restored universal humanity. But when she embraced the young girl and her friends—because the 13-year-old came to the stage with her entire entourage—their radical imaginations had already been stretched in unforeseen ways. The young girl's name is Zulaikha Patel, and this Black girl with a Gujarati Muslim name reminds us of the confluence of oceanic histories that converge in South Africa's racial formation, opening surprising twists in the entangled geographies of Black and subaltern radical traditions. Her raised fist twists a young girl's call for spaciousness into a planetary demand.

References
Ambedkar B R (1956) "Buddha or Karl Marx." Babasabeb Ambedkar International Association for Education, Japan. http://velivada.com/wp-content/uploads/2017/07/buddha-or-karl-marx-book-in-english.pdf (last accessed 7 November 2018)
Chari S (2017a) Detritus, difference, politics. *Somatosphere* 30 October. http://somatosphere.net/2017/10/detritus.html (last accessed 3 December 2018)

Chari S (2017b) The Blues and the damned: (Black) life that survives capital and biopolitics. *Critical African Studies* 9(2):152–173

Coronil F (1996) Beyond occidentalism: Towards non-imperial geohistorical categories. *Cultural Anthropology* 11(1):51–87

Davis A Y (1998) *Blues Legacies and Black Feminism: Gertrude "Ma" Rainey, Bessie Smith, and Billie Holiday*. New York: Vintage Books

Fanon F (1961) *Les damnés de la terre*. Paris: François Maspero

Fanon (1967) *The Wretched of the Earth* (trans C Farrington). London: Penguin

Gilmore R W (2007) *Golden Gulag: Prisons, Surplus, Crisis, and Opposition in Globalizing California*. Berkeley: University of California Press

Gilmore R W (2008) Forgotten places and the seeds of grassroots planning. In C Hale (ed) *Engaging Contradictions: Theory, Politics, and Methods of Activist Scholarship* (pp 31–61). Berkeley: University of California Press

Gilmore R W (2017) Abolition geography and the problem of innocence. In G T Johnson and A Lubin (eds) *Futures of Black Radicalism*. London: Verso

Glissant É (2010) *Poetics of Relation*. Ann Arbor: University of Michigan Press

Harvey D (2007 [1982]) *The Limits to Capital*. London: Verso

Hunter M (2019) *Race for Education: Gender, White Tone, and Schooling in South Africa*. Cambridge: Cambridge University Press

McKittrick K (2006) *Demonic Grounds: Black Women and the Cartographies of Struggle*. Minneapolis: University of Minnesota Press

McKittrick K and Woods C (eds) (2007) *Black Geographies and the Politics of Place*. Toronto: Between the Lines

Moten F (2003) *In the Break: The Aesthetics of the Black Radical Tradition*. Minneapolis: University of Minnesota Press

Robinson C (2000 [1983]) *Black Marxism: The Making of the Black Radical Tradition*. Chapel Hill: University of North Carolina Press

Sekyi-Out A (1996) *Fanon's Dialectic of Experience*. Cambridge: Harvard University Press

Skaria A (2015) Ambedkar, Marx, and the Buddhist question. *South Asia: Journal of South Asian Studies* 38(3):450–465

Vergès F (1999) *Monsters and Revolutionaries: Colonial Family Romance and Metissage*. Durham: Duke University Press

Woods C (1998) *Development Arrested: Race, Power, and the Blues in the Mississippi Delta*. London: Verso

Economic Democracy

Andrew Cumbers

Adam Smith Business School, University of Glasgow, Glasgow, UK;
andrew.cumbers@glasgow.ac.uk

Making Space for Economic Democracy
Economic Democracy as Labour Freedom

Economic democracy is often thought of in collective terms as the realm of collective bargaining or about giving workers collective ownership of the means of production. These are important ingredients but reflect a rather particularist and "workerist" (Cleaver 2008) sense of labour, work, and the economy. In embracing a broader and more pluralistic perspective, a project of radical economic democracy must take the individual as its starting point. One of the shared concerns of both Marx and the more radical liberals such as John Stuart Mill (Ellerman 1992) was to give individuals ownership of their labour, with the implication that it is they who choose how it is used, in opposition to the diverse servitudes of slavery, feudalism, or capitalism.[1] For Marx, this liberation of the individual worker is about giving her the power and control over how she uses her labour, thereby overcoming the alienation of the capitalist labour process (see Megill 2002).

In the context of massive industrialisation in the 19th century, it is not surprising that the emergent working class becomes the subject for revolutionary transformation. But in the very different world of the 21st century—of ecological crises, massive wealth inequalities, a growing precariat, and the marginalisation of many from the contemporary labour process—the project of radical economic democracy needs a degree of reformulation. Starting with providing the individual with ownership and control of their labour—in pursuit of their own reproduction and flourishing—is fundamental.[2]

Framing economic democracy thus links to a political agenda around the commons or "commoning" (De Angelis 2016) rather than the clarion call of earlier generations to provide (largely industrial and male) workers with the fruits of their labour. A focus upon individuals operating in "free exchange" with others about how they use and organise their labour, as well as shared resources, in a sustainable fashion, is key to achieving the social and ecological transformation required. It also poses the question of the individual economic rights of others in the community who are unable or no longer have the capacities to work. How might their ownership rights to flourish be addressed and what mechanisms (e.g. a basic living income, cooperative stakeholding) might serve these purposes?

This positive sense of individual economic freedoms articulated here contrasts sharply with the negative freedoms of Hayek and the neoliberal tendency (Burczak 2006). Where the latter views freedom as the right of elites to appropriate the labour of others, a focus on the "economic and property rights" of labour

Keywords in Radical Geography: Antipode at 50, First Edition. Edited by the *Antipode* Editorial Collective.
© 2019 The Authors/Antipode Foundation Ltd. Published 2019 by John Wiley & Sons Ltd.

leads in a very different direction; that of autonomous self-government for individuals, families, and communities. Moreover, the principle of self-government of one's own labour to meet social needs—use value over exchange value—can only be realised through collective projects and working with others. One of Marx's fundamental insights was to see labour as a collective social product. We can only survive as a family, community, city, or even planet through our cooperation with others in the management of resources (including labour).

What might seem counterintuitive at first—but has a strong intrinsic logic—is that exercising individual labour rights in a way that also protects the rights of others from exploitation and alienation quickly leads to collective forms of ownership, in opposition to private and corporate[3] forms. Securing the individual's right to participation in decisions about their own labour—and the inescapable point that follows that all individuals have this right (Dahl 1985)—can only be achieved through democratic and cooperative means. In practice, this is best achieved through diverse forms of public and cooperative ownership.

Forging Economic Democracy In, Against, and Beyond the State

Addressing the current ecological, political, and economic crises that confront us involves reclaiming space and place from capitalist appropriation and forging new collective organisations, institutions, and identities that can transform economic practices. The resurgence of an agenda around the global commons is critical to this task, with its insistence on carving out new spaces that can reclaim resources, work, and social being for collective and socially useful purposes in environmentally sustainable ways (De Angelis 2016).

A critical question arises, however. How do we get there? Can a transformative economic and social project be achieved outside of any engagement with the state as many commons proponents argue? While much is made of radical autonomous projects at the local scale that work at the interstices of capital (Holloway 2010), it is doubtful that more transformative systemic change at higher scales can be achieved without actively reclaiming commons from the state. A sobering point is that the deepening inequalities and crises that we face are not leading to effective Left mobilisations, but rather a resurgence of right-wing populism and even fascism. In parallel it is notable that, as Gramsci (1971) long ago warned, advanced forms of capitalism lead to a deepening of the relations between capital, the state, and civil society. During periods of crisis, political, and economic elites often strengthen their grip on power, rather than being open to challenges from below (Mirowski 2013). The ability to combine coercive state powers with dominant metanarratives (e.g. austerity, taking back control, anti-immigrant rhetoric, Islamophobia) being key to preserving elite rule and wider public support for the status quo.

Forging a radical economic democracy requires creating spaces for labour agency in opposition to, as much as independent from, capital "in, against, and beyond the state" (Angel 2017; Cumbers 2015). This also involves "bringing a realistic everyday politics of social reproduction" (Pitts and Dinerstein 2017:430) to social struggles in terms of reclaiming spheres of work and life for collective projects in opposition to elite commodification processes. This can involve new forms of working in common, such as those typical in community gardening projects which attempt to reclaim urban spaces from property-based speculation, commodification and gentrification in cities (Crossan et al. 2016; Cumbers et al. 2018b). These are never completely autonomous from incorporation into dominant agendas, and, as such, need constant struggle creating and recreating alternative economic identities and practices. But neither can they be innocent of broader spatial governance processes and structures, needing strategies for upscaling and indeed scaling out if they are to have broader transformative effects (MacKinnon 2011). This involves making claims and advances on state institutions and structures, such as city planning boards, land registry regimes, and even changing broader regulatory regimes through parliamentary activity.

Reclaiming Space for Public Ownership
Public ownership is critical to a project of radical economic democracy. So much of the neoliberal attack on collective institutions has been targeted at existing public institutions at all scales where both direct privatisation of resources and assets and increasingly innovative mechanisms for allowing rent seeking and private profit to flourish in public–private ventures where exchange value increasingly expunges use value. The financialisation of basic service sectors like water and energy further alienate, marketising resources for selfish elites in the here and now, rather than sustaining them for future generations. In response, new possibilities for alternative, democratic, and participatory modes of governance need articulating.

The re-emergence of new forms of public ownership at local and national levels marks a significant moment in the emerging contours of a post-neoliberal order as the latter's contradictions become more socially and ecologically urgent. Some have tentatively identified the global trend towards remunicipalisation (see Kishimoto and Petitjean 2017) as part of a Polanyian double-movement of social and state re-regulation of key strategic sectors of the economy (e.g. water, electricity, transport, waste) in the wake of the failings and contradictions of privatisation and marketisation (Hall et al. 2013). Although history and Polanyi himself reminds us that such a double-movement can be progressive, it also can turn malign in its implications for society and democracy (Polanyi 1944). Notably, it is the authoritarian, immigrant-hating regime of Viktor Orbán that leads the way with renationalisation programmes in Hungary while the superficially "liberal" EU continues to impose new privatisations as part of its fiscal discipline in Greece and elsewhere.

But a refashioned economic democracy around public ownership can achieve two vital things: it can challenge the economic rationality of capital and private appropriation of labour, land, resources, and much else, while also advocating radical and progressive alternatives to earlier forms of flawed state-led projects. Recent experiences against privatisation in Latin America are apposite here. In 2007, the Peruvian city of Huancayo was faced with the urgent requirement to modernise its local sanitation system which was becoming unfit for purpose. The preferred multi-scalar neoliberal option by the national government was a public–private partnership (involving the German government and the Inter-American Development Bank), but following a grassroots protest and mobilisation, the city opted for its own trans-local public–public partnership involving local NGOs and technical assistance from another innovative municipal enterprise, ABSA of Argentina.[4] Such creative spatial strategies help reclaim local public spaces for social ends over exchange value while also enlisting broader support and participation from both the state and civil society.

Remunicipalisation campaigns in the energy sector in Germany—notably in Berlin and Hamburg—have attracted much attention because of the way that diverse radical grassroots coalitions have mobilised against privatisation, with varying degrees of success (Becker et al. 2015). To some extent, these initiatives can be viewed as articulating alternative social and ecological visions in line with "right to the city" movements (Beveridge and Naumann 2014). But they were also partly enabled by the particularities of German federal state and constitutional structures that provide opportunity spaces for social movement actors to reshape public institutions. In neither Berlin nor Hamburg has the outcome (yet) been a transformation to the kinds of participatory and deliberative public organisations that we would wish for, but the mobilisations have renewed grassroots agency while also contesting dominant state logics, framing alternative discourses around social and ecological justice.

Beyond these specific examples, radical geography needs to fashion a new spatial architecture around public and collective ownership and the wider goal of economic democracy if we are to realise the full emancipation of our labours. In the 20th century, the dominant Left traditions of socialism and social democracy tended to have highly verticalist and nation-centric spatial imaginaries where the forms of ownership were heavily centralised and top-down, often eviscerating older localist forms of mutualism and municipal socialism. The adherence to what John O'Neill has described as "a Cartesian rationalism and the technocratic conception of planning" (2006:67) resulted in the autocratic imposition of centrally imposed state projects with neat geometries onto messy, disordered "on the ground" economic realities. Some autonomous writers (e.g. Holloway 2010) lean in the opposite direction, celebrating a trans-local commons, evoking moments and fragments of commons without much sense of the dynamics required to achieve more transformative systemic change at higher spatial scales. An urgent task for radical geography in the 21st century is surely to navigate between these opposing tendencies with a spatial politics that is sensitised to local autonomy, individual empowerment, and decentred economic decision-making while still being alert to broader responsibilities to social and ecological justice. This requires institutional forms and arrangements around diverse forms of collective ownership, in and outside the state, rather than essentialising one or the other.

Endnotes

[1] These ideas are developed in much greater depth in Cumbers et al. (2018a).
[2] From a Marxist perspective, some of the most compelling arguments here from those working in the autonomous tradition who write against "workerist" and "immaterialist" accounts (e.g. Cleaver 2008; Dinerstein 2015; Pitts and Dinerstein 2017).
[3] Corporate forms of capitalism have over time been sanctified by states as "individualised" forms of property rights of course by national and supranational state bodies such as the EU and WTO.

[4] ABSA was formed as a result of a remunicipalisation struggle against a foreign consortium involving the now defunct and disgraced corporation Enron in 2001. It is an interesting hybrid, part owned by the water sector trade union and part owned by the provincial government of Buenos Aires (Kishimoto et al. 2015).

References

Angel J (2017) Towards an energy politics in-against-and-beyond the state: Berlin's struggle for energy democracy. *Antipode* 49(3):557–576

Becker S, Beveridge R and Naumann M (2015) Remunicipalization in German cities: Contesting neoliberalism and reimagining urban governance? *Space and Polity* 19(1):76–90

Beveridge R and Naumann M (2014) Global norms, local contestation: Privatisation and de/politicisation in Berlin. *Policy and Politics* 42(2):275–291

Burczak T A (2006) *Socialism After Hayek*. Ann Arbor: University of Michigan Press

Cleaver H (2008) Deep currents rising: Some notes on the global challenge to capitalism. In W Bonefeld (ed) *Subverting the Present, Imagining the Future: Insurrection, Movement, Commons* (pp 127–160). New York: Autonomedia

Crossan J, Cumbers A, McMaster R and Shaw D (2016) Contesting neoliberal urbanism in Glasgow's community gardens: The practice of DIY citizenship. *Antipode* 48(4):937–955

Cumbers A (2015) Constructing a global commons in, against, and beyond the state. *Space and Polity* 19(1):62–75

Cumbers A, McMaster R, Cabaco S and White M J (2018a) "Reconfiguring Economic Democracy: Collective Agency, Individual Economic Freedom, and Public Participation." Working Paper 1, Economic Democracy Project, University of Glasgow

Cumbers A, Shaw D, Crossan J and McMaster R (2018b) The work of community gardens: Reclaiming place for community in the city. *Work, Employment, and Society* 32(1):133–149

Dahl R (1985) *A Preface to Economic Democracy*. Oakland: University of California Press

De Angelis M (2016) *Omnia Sunt Communia: On the Commons and the Transformation to Postcapitalism*. London: Zed

Dinerstein A (2015) *The Politics of Autonomy in Latin America: The Art of Organising Hope*. London: Palgrave Macmillan

Ellerman D (1992) *Property and Contract in Economics: The Case for Economic Democracy*. Oxford: Blackwell

Gramsci A (1971 [1929–1935]) *Selections from the Prison Notebooks* (eds and trans Q Hoare and G Nowell Smith). New York: International

Hall D, Lobina E and Terhorst P (2013) Re-municipalisation in the early 21st century: Water in France and energy in Germany. *International Review of Applied Economics* 27(2):193–214

Holloway J (2010) *Crack Capitalism*. London: Pluto

Kishimoto S, Lobina E and Petitjean O (2015) *Our Public Water Future: The Global Experience with Remunicipalisation*. Amsterdam: Transnational Institute

Kishimoto S and Petitjean O (eds) (2017) *Reclaiming Public Services: How Cities and Citizens are Turning Back Privatisation*. Amsterdam: Transnational Institute

MacKinnon D (2011) Reconstructing scale: Towards a new scalar politics. *Progress in Human Geography* 35(1):21–36

Megill A (2002) *Karl Marx: The Burden of Reason (Why Marx Rejected Politics and the Market)*. Lanham: Rowman and Littlefield

Mirowski P (2013) *Never Let a Serious Crisis Go to Waste: How Neoliberalism Survived the Financial Meltdown*. London: Verso

O'Neill J (2006) Knowledge, planning, and markets: A missing chapter in the socialist calculation debate. *Economics and Philosophy* 22(1):55–78

Pitts F H and Dinerstein A C (2017) Corbynism's conveyor belt of ideas: Postcapitalism and the politics of social reproduction. *Capital and Class* 41(3):423–434

Polanyi K (1994) *The Great Transformation: The Political and Economic Origins of our Time*. Boston: Beacon Press

Emotions

Kye Askins

School of Geographical and Earth Sciences, University of Glasgow, Glasgow, UK;
kye.askins@glasgow.ac.uk

The truth will set you free, but first it will piss you off. (Steinem 1998)

To love well is the task in all meaningful relationships, not just romantic bonds. (hooks 2000)

On *Antipode*'s 40th anniversary, the then editors wrote: "We're in the midst of some exceptionally challenging, complex and momentous changes to the global economy, polity, society and ecology". Emphasising *Antipode*'s foundational remit for progressive change, they argued that "those of us who work with ideas and books, abstractions and words, among the sundry tools of the academic trade ... are faced with the task of using them to engage with the world in progressive ways" (Castree et al. 2010:2). *The point is to change it.*

Ten years on, the challenges are arguably more acute than ever. This brief intervention, embedded in theory and passion, argues that critical geographers/scholars cannot set about such engagement without *feeling*. Oppression and injustice are deeply emotional; tackling them likewise.

So, I take my remit here to ask: "How does it feel?" What are the radical logics of including and involving emotions?

How does it feel to be the target of racist/sexist/homophobic/ablist abuse? How does it feel to be dispossessed/homeless? How does it feel when you can't provide food, water, clothing for yourself/children/family? How does it feel to be trafficked? How does it feel to live under a political regime that violently curtails human rights? How does it feel to witness (directly or indirectly) another's suffering? How does it feel to research, teach and write—using our sundry tools—about these issues?

Indeed, the last decade has seen a significant increase in work exploring the role of emotions and affect in how social and spatial relations unfold. "How does it feel?" is gaining traction as a valid research focus, with how people feel (as well as think and do) politics becoming established in social movement and related social science circles. However, emotionality is still largely ignored, undermined, dis-valued in Enlightenment, masculinist scholarship more broadly: emotions and the body remain normatively constructed as the binary shadow to Rationality and Mind.

Yet emotions are central. Governments are manipulating people's fears and anxieties to re/produce enmity and otherness between marginalised groups, such that the status quo and the "1%" remain unchallenged. Dominant media discourses tell certain stories in sensational ways, to sew and spread anger and

Keywords in Radical Geography: Antipode at 50, First Edition. Edited by the *Antipode* Editorial Collective.
© 2019 The Authors/Antipode Foundation Ltd. Published 2019 by John Wiley & Sons Ltd.

resentment among their readership, and reiterate powerful myths and interests. Corporate actors (from international conglomerates to local entrepreneurs) work hard to heighten desire for their products and services. Structures, industries and geographies have been carefully honed to circulate affects and emotions for political, capitalist, authoritarian ambitions, around the world and at all scales.

When public feelings "run high", then, these are not individual sentiments but vital forces (used) in mutually co-constructing consent for state-sponsored military-carceral expansion and the retreat from public welfare, thus a radical project must grapple with emotions and affect, to address "the embodied and unconscious dimensions of oppression" (Ioanide 2018:3). Emotions produce real effects, with significant and often severe consequences, yet because we (as academics, as society) struggle to capture them in reason and evidence, they are seldom understood or treated as economies and polities (Ahmed 2004).

Thinking Emotionally

If utilising ideas and books, abstractions and words to resist/enable resistance to the complex relations and geographies of power is central to the task of radical scholarship, *a critical question and nagging feeling*, for me, is how to think about and with emotions. Understanding the world rationally only goes so far in progressive change for sustainable justice: we need to fold joy and compassion, hate and envy into these ideas and books, lectures and seminars.

Analytical thinking *is* vital. In emphasising the emotional, rationality is neither lost nor subsumed. Feminism has long argued for the consideration of subject positions in academic endeavour, critically reflecting on how bodies are caught up in knowledge production, through relations in-and-beyond the field: the embodied and emotional as always already enmeshed in rational apprehension of one another through ideological and socially constructed differences. Thinking/theory/ideas are imbued through feeling/affect/emotion, and vice versa. Body-and-mind are interconnected, thus radical analysis involves both.

Certainly, there is *desire* among radical geographers to work towards justice, beyond rational response. There is outrage! There is anger, upset, dismay ... and hope, care, concern, at times elation. Moreover, the role of emotions in motivating, *moving* people in thought and to action, activism and resistance is well established (e.g. Bosco 2007). Radical scholarship emphasises the importance of building, maintaining, extending solidarities between diverse social movements, scholars and publics. Such work must consider how solidarities are materially enacted (through provision of money, food, shelter, etc.); ideological (embedded in and sustained through ideas and words); and also produced through emotional connection. The analytic utility and radical potential of emotional geographies is in its attention to the role of emotions in social and spatial relations, and *how they do different kinds of work in different contexts*. Anger may be processed through violence or action for justice; grief may draw sympathy, empathy and active support, or prompt exclusionary behaviours. Paying attention to feelings as dynamic, situated in and relational across space, social practices and politics (Smith et al. 2009) deepens understanding of emotional connections between activists as *so central to solidarity* that they "have become the targets of both social movement strategies for growth and police strategies for social control" (Clough 2012:1667).[1]

Further, thinking with/through/about emotions demands a critical (re)orientation to the spatiality of politics, in which the everyday is intertwined through a range of scales. This resonates with work in feminist geopolitics, regarding the ways in which intimacies and politics are intertwined across diverse sites, highlighting relations of power acting on/through bodies, and how emotions, attitudes and comportments simultaneously enhance and/or constrain particular geopolitical subjectivities. Emotions, then, can't be treated through an analytical lens that contains, separates and limits these unruly spectres to an apolitical everyday. Indeed, neoliberal attempts to financialise feelings and affects precisely alert us to their multi-scalar significance, how emotions are envisaged, produced, circulated in capitalist politics/policies (Swyngedouw 2010).

Conceptualising mundane "micro-politics of emotion" as mutually co-productive of more formal political structures/systems critically speaks to a geography of responsibility to multiple others across scale, central to efforts for more just and sustainable worlds. Recent critical debates draw out scholarly and wider responsibilities to equality precisely through the relational geographies of individuals and communities, species and things, and land-, sea- and air-scapes (Bawaka Country et al. 2016). Feminist, critical race, Indigenous and queer scholarships are extending theories regarding body-and-mind interconnectivity to more-than-human concepts that encompass hybrid, plural, becoming bodies as pivotal in challenging and shifting oppressions of all kinds, open to the sensorial, embodied, fleshy and felt. Richa Nagar (in Chowdhury et al. 2016:1810) argues for "a richly chaotic terrain for undoing all cannons and for enabling an ever-contested, embodied, tentative, and unfolding cocreation with differentially situated knowledges and experience" to challenge hegemonic power. Thinking emotionally has a vital role in such progressive approaches to destabilising normative, exclusionary structures and discourses.

Working Emotionally

Geographers tell robustly researched stories of people and place, which are part of shaping those lives and worlds. Storytellers can empower geographical imaginations to envision how the world could be, and should take seriously the affective power of those stories and their circulations. Again, this is not as an alternative to theoretical work; rather to enrich and deepen research, writing, teaching, public engagement and other academic endeavours. Similar to how radical scholarship understands *both* theory *and* practice as implicitly political (knowledge and its production as political), *both* rational *and* emotional concerns are important for engaging more fully in and with the world. The discipline of radical geography must be relevant to those people whose distress, suffering, fury and political commitments are the material of academic interest. To speak back to such emotive positions only with theory seems ... partial.

Normative academic writing, conference presentation, knowledge circulation has long been entrenched in Enlightenment, masculinist and colonial structures of exclusion. Working emotionally requires resisting analytic closure, not least since emotions are socio-culturally constructed and experienced: again, the analytic utility of emotional geographies is in attending to the role of emotionality in social and spatial relations and political economies, not defining/understanding any one emotion (there is no reductive, universal "given" for anger/love/jealousy). This "richly chaotic" (Nagar in Chowdhury et al. 2016) refusal to censure and conclude remains marginalised. How does that feel, for the many whose stories/thinking/emotions "do not fit"? It is possible to remain rigorous and develop scholarly debates through *both* conceptual framing *and* with passion, trepidation, love and care: moreover, working emotionally is a radical movement for progressive change. Loss and trauma, violence and persecution are held and remembered through bodies and what they endure; vulnerability and precariousness ignite both sub- and conscious forms of knowing and doing, and "affective memories can serve as a political horizon" (Salih 2016). Such accounts are being written and told; the more radical of them in accessible language for, collaboratively with, and openly accessible to those whose stories they tell.[2]

I think, and my gut tells me, that it is critical to research and tell the emotional dimensions of injustice and exclusion. It is very difficult to counter hate-driven oppression with rationales, with arguments based on quantitative evidence. These logics are unmoored in an ocean of state- and corporate-sponsored sexist, racist, homophobic, ablist discourses that affectively fuel the hatred. I've experienced outright, and outraged, denial of asylum claims and refugee rights among politicians, policy-makers and individuals, when presented with statistics on inequality and legislation on human rights, the latter both outweighed by hegemonic fears, greed and stigma ... But I've also seen the affective and effective drip of emotive stories, with their potential to touch *"hearts and* minds", making a difference through education, research, community engagement.

Moreover, if scholarship only tells stories of inequality and despair, we risk re/making a world for academic colleagues, research participants, students and publics that emphasises the worst of humanity, and shuts down progressive change. "Critical thinking without hope is cynicism. Hope without critical thinking is naïveté" (Popova 2015). Countering pessimism in this "era of crisis" is a radical act in itself; critical scholarship can offer thoughtful and practical interventions that simultaneously draw on hope, agency, and enact transformative change across classrooms,[3] research, public engagement activities, and personal encounters as locally embedded citizens (Fuller and Askins 2010; Pickerill 2008; Van Wijnendaele 2013).

To this end, the discipline can usefully combine radical logics with an ethics of care: *caring-with* as a political project crucial to thinking, acting and becoming in more sustainable present-future worlds. Risking vulnerability, as an anti-hegemonic strategy, radical geography can further step away from "the objective" view and shift closer towards an academy that cares-with and feels-with others, recognising and valuing interdependence as an ethic vital to both care and justice (Askins and Blazek 2016). Grace Lee Boggs (2012) speaks poignantly on the broader need to move beyond revolution as binary opposition (cycles of left/right swings in which *power over* remains reductive)[4] and reimagine revolutionary politics *as evolutionary*, a project of transformation integrally connected to global and historical contexts and simultaneously intimately embedded in the individual soul.

I've witnessed heart-melting moments of generosity and compassion, between strangers of all manner of difference, who, in particular moments and places, feel compelled to reach out, connect and support each other. Such caring gestures are always powerful, overcoming hegemonic fears and stigmas to produce new relations and possibilities, already enacting a politics of engagement and "emotional citizenry", and anticipating how individual relations may enable collective change (Askins 2016). These gestures, those people, such care always floor me ... make me laugh, make me cry ... and *move me* to relay their radical potential in teaching, writing, everyday conversations. How does it feel? *The point is that it does.*

Endnotes

[1] Similarly, emotions are important to building solidarity among radical academic endeavours, and simultaneously a target of neoliberal university control tactics, as scholars are pitted against each other in competitive processes that produce fear and anxiety around potential precarity in our own working lives.
[2] There are issues here regarding the co-production of research and collective writing, beyond the scope of this piece (see Chowdhury et al. 2016).
[3] Radical pedagogies are crucial, in what, how and where we "teach" (see Springer et al. 2016).
[4] This resonates with anarchist geographies, see Springer (2013) and an *Antipode* Special Issue (http://142.207.145.31/index.php/acme/issue/view/72).

References

Ahmed S (2004) *The Cultural Politics of Emotions*. Edinburgh: Edinburgh University Press
Askins K (2016) Emotional citizenry: Everyday geographies of befriending, belonging, and intercultural encounter. *Transactions of the Institute of British Geographers* 41(4):515–527
Askins K and Blazek M (2016) Feeling our way: Academia, emotions, and a politics of care. *Social and Cultural Geography* 18(8):1086–1105
Bawaka Country, Wright S, Suchet-Pearson S, Lloyd K, Burarrwanga L, Ganambarr R, Ganambarr-Stubbs M, Ganambarr B, Maymuru D and Sweeney J (2016) Co-becoming Bawaka: Towards a relational understanding of place/space. *Progress in Human Geography* 40(4):455–475
Boggs G L (2012) *The Next American Revolution: Sustainable Activism for the 21st Century*. Berkeley: University of California Press
Bosco F (2007) Emotions that build networks: Geographies of human rights movements in Argentina and beyond. *Tijdschrift voor Economische en Sociale Geografie* 98(5):545–563
Castree N, Chatterton P, Heynen N, Larner W and Wright M W (eds) (2010) *The Point Is To Change It: Geographies of Hope and Survival in an Age of Crisis*. Oxford: Wiley-Blackwell
Chowdhury E H, Pulido L, Heynen N, Rini L, Wainwright J, Inayatullah N and Nagar R (2016) Book review forum—*Muddying the Waters: Co-authoring Feminisms Across Scholarship and Activism*. *Gender, Place, and Culture* 23(12):1800–1812
Clough N (2012) Emotion at the center of radical politics: On the affective structures of rebellion and control. *Antipode* 44(5):1667–1686

Fuller D and Askins K (2010) The (dis)comforting rise of public geographies: A public conversation. *Antipode* 39(4):579–601

hooks b (2000) *All About Love: New Visions*. New York: HarperCollins

Ioanide P (2018) *The Emotional Politics of Racism: How Feelings Trump Facts in an Era of Colorblindness*. Redwood City: Stanford University Press

Pickerill J (2008) The surprising sense of hope. *Antipode* 40(3):482–487

Popova M (2015) Hope, cynicism, and the stories we tell ourselves. *Brain Pickings* 9 February. https://www.brainpickings.org/2015/02/09/hope-cynicism/ (last accessed 17 April 2018)

Salih R (2016) Bodies that walk, bodies that talk, bodies that love: Palestinian women refugees, affectivity, and the politics of the ordinary. *Antipode* 49(3):742–760

Smith M, Davidson J, Cameron L and Bondi L (2009) *Emotion, Place, and Culture*. Farnham: Ashgate

Springer S (2013) Anarchism and geography: A brief genealogy of anarchist geographies. *Geography Compass* 7(1):46–60

Springer S, de Souza M L and White R J (eds) (2016) *The Radicalization of Pedagogy: Anarchism, Geography, and the Spirit of Revolt*. London: Rowman & Littlefield

Steinem G (1998) "Herstory." Keynote address to Stanford University

Swyngedouw E (2010) The communist hypothesis and revolutionary capitalisms: Communist geographies for the 21st century. In N Castree, P Chatterton, N Heynen, W Larner and M W Wright (eds) *The Point Is To Change It: Geographies of Hope and Survival in an Age of Crisis* (pp 298–319). Oxford: Wiley-Blackwell

Van Wijnendaele B (2013) The politics of emotion in participatory processes of empowerment and change. *Antipode* 46(1):266–282

Enough

Natalie Oswin

Department of Geography, McGill University, Montreal, QC, Canada;
natalie.oswin@mcgill.ca

During the national March For Our Lives that the student survivors of the mass shooting at Marjory Stoneman Douglas High School in Florida organised in March 2018, activist Emma González stated, " … in this case if you actively do nothing, people continually end up dead, so it's time to start doing something" (Grinberg and Muaddi 2018). Meanwhile, protestors at simultaneous rallies across the country chanted, "Enough is enough!".

Black Lives Matter—Toronto used its position as "Honoured Group" at the July 2016 Toronto Pride Parade, one of the largest such events in the world, to stage a sit-in and demand funding commitments for queer people of colour organisations, the removal of police floats from the event, and more. When Toronto Pride announced in 2017 that it would accede to these demands, new board member Akio Maroon stated, "I think the membership just said, you know what? Enough is enough" (Simmons 2017).

These are just two examples from Canada (where I live and work) and the United States (the behemoth next door) of recent moves to say "enough". Other examples from just these two national contexts include the Dakota Pipeline protests, the Idle No More movement, the Ferguson, Missouri, protests, the Me Too movement, various teacher's strikes, and the high-profile resignations of commissioners from Canada's National Inquiry into Missing and Murdered Indigenous Women and Girls. In these instances (and many more in Canada, the US, and elsewhere), people are powerfully standing up and insisting on the need for progressive action. There is a spreading exhaustion with the status quo, an expanding desire on the Left for radical rather than incremental change.

Further, many of these examples evidence the now widespread recognition in some activist worlds that no social justice issue is efficaciously approached as a single issue. Consider gun violence, the issue at the heart of the March For Our Lives. It is absolutely a political-economic issue, as the NRA and its allies in government put profits ahead of people. It is also clearly a race, gender, and sexuality issue, as people of colour, women, and sexual and gender minorities are the disproportionate victims of gun violence. In other words, economic inequality, toxic masculinity, settler colonialism, misogyny, homophobia, transphobia, and racism all sustain the USA's position as a world leader in gun deaths. Further, as the Black Lives Matter intervention at Toronto Pride shows, our putatively progressive movements need intersectional analyses too. LGBT issues are about much more than the policing of a heterosexual–homosexual binary. They are about hetero- and cis-normativity, which are sustained by intertwined race, settler, class, gender, and sexual norms that make specific forms of heterosexual formations and cis-gender embodiments seem right.

Radical geographers are of course extremely interested in these sorts of hearten-ing, necessary, and long overdue efforts to speak truth to power. Moreover, through their scholarship, teaching, and activism, they are doubtless already con-tributing in all kinds of ways to the creation of a political field in which these sorts of strong and intersectional actions occur. To do so fully, effectively, and carefully, though, many recognise that we need our own moment of reckoning. And this moment has been coming for quite some time.

Throughout the 1990s, during the "cultural turn" that swept across the human-ities and social sciences, an unprecedentedly diverse wave of critical scholarship shook the foundations of the discipline of geography in Anglo-America. In the 1950s and 1960s, positivist spatial science was the dominant game in town. Then, in the 1970s and 1980s, Marxist and humanistic geographies usefully com-plicated the field, bringing concerns with political relevance and social inequities (in the former case) and human agency (in the latter case) to the fore. But these bodies of work were buoyed by disturbingly limited analyses of embodiment, and thus they too were forcefully challenged when "cultural geography" was re-invented in the 1990s. Inspired in good measure by the work coming out of the Centre for Contemporary Cultural Studies in Birmingham (UK) and by the rise of identity politics and environmental movements, geographers across various sub-disciplines began rethinking culture as plural, heterogeneous, and shot through with power relations. Thus, a shift away from a narrow emphasis on class-based differences and presumptions of universal subject-hood began in earnest within the discipline, as geographers in large numbers turned to a wide array of concep-tual frameworks—most notably poststructuralist, critical race, queer, feminist, postcolonial, and psychoanalytic theories—for insight and inspiration. Conse-quently, positionality, the politics of knowledge production, the permeable boundaries between the cultural, the economic, the social, and the political, and the imbrication of discourse and materiality have been key themes for many geographers ever since.

At the peak of the "cultural turn", Cindi Katz captured its potential well. She wrote: "despite years of feminist, postcolonial, queer, and antiracist critique, and the rich, different productions of knowledge offered from these quarters, much social theory remains largely impervious to this work" (1996:488). She continued, powerfully:

> ... we have to learn to be genuinely cartographic—mapping a politics from our situ-ated positions, but also making usable and shared maps of the worlds we inhabit col-lectively. Gender is not class and class is not race; and the maps of their politics are not homologous. Yet we are lost—metaphorically and materially—if we think they are of separate worlds. Classed, sexed, raced, and gendered, we can produce renegade cartographies at once situated, fluid, and incorporative. (Katz 1996:495)

She argued for the need to grapple with the ways in which the "theoretical twists and turns" of the 1990s in geography were "as much about power and authority as about the production of theory and the constitution of knowledge" (1996:487). Katz also implored "minoritarian" theorists to refuse their marginalisation. And many have done so, both before and after Katz wrote these words. See, for instance, Bell and Valentine (1995), Blunt and McEwan (2002), Doan (2007), Kobayashi and Peake (1994), Mahtani (2006), McKittrick (2006), Nast (1994), Pulido (2002), Rose (1993), Ruddick (1996), Sanders (1990), and Shaw et al. (2006) for potent arguments for the need to take gender, race, Indigeneity, and sexuality much more seriously in geography.

These and many more interventions have changed geography for the better such that today there are thriving literatures on critical race, postcolonial, Indigenous, queer, trans, and feminist geographies. Yet, the struggle continues. Epistemic narrowness and a lack of diversity mar the discipline, in both its mainstream *and* its "critical" or "radical" iterations, into the present. As Katz (2017) recently put it: "Geography is less white, less masculinist, less straight than it was in 1989, but the erasures, occlusions, mansplaining, and minimizing go on ...".

Indeed. Consider the debates over the "planetary urbanization" thesis that are currently taking place in geography and the cognate field of urban studies. This Marxist/political economy approach has much to offer. Yet its architects largely ignore decades of critical race, postcolonial, Indigenous, feminist, queer, and trans geographical and urban studies scholarship, and thus it has come in for much critique (see, for instance, Buckley and Strauss 2016; Derickson 2018; McLean 2018; Oswin 2018; Reddy 2018; Roy 2016). It is frustrating that there was a need to make these critiques more than a quarter century after the "cultural turn". And it is especially frustrating that the critiques have been characterised by some as "dismissive caricatures" (Brenner 2018) and as "ad hominem attacks" (Keil 2018). Comparing these responses to David Harvey's (1992) well known dismissal of the critiques that feminist scholars Rosalyn Deutsche (1991) and Doreen Massey (1991) made of his book *The Condition of Postmodernity*, the sense of déjà vu is strong.

And the continued resistance to calls to diversify geographical approaches that is evident in the current planetary urbanisation debates is by no means exceptional. Thus scholars, as they did in the 1990s and early 2000s, have had to keep making the case for bringing critical race, postcolonial, Indigenous, feminist, queer, and trans approaches into geography. Minelle Mahtani critiques the field of social and cultural geography for what she calls "the ongoing production of ... toxic geographies, or emotionally toxic material spaces, for geographers of colour" (2014:360). Camilla Hawthorne and Brittany Meché (2016) and Sharlene Mollett and Caroline Faria (2018) also call out the enduring overwhelming whiteness of geography. Sarah Hunt points out that, within geography, "Indigenous knowledge is rarely seen as legitimate on its own terms, but must be negotiated in relation to pre-established modes of enquiry" (2014:29; see also Daigle forthcoming). Browne et al. note that "geographical enquiry has yet to explore the lives and experiences of people, including trans people, that trouble and call into question ... hegemonic, normative [gender] binaries" (2010:573). And I and others have long argued for the need for critical geography to add "sexuality" to the "race, class, gender" trinity, and for queer geography to think about sexuality in intersectional terms, but there has not been sufficient movement on either of these fronts (see Catungal 2017; Oswin 2008).

To echo the activists with whom I began, those who want action instead of the same old debates, enough is enough. We have enough evidence. We have enough testimonials. It is time to do geography differently.

Enough with the gatekeeping. Enough with the micro- and macro-aggressions that "minority" scholars experience daily. Enough with being called "minority" scholars. Enough with ignoring epistemological differences through recourse to "excellence". Enough with discourses of "merit" that obscure the unlevel playing field that gives many of us unfair advantages. Enough with using communities for vanity projects that fuel individual careers. Enough telling us to cite your canon while ignoring ours. Even with the canon. Enough with forgetting the violence and exploitation upon which our discipline is built.

It is time to listen and support those who literally have skin in the game. We must diversify our discipline, through hiring and student recruitment practices, through panel composition, through editorial practices, through the assembly of research teams, and through the adjustment of research and citation practices.

I am not naïve. I know that many will scoff at this call, because the problematic politics and practices of geographic knowledge production are deeply entrenched. They are, in other others, not the result of some people just not getting it and needing to be convinced, but the result of diffuse, complex, and deeply grooved conceptual and personal pathways. I also know that many others will nod along but will be unable to work toward actual change due to the constraints of precarious institutional positions. So, it is up to those of us who have the drive and ability to heed this call to stand up, with humility and a willingness to be uncomfortable, and push harder than ever before. The politics of the institution may seem tedious and tiring and insulated from everyday life, but as "critical" and "radical" geographers we research and teach on life and death issues every day, and if we are to maximise our solidarities outside our institutions we must change them from within.

References

Bell D and Valentine G (eds) (1995) *Mapping Desire: Geographies of Sexualities*. London: Routledge

Blunt A and McEwan C (2002) *Postcolonial Geographies*. London: Continuum

Brenner N (2018) Debating planetary urbanization: For an engaged pluralism. *Environment and Planning D: Society and Space* 36(3):570–590

Browne K, Nash C and Hines S (2010) Towards trans geographies. *Gender, Place, and Culture* 17(5):573–577

Buckley M and Strauss K (2016) With, against, and beyond Lefebvre: Planetary urbanization and epistemic plurality. *Environment and Planning D: Society and Space* 34(4):617–636

Catungal J P (2017) Toward queer(er) futures: Proliferating the "sexual" in Filipinx Canadian sexuality studies. In R Diaz, M Largo and F Pino (eds) *Diasporic Intimacies: Queer Filipinos and Canadian Imaginaries* (pp 23–40). Chicago: Northwestern University Press

Daigle M (forthcoming) The spectacle of reconciliation and hollow gestures of recognition: On (the) unsettling responsibilities to Indigenous peoples. *Environment and Planning D: Society and Space*

Derickson K (2018) Masters of the universe. *Environment and Planning D: Society and Space* 36(3):556–562

Deutsche R (1991) Boys town. *Environment and Planning D: Society and Space* 9(1):5–30

Doan P (2007) Queers in the American city: Transgendered perceptions of urban space. *Gender, Place, and Culture* 14(1):57–74

Grinberg E and Muaddi N (2018) How the Parkland students pulled off a massive national protest in only five weeks. *CNN* 26 March. https://edition.cnn.com/2018/03/26/us/march-for-our-lives/index.html (last accessed 20 August 2018)

Harvey D (1992) Postmodern morality plays. *Antipode* 24(4):300–326

Hawthorne C and Meché B (2016) Making room for black feminist praxis in geography. *SocietyandSpace.org* 30 September. http://societyandspace.org/2016/09/30/making-room-for-black-feminist-praxis-in-geography/ (last accessed 20 August 2018)

Hunt S (2014) Ontologies of Indigeneity: The politics of embodying a concept. *Cultural Geographies* 21(1):27–32

Katz C (1996) Toward minor theory. *Environment and Planning D: Society and Space* 14(4):487–499

Katz C (2017) On rocking "the project": The beat goes on. *AntipodeFoundation.org* 23 October. https://antipodefoundation.org/2017/10/23/on-being-outside-the-project/ (last accessed 20 August 2018)

Keil R (2018) The empty shell of the planetary: Re-rooting the urban in the experience of the urbanites. *Urban Geography*. https://doi.org/10.1080/02723638.2018.1451018

Kobayashi A and Peake L (1994) Unnatural discourse: "Race" and gender in geography. *Gender, Place, and Culture* 1(2):225–243

Mahtani M (2006) Challenging the ivory tower: Proposing anti-racist geographies in the academy. *Gender, Place, and Culture* 13(1):21–25

Mahtani M (2014) Toxic geographies: Absences in critical race thought and practice in social and cultural geography. *Social and Cultural Geography* 15(4):359–367

Massey D (1991) Flexible sexism. *Environment and Planning D: Society and Space* 9(1):31–57

McKittrick K (2006) *Demonic Grounds: Black Women and the Cartographies of Struggle*. Minneapolis: University of Minnesota Press

McLean H (2018) In praise of chaotic research pathways: A feminist response to planetary urbanization. *Environment and Planning D: Society and Space* 36(3):547–555

Mollett S and Faria C (2018) The spatialities of intersectional thinking: Fashioning feminist geographic futures. *Gender, Place, and Culture* 25(4):565–577

Nast H (1994) Women in the field: Critical feminist methodologies and theoretical perspectives. *The Professional Geographer* 46(1):54–66

Oswin N (2008) Critical geographies and the uses of sexuality: Deconstructing queer space. *Progress in Human Geography* 32(1):89–103

Oswin N (2018) Planetary urbanization: A view from outside. *Environment and Planning D: Society and Space* 36(3):540–546

Pulido L (2002) Reflections on a White discipline. *The Professional Geographer* 54(1):42–49

Reddy R (2018) The urban under erasure: Towards a postcolonial critique of planetary urbanization. *Environment and Planning D: Society and Space* 36(3):529–539

Rose G (1993) *Feminism and Geography: The Limits of Geographical Knowledge*. Cambridge: Polity

Roy A (2016) What is urban about critical urban theory? *Urban Geography* 37(6):810–823

Ruddick S (1996) Constructing difference in public spaces: Race, class, and gender as interlocking systems. *Urban Geography* 17(2):132–151

Sanders R (1990) Integrating race and ethnicity into geographic gender studies. *The Professional Geographer* 42(2):228–231

Shaw W S, Herman R D K and Dobbs R (2006) Encountering Indigeneity: Re-imagining and decolonizing geography. *Geografiska Annaler: Series B, Human Geography* 88(3):267–276

Simmons T (2017) "Enough is enough": Pride Toronto board member explains decision to ban police from parade. *CBC News* 23 January. http://www.cbc.ca/news/canada/toronto/pride-board-member-response-1.3947820 (last accessed 20 August 2018)

Experimentations

Jenny Pickerill

Department of Geography, University of Sheffield, Sheffield, UK;
j.m.pickerill@sheffield.ac.uk

Being experimental and experimenting in other ways of being is a central tenet of radical geography (Bell and Valentine 1995; Gibson-Graham 2006; Harvey 2000). What constitutes experimental geography is contested and unlimited, and includes experimentation in methods, knowledge, practices, representations and with interdisciplinarity (Kullman 2013). Yet across the many forms of experimentation is a common emphasis on venturing into the unknown, creativity, openness and on facilitating change (Gross 2010; Kerr 2008; Paglen 2008). It is a "conscious mode of intervention ... a project to make others conscious of their 'world-making'" (Last 2012:709) that seeks change beyond academia. It is a process of "learning by doing" (Caprotti and Cowley 2017). There is a joy in this engagement with the unknown, a creative risk, a hope, a play with surprise and a release from the strictures of conventional academic knowledge production.

There is often a distinction made between geographers conducting experiments or staging experimental interventions (experimental geographies), and the geographical analysis of already existing sites of experimentation (geographies of experiments) (Kullman 2013). For example, geographers working on experimental urbanism (Bulkeley et al. 2018; Evans 2011; Evans et al. 2016) have often not set up these experiments themselves, but use them to explore the possibilities of alternative urban futures. Alternatively, creative cultural geographers experiment in collaborating with artists in new methods of representation and intervention (Hawkins 2011). This distinction, however, is troubled by the way many geographers actively participate in a variety of experiments while not being the sole or lead author, and such experiments rarely being established primarily as an academic moment, for example, Paul Chatterton's role as a co-founder of the LILAC eco-community in Leeds, England. Indeed, experimentation is often an explicitly more collaborative and responsive way of working, of being co-experimenters.

While some reject any focus on useful or productive outcomes from such experimentation, or argue that such outcomes will always remain unmeasurable, for many geographers the point of experimentation is to produce a noticeable change, even if the form of such change will initially be unknown and uncertain. Despite openness and surprise being an important element of experiment, without some clear intent there is "a danger of not arriving at anything at all" (Last 2012:716).

Experimentation has an uneasy history in geography, which is seemingly ignored and troublingly repeated by some contemporary geographers. There are worrying links between experimentation, exploration and exploitation in the

Keywords in Radical Geography: Antipode at 50, First Edition. Edited by the *Antipode* Editorial Collective.
© 2019 The Authors/Antipode Foundation Ltd. Published 2019 by John Wiley & Sons Ltd.

colonial past of the discipline. But the quest to experiment, for example in exploration of urban hidden spaces, has led to accusations of unreconstructed notions of the white male macho explorer (Mott and Roberts 2014). Equally there are "dark" experimental spaces, such as refugee camps or concentration camps, and problematic notions of "testing" subjects in such spaces (Kullman 2013; Last 2012).

If there is an interest by geographers in experimental interventions that enact societal and environmental change, then there has also been a particular focus on certain types of experimentations. Geographers have a fascination with how money, food, transport, energy and housing can be collectively done differently, with more attention to reducing environmental impact and facilitating social change. These have been explored through urban labs, community economies, community bicycle repair shops, food growing networks, community gardens, land trusts, co-housing and autonomous geography projects (Carlsson and Manning 2010). These experiments are examined through attention to their new governance structures, planning processes, household infrastructures, collaborations, and social practices, which require "attention to the full range of bodies, texts and practices that constitute spaces of experimentation" and how they weave together (Powell and Vasudevan 2007:1790).

Common to these experimentations is an emphasis on the neighbourhood scale, the everyday, personal and collective action, often with the explicit intention that should such experimentations work then they can be "scaled up". But experimentations are not limited to these often-urban neighbourhood places; they are also in online spaces (activist media), transient moments (music festivals and protest camps), and long-running rural spaces (eco-communities). Much geographical research has focused on small-scale alternative bundles of practices (such as local food growing schemes) that signal post-capitalist possibilities (Gibson-Graham 2006). Yet research into eco-communities calls for greater attention to be paid to the interdependencies and interrelationships between the social, material, political, economic and environmental activities of these alternatives, and how they interconnect between places and scales (Pickerill 2016).

Experimentations are not necessarily temporary nor spatially bounded; experiments overflow into their surroundings creating "complex cartographies" such that "an experimental intervention is necessarily a temporal-spatial one" (Davies 2010:668), and obsolete experiments will be replaced by new ones while others elsewhere will endure. Not only will these experimentations be embodied differently but "the body becomes an experimental site" (ibid.) where, for example, alternative food practices are engaged with. Experimentation is not limited to organised defined experiments, but encompasses the ongoing processes of experimenting with new ways of daily life, be that in response to changing personal circumstances or crises (Kullman 2013). Crucially, there will be a multiplicity in how such experimentation is generated, experienced, understood and contested.

While many of these experimentations involve material changes, such as developing new infrastructures for growing and sharing food, they are social as much as material, often requiring changes to social practices, social relations and social expectations. Change often needs to be collective; a process of encouraging each other in shifting cultural expectations of what is acceptable. Understanding these social dynamics, particularly how people adapt and reconfigure what they deem acceptable and appropriate, is central to enabling transformative change. Indeed Caprotti and Cowley (2017) argue that there needs to be more focus on these subjects of experimentation.

With any experimentation it is vital to ask "who or what is really being transformed, and to what ends?" (Last 2012:710). Given the history of experimentation in geography there remains a risk that experimentation is about the privileged, heroic experimenter. This is evident in the experimental work of eco-communities, most of which have a diversity problem and struggle to reach beyond a highly educated, white, able-bodied cohort. Chitewere's analysis of EcoVillage at Ithaca (EVI) detailed how the class and identity of its residents' acts to exclude differentiated others. Only those "with both the economic and social capital" (2018:140), who have money but only work part-time or from home can join EVI. Seemingly subtle assumptions about participation, common values, lifestyle and food choices shape who gets to be part of eco-community experiments. There are important questions here about power, privilege and participation.

Yet some forms of experimentation, especially in cities, seek to turn all citizens into experimental subjects. Datta's (2018) work on smart cities in India contends that urbanism in the global South can violently exclude subaltern others who are deemed to not fit into the experimental future city. Experimentation here is a form of neoliberal expansion rather than a politically progressive intervention for social change.

Even seemingly politically progressive experimentations can actually be built on troubling exclusions. What might initially appear as alternative forms of transformation can be built on neoliberal rationalities, reproducing neoliberal conditions that undermine their radical potential. Argüelles et al. summarise such rationalities as a focus on individual responsibility rather than calling for State intervention, which in turn "might help to legitimize neoliberal attempts of disposing the State from its economic and societal functions" (2017:38). This ability to retreat from the State is reliant on the "privileged progressive whiteness that permeates" (Argüelles et al. 2017:40) these experiments, an environmental and social privilege that enables such individuals to self-provide, self-organise and improve their quality of life. Absence of a critical analysis of privilege and power in such experiments means the broader political possibilities of transformative change are limited. There is a need, then, to be vigilant to the *politics* of experiments (Powell and Vasudevan 2007) and to their *justice* (Caprotti and Cowley 2017).

It is vital that geographers are vigilant in their construction of and/or support for such experimentation. There is a tension here in the scale and temporality of what such experiments are trying to achieve, whether they are momentary interventions that offer temporary (and ultimately inadequate) solutions to "solvable" problems, or whether they are tackling larger structural issues for which solutions are less immediately likely but through which, in the long run, more transformative and inclusive changes might be realised. There is also a tension in how crisis can appear to offer an opportunity for experimental innovation, given that crisis can often be politically manufactured to facilitate regressive change.

Giuseppe Feola (in his project "Societal Transformation to Sustainability through the Unmaking of Capitalism?") argues that geographers have not paid enough attention to the need to make space—symbolically, materially, temporally and spatially—for experimentation. Using the example of the need for space to experiment in degrowth, Feola explores the necessity to unmake, deconstruct, destabilise or displace existing capitalist socio-ecological configurations. Without adequate attention to processes of decay then the dynamics identified by Argüelles et al. (2017), of apparent alternatives continuing to be built within neoliberal rationalities, are likely to continue. Unmaking does not, however, require the erasure of all that has gone before; rather it is a creative and generative process alongside acts of refusal, a material as well as a social unmaking. This is a dynamic open-ended process of destruction alongside creation, mirroring Davies' (2010) articulation of the spatial-temporality of experimentations folding into and over one another. This process will likely be messy, at times personally contradictory, hidden rather than overt, but always generative.

Using this notion of unmaking highlights the need to understand better the failure of experimentations. Despite experimentation often being about understanding whether something works, is successful and has the potential to facilitate social change, the failure of experiments is too rarely discussed. Geographers have started to more openly talk about failure, and have sought to recast failure as a necessary part of the research process, arguing that discussing such failures is enriching and ultimately productive (Harrowell et al. 2018). Failure is integral to the process of learning, reflecting and improving. Yet disclosing failure remains an inherently risky act in the contemporary neoliberal university and is often avoided (Klocker 2015). Indeed many geographers have tended to identify the positive possibilities for transformation in experimentations such as eco-communities and alternative food networks, rather than their limitations.

What is powerful about experimentations is their messy, unfinished, fluid and open nature, their unbounded, overflowing implications, but only if we also engage in what is not achieved, undone or remade. This involves staying with the mess, working with experimentations over longer timeframes and critically reflecting on failure, who is included, who and what is being transformed, and the experimental subjects.

Experimentations are multi-level and multi-dimensional, and they have particular spatialities and temporalities. Caprotti and Cowley (2017) argue that experimentations require some element of structure, of boundedness, a clear intent, beginning and end, and Kullman suggests that experiments need to be "controlled enough to hold together" (2013:885). Yet geographers should also expand the notion of experimentation to include more open-ended, fluid interventions in "world-making" which might be better positioned to tackle larger structural issues and therefore more transformative change. Experimentation allows the opportunity for change, a hopefulness in an uncertain world, but we must remain vigilant in ensuring such hope is reflected in practice and is not yet another form of neoliberal appropriation.

References

Argüelles L, Anguelovski I and Dinnie E (2017) Power and privilege in alternative civic practices: Examining imaginaries of change and embedded rationalities in community economies. *Geoforum* 86:30–41

Bell D and Valentine G (eds) (1995) *Mapping Desire: Geographies of Sexualities*. London: Routledge

Bulkeley H, Marvin S, Voytenko Palgan Y, McCormick K, Breitfuss-Loidl M, Mai L, von Wirth T and Frantzeskaki N (2018) Urban living laboratories: Conducting the experimental city? *European Urban and Regional Studies* https://doi.org/10.1177/0969776418787222

Caprotti F and Cowley R (2017) Interrogating urban experiments. *Urban Geography* 38 (9):1441–1450

Carlsson C and Manning F (2010) Nowtopia: Strategic exodus? *Antipode* 42(4):924–953

Chitewere T (2018) *Sustainable Community and Green Lifestyles*. London: Routledge

Datta A (2018) The "digital turn" in postcolonial urbanism: Smart citizenship in the making of India's 100 smart cities. *Transactions of the Institute of British Geographers* https://doi.org/10.1111/tran.12225

Davies G (2010) Where do experiments end? *Geoforum* 41:667–670

Evans J, Karvonen A and Raven R (eds) (2016) *The Experimental City*. London: Routledge

Evans J P (2011) Resilience, ecology and adaptation in the experimental city. *Transactions of the Institute of British Geographers* 36(2):223–237

Gibson-Graham J K (2006) *A Postcapitalist Politics*. Minneapolis: University of Minnesota Press

Gross M (2010) *Ignorance and Surprise: Science, Society, and Ecological Design*. Cambridge: MIT Press

Harrowell E, Davies T and Disney T (2018) Making space for failure in geographic research. *The Professional Geographer* 70(2):230–238

Harvey D (2000) *Spaces of Hope*. Berkeley: University of California Press

Hawkins H (2011) Dialogues and doings: Sketching the relationships between geography and art. *Geography Compass* 5(7):464–478

Kerr I (2008) Research and development. In N Thompson (ed) *Experimental Geography* (pp 63–77). New York: Melville House

Klocker N (2015) Participatory action research: The distress of (not) making a difference. *Emotion, Space, and Society* 17:37–44

Kullman K (2013) Geographies of experiment/experimental geographies: A rough guide. *Geography Compass* 7(12):879–894

Last A (2012) Experimental geographies. *Geography Compass* 6(12):706–724

Mott C and Roberts S M (2014) Not everyone has (the) balls: Urban exploration and the persistence of masculinist geography. *Antipode* 46(1):229–245

Paglen T (2008) Experimental geography: From cultural production to the production of space. In N Thompson (ed) *Experimental Geography* (pp 27–33). New York: Melville House

Pickerill J (2016) *Eco-Homes: People, Place, and Politics*. London: Zed

Powell R C and Vasudevan A (2007) Geographies of experiment. *Environment and Planning A* 39(8):1790–1793

Fieldwork

Kiran Asher[iD]

Women, Gender, Sexuality Studies, University of Massachusetts Amherst, Amherst, MA, USA;
kasher@umass.edu

Radical geography aims to know the world and change it for the better. These two goals are linked to Enlightenment modernity and science. However, unlike mainstream science, in radical geography both knowledge production and struggles for justice are political, power-laden processes.[1] Furthermore, critical geographers are learning from feminist, anti-colonialists and others committed to radical social change to acknowledge that uneven capitalist development is premised on exploiting colonialised/raced, gendered, sexualised and non-human Others. Thus, addressing persistent economic and class inequities necessarily means grappling with concomitant social, sexual and environmental violence. Changing the world for the better then not only requires imagining different relationships between humans, but also between humans and nature.

The task of imagining and constructing just worlds requires fieldwork. In Ecology and Political Science, the disciplines in which I was formally trained, fieldwork meant going somewhere (the "field") to collect verifiable, generalisable data about specific aspects of the environment, society, space, or politics. But extended research on ecology, economic development, and social movements revealed that such positivist approaches did not account for the messy connections between people, places, and non-human denizens of this planet. I also made the uncomfortable discovery that my research questions, categories, and field sites were overdetermined by history, politics, and my own identity as a "third world woman" (Asher 2017; Katz 1994). By "fieldwork", then, I do not refer to a method of knowing something, or the means of taking action for scientific or political ends. Rather I contend that fieldwork entails fundamentally interrogating the work done within fields of inquiry, including radical geography, to produce legible and legitimate objects and subjects of knowledge and action.

Such work was gathering pace in the 1990s. Scholars from a range of disciplines—anthropology, environmental history, geography, sociology—contested the technical, apolitical, and ahistorical approaches of positivist, empiricist sciences. In conjunction with colleagues from the humanities, and drawing on post-structural and post-colonial critiques of modernity and the Enlightenment, they opened up conversations about the meanings, production, affects, and effects of scientific knowledge.[2] As a field biologist, I found my way into these conversations with Donna Haraway's (1989) *Primate Visions*, and then via feminist writing.

Feminists questioned the masculinist and essentialist assumptions pervading fields from anthropology to zoology in order to focus attention on the partial

Keywords in Radical Geography: Antipode at 50, First Edition. Edited by the *Antipode* Editorial Collective.
© 2019 The Authors/Antipode Foundation Ltd. Published 2019 by John Wiley & Sons Ltd.

nature of evidence, and the unpredictable and contingent nature of research. They also examined how the foundational categories and dualisms of the Enlightenment (nature–culture, object–subject, feminine–masculine, sex–gender, colony–nation, knowledge–praxis, and more) were constituted as a result of power, representation, and political economy. Feminist anthropologists and geographers scrutinised how ethnography and cartography assumed that "cultures" and "fields" existed a priori, rather than as products of uneven relations between peoples and places. Feminist and other geographers also denounced the discipline's links to the military and the territorial and imperialist imperatives of states. Marxist feminists traced how the uneven relations of capitalist production piggyback on patriarchal, racist and imperialist structures, and how social production necessarily but silently depends on social reproduction (Federici 2012; Katz 2001b; Mies 1982; Wright 2006). Applied to the Marxist variant of modernity, this means rethinking the analytical parameters of class and the political promise of revolutionary change.

Since the 1990s, various feminist, post-colonial, transnational, "of colour", black, queer, decolonial, post-humanist, and other critical perspectives have reframed debates about science, the nature of subjectivity, domination, and resistance; and posited new forms of radical politics. Despite this work, empiricism continues to haunt geography, and positivist methods (strongly influenced by "evidence-based science") have re-emerged in the 21st century. As I face the challenging task of rethinking the fields and work of social and environmental justice, I repeatedly return to the realisation that the ethical imperative to act must be accompanied by persistent critique. I learn that grappling with the pitfalls of essentialist and oppositional thinking involves the impossible necessity of depending on the very categories of my critiques, for example, "third world women" and "nature".

Once again feminist politics and theory offer an illustration and a key lesson. Western feminists distanced themselves from the "women=nature" connection as they sought access to subjectivity and citizenship by claiming that they too belonged to "culture/reason". In later moments of feminist politics and theorising, attempts to make third world or non-western women visible and retrieve their heretofore hidden agency and voice sought to highlight their traditional or cultural knowledge about the natural world. Some versions of this are about retrieving the speech of subaltern women. In her discussion of the debates around women, gender and development, Kriemild Saunders (2002:13) illustrates the problems with both positions:

> Rather than thinking the impoverished Third World Woman as a sovereign, having a privileged insight into development processes, it may be more appropriate and cogent to see her positioning as a symptom of the over-determined effects and resistances to multiple oppressions and exploitative processes. This delineation definitively takes away the authority of a sovereign, revolutionary subject.

Neither claiming connections to nature nor distancing oneself from it are unproblematic options, and one needs to think about the vulnerability of all positions. These are key lessons for those of us foregrounding or formulating other (non-

Eurocentric, radical, post-humanist, subaltern are some contingent descriptors) ways of knowing but wishing to avoid nativist thinking.

To build on these lessons, I turn to Gayatri Spivak's anti-disciplinary texts to grapple specifically with two key issues of fieldwork: first, the ethics and politics of representation (present in her call "to learn to learn from below" and to "field-work" itself); and second planetarity—a concept that asks us to think alterity beyond the human. In the afterword to *Imaginary Maps*, her translation of Mahasweta Devi's short stories about tribals in India, she notes:

> I have no doubt that we must *learn* to learn from the original practical ecological philosophers of the world, through slow, attentive, mind-changing (on both sides), ethical singularity that deserves the name of "love"—to supplement necessary collective efforts to change laws, modes of production, systems of education and health care ... Indeed, in the general predicament today, such a supplementation must become the relationship between the silent gift of the subaltern and the thunderous imperative of the Enlightenment to "the public use of Reason", however hopeless that undertaking might seem. One filling the other's gap. (Spivak 1995:201)

Spivak's call to learn from below, which she calls "fieldwork" in a 2002 interview with Jenny Sharpe (Sharpe and Spivak 2002) is very different from the retrieval of subaltern knowledge or subjectivity. When focusing on other knowledges, Spivak's work provides critical methodologies to grasp how our desire for "other" alternatives is also bound up in political economy. Spivak's approach is deconstructive, anti-positivist, and focuses on understanding those ambiguous desires, analyses, and actions. The much-misunderstood argument in her "Can the subaltern speak?" (1988) is that those who wish to make the subaltern speak must locate themselves in the international division of labour.

Spivak (1999) calls for a historico-political perspective to trace the erasure and mobilisation of "culture" in dominant narratives, including those of the state, nationalism, and development. Her methodology entails a scathing critique of capitalist development and "supplements Marxism" in service of feminist, anti-racist and anti-colonial efforts. This "supplementing" neither offers a corrective to Marxism nor rejects it. Rather it works with Marx's thought to trace the insertion of rural communities, especially third world women, into the circuits of global capitalism.

Spivak draws attention to non-Eurocentric ecological movements and to "planetarity"—a term she uses for being concerned about nature and ecology beyond human self-interest or functionality. Once again a caution that this is not about recovering subaltern practices. She notes:

> In order for planetarity to be used for feminist utopias or ecological justice or whatever, you would have to put it in the value form, and I use the term value form in the original Marxian, not Marxist, sense. Marxists have either given it up or are confused about it, one reason being that the English translation of that simple sentence in Marx describing value has right from the start been wrong. Marx writes that it is *inhaltlos und einfach*, "contentless and simple". Why contentless? Because it allows the use of a form. All the English translations are "slight and simple" or "slight in content". How could they mistranslate a word like *inhaltlos* in which the *-los* is cognate to English

"-less"? The only answer is that they didn't understand what Marx was trying to say. Take the example of a bottle of water where you have the ingredients listed and assigned percentages. That is water put in the value form. Because the value form ... is what makes commensurability possible. So by putting a certain percentage on this ingredient you make this water commensurable with roast beef, say. You can compare. That's all it is. If you want planetarity to travel to ecological justice, or utopian feminism, or whatever, you have to put planetarity in the value form, and its unmotivated reminding task—of an epistemological gap—evaporates. Marx was inviting us to understand and use the pharmakonic potential of quantification through an appreciation of the value form. Planetarity is elsewhere, always, from finding a measure. (Spivak 2011:61–62)

Spivak repeatedly returns us to non-passages and aporias. She rereads Marx to trace the work and expansion of capital, and how it draws more and more of the world into its orbit by turning all kinds of products and knowledge, including the "Indigenous" variety, into the value form. But capitalist development cannot simply be replaced by Indigenous alternatives. Nor can it be accepted, given that it is unjust by its very nature. Marx recognised this, but his thinking was limited by his time and culture. It is imperative to supplement Marxism and re-engage his writings, including about socialism as an alternative to capitalism. It is equally imperative to think beyond economic globalisation to "re-imagine the planet" (Spivak 2012:335). It is for these tasks that Spivak invites us to think ethically with the responsibility-based livelihoods of aborigines—and through them rethink the relations between "culture" and "nature".

My fieldwork then is my homework, and its challenge is to articulate why and how the grammars of feminism, anti-colonialism, and Marxism in a transnational frame are crucial to struggles for environmental and social justice. Such work is also at the heart of radical geography. Its challenges are as large as its goals are ambitious: to change the world and suture the many severed relations between nature-cultures through methodological, epistemic and actual multi-lingualism.

Endnotes

[1] Of course as this volume outlines, there are vast debates about the parameters and genealogies of radical geography and those of the notion of "justice".

[2] This too is an extensive literature of which the following are a limited sample (see Bourdieu 1993; Cronon 1992; Hall et al. 1996; Haraway 1988; Katz 2001a; Staeheli and Lawson 1994; Vishweswaran 1997; D.L. Wolf 1996; E.R. Wolf 1982; see also, more recently, Bryan and Wood 2015; Spivak 2014; Wainwright 2013).

References

Asher K (2017) Thinking fragments: Adisciplinary reflections on feminism and environmental justice. *Catalyst: Feminism, Theory, Technoscience* 3(2):1–28

Bourdieu P (1993) *The Field of Cultural Production*. New York: Columbia University Press

Bryan J and Wood D (2015) *Weaponizing Maps: Indigenous Peoples and Counterinsurgency in the Americas*. New York: Guilford Press

Cronon W (1992) *Nature's Metropolis: Chicago and the Great West*. New York: W.W. Norton

Federici S (2012) *Revolution at Point Zero: Housework, Reproduction, and Feminist Struggle*. Oakland: PM Press

Hall S, Held D, Hubert D and Thompson K (eds) (1996) *Modernity: An Introduction to Modern Societies.* Oxford: Blackwell

Haraway D (1988) Situated knowledges: The science question in feminism and the privilege of partial perspective. *Feminist Studies* 14(3):575–599

Haraway D (1989) *Primate Visions: Gender, Race, and Nature in the World of Modern Science.* New York: Routledge

Katz C (1994) Playing with field: Questions of fieldwork in geography. *The Professional Geographer* 46(1):67–72

Katz C (2001a) On the grounds of globalization: A topography for feminist political engagement. *Signs* 26(4):1213–1234

Katz C (2001b) Vagabond capitalism and the necessity of social reproduction. *Antipode* 33 (4):709–728

Mies M (1982) *Lace Makers of Narsapur: Indian Housewives Produce for the World Market.* London: Zed Press

Saunders K (2002) Towards a deconstructive post-development criticism. In K Saunders (ed) *Feminist Post-Development Thought: Rethinking Modernity, Post-Colonialism, and Representation* (pp 1–38). London: Zed

Sharpe J and Spivak G C (2002) A conversation with Gayatri Chakravarty Spivak: Politics and the imagination. *Signs* 28(2):609–624

Spivak G C (1988) Can the subaltern speak? In C Nelson and L Grossberg (eds) *Marxism and the Interpretation of Culture* (pp 271–313). Chicago: University of Illinois Press

Spivak G C (1995) *Imaginary Maps: Three Stories by Mahasweta Devi.* London: Routledge

Spivak G C (1999) *A Critique of Postcolonial Reason: Toward a History of the Vanishing Present.* Cambridge: Harvard University Press

Spivak G C (2011) Love: A conversation (Gayatri Chakravorty Spivak with Serene Jones, Catherine Keller, Kwok Pui-Lan and Stephen D. Moore). In S D Moore and M Rivera (eds) *Planetary Loves: Spivak, Postcoloniality, and Theology* (pp 55–78). New York: Fordham University Press

Spivak G C (2012) *An Aesthetic Education in the Era of Globalization.* Cambridge: Harvard University Press

Spivak G C (2014) Scattered speculations on geography. *Antipode* 46(1):1–12

Staeheli L A and Lawson V (1994) A discussion of "Women in the field": The politics of feminist fieldwork. *The Professional Geographer* 46(1):96–102

Vishweswaran K (1997) Histories of feminist ethnography. *Annual Review of Anthropology* 26:591–621

Wainwright J (2013) *Geopiracy: Oaxaca, Militant Empiricism, and Geographical Thought.* New York: Palgrave Macmillan

Wolf D L (ed) (1996) *Feminist Dilemmas in Fieldwork.* Boulder: Westview

Wolf E R (1986) *Europe and the People Without History.* Berkeley: University of California Press

Wright M W (2006) *Disposable Women and Other Myths of Global Capitalism.* London: Routledge

Fracking

Bruce Braun

Department of Geography, Environment, and Society, University of Minnesota, Minneapolis, MN, USA;
braun038@umn.edu

In the years after 2008, rural residents of western North Dakota complained that the dark night sky, to which they were accustomed, had disappeared. The culprit —the flaring of unwanted gas from oil wells—was a side effect of the rush to capture oil rents in the region, as new technological innovations and high oil prices made "tight oil" plays across the western United States commercially viable. Indeed, the rapid spread of "fracking"—a contentious method of introducing fractures to rock formations to release oil and gas—became a flashpoint for social and environmental conflict across the US in the early 2000s as new resource frontiers were opened and old ones reactivated, disrupting communities, introducing new environmental risks, and intensifying struggles over land and resources. But fracking was not just a concern for local communities and Indigenous groups in the US, a few of which embraced the economic development that it promised to bring. For fracking's atmospheric effects were not limited to the disappearance of the dark night sky. In places like North Dakota, gas flaring and well emissions added more methane to the atmosphere. And, as oil and gas production soared and global oil prices fell, new energy infrastructure locked in fossil fuel use for the foreseeable future, threatening to further accelerate global climate change and, with it, devastation to distant communities vulnerable to rising seas or extreme weather.

Fracking, it turns out, is much more than a new oilfield technology, and much more than a local issue, for it stitches together the earth and the sky, the local and the global, the geological and the political, and the past and future. Dig into fracking and you are invariably led in a thousand directions—into the nature of capitalism, the social relations of technology, the future(s) of humanity, histories and politics of settler societies, even the spectre of mass extinction. Fracking allows for many stories. I focus on three: fracking as a spatial and geophysical manifestation of "fossil capital" that may give the lie to the idea that "cheap nature" is near its end; fracking as a geo-technical assemblage that challenges how we write political economy and what we include in its accounts; and fracking as a site in which geological and political pasts and futures are produced and contested, including the violent histories and contested futures of settler colonialism. Taken together, we can begin to see fracking as a contentious political site in which worlds are made and unmade, locally and globally.

What does it mean to say that fracking is a manifestation of fossil capital? For Andreas Malm (2016) adding the adjective "fossil" to "capital" is essential to understand the true nature of capitalist globalisation (in both senses of the term

Keywords in Radical Geography: Antipode at 50, First Edition. Edited by the *Antipode* Editorial Collective.
© 2019 The Authors/Antipode Foundation Ltd. Published 2019 by John Wiley & Sons Ltd.

"nature"), namely, that the transformation of fossil fuels into atmospheric carbon dioxide is today *intrinsically linked* to capital accumulation. Malm dates this tight coupling to a shift from "flow" and "animate" energy to "stock" energy that occurred in the 1820s, in part a response to labour militancy at the mill and factory and, more generally, an effect of capital's continuous need to capture relative surplus value, which increasingly required high-energy density fuels to drive its labour-saving machines. Coal, and later oil and gas, Malm argues, also liberated production from the spatial and political constraints of water power, allowing production to move to rapidly growing cities where labour was plentiful and cheap, Manchester in the 1800s, China today. Hence, in production and circulation alike, fossil fuels are today the "general lever for surplus-value production" (Malm 2016:288).

Fracking is thus driven by the demand for ever more energy by globalising capital, part of its "inner logic". But for some, the turn to fracking and the search for new "unconventional" sources of oil and gas also portends the end of "cheap" nature and with it the end of capitalism as we know it, as costs of production rise and pollution and climate change produce historical natures "hostile to capital accumulation" and "temporarily fixed (if at all) only through increasingly costly and toxic strategies" (Moore 2015:98). By this view, the search for unconventional oil marks the *apogee* of capitalism, that moment of inflection when the ability to extract relative surplus value begins to lose steam. But is cheap nature really nearing an end? Arguably such stories fail to account for the ability of capital to achieve ever new efficiencies, and open ever new territories for appropriation. Since the crash of oil prices in 2014—a result of the *overproduction* of fossil fuels that accompanied the fracking boom—the production cost for a barrel of North Dakota oil has decreased by 30%, as advances in well drilling, automation, and the completion of the Dakota Access pipeline has led to dramatic reductions in labour costs in both production and circulation. These efficiencies have spread across the industry, enabling more oil fields across the continent to be commercially viable. Far from heralding the end of capitalism, fracking reminds us that cheap nature is not a fixed stock that is used up, but rather something that capital continuously produces anew. Not only has it not been exhausted, it may not be any time soon. Nor is it clear that pollution and climate change inject a fatal "negative value" into capitalism, rather than new opportunities for accumulation. Indeed, viewed politically and ecologically, the fantasy that fossil capital has natural limits that will usher in capitalism's end may be the most dangerous dream of all.

But fracking as the latest chapter in the history of fossil capital is not the *only* story to tell. As a geo-technical apparatus, fracking also challenges what we include within the explanatory frames of political economy. Here the example of North Dakota is again instructive. By the 1990s, North Dakota had already experienced two modest and short-lived oil booms, as high energy prices driven by geopolitical instability encouraged exploitation of the region's modest and relatively expensive conventional oil reserves. When oil prices fell, investment dried up and producers shifted to more profitable plays, bringing the boom to an end. To understand why the current oil boom is much larger and has lasted much longer, we need to understand the petroleum system buried far below the state's rangelands and wheat fields, and the technologies developed to exploit it. In other words, we need to think *geologically*.

Schematically, a petroleum system consists of a source rock in which oil and gas originates, a permeable reservoir rock into which the oil and gas migrates, and an impermeable seal which limits the upward migration of hydrocarbons toward the earth's surface. Often this is covered by a thick overburden that must be removed or drilled through. In conventional oil and gas wells, recovery of these fossil fuels is rather straight-forward and relatively inexpensive—one drills through the seal and the oil and gas are released to the surface. Furthermore, as oil and gas are extracted, more oil and gas (and water) migrate to the borehole, resulting in a continuous flow that can last for decades. "Tight" oil is very differ- ent. Here, oil and gas are trapped in porous but *impermeable* rock. This is the case in North Dakota, a result of geological processes far back in the Devonian period, when what is now "North Dakota" on US maps was located near the equator and covered with shallow, algae-rich seas. As fine sediment was swept into the sea, it formed layers which, under the weight of subsequent sediment, com- pressed and heated into mudrock (shale) in which the trapped algae literally cooked into oil and gas. Crucially, shale is impermeable, its small pores discon- nected or the pathways between them too narrow for oil and gas to move.

In *Fossil Capital*, Andreas Malm (2016:287) writes that "capital recognizes no boundary in nature ... solely concerned with the expansion of abstract value, it can drain nature of biophysical resources without really noticing what is in there". But is this really so? Is capital blind to the *materiality* of the biophysical world? Or is it political economy—our analysis of how capitalism works—that tends to be so? True, capital often makes every effort to not really "notice what is there" when it comes to such things as waste. But arguably in the process of extraction, capital must attend very *closely* to what is there. Indeed, as we can see in the case of fracking, the affordances and dynamics of geological formations matter for *how* capital circulates, and for the social, political and technical worlds that form around it. Fracking is a method to make impermeable rock permeable, the details of which are relatively simple, although technically advanced. First, a well is drilled and explosives set off deep underground to introduce small fractures into the surrounding rock. Then, water filled with "proppants" is pumped into the wells at high pressure, widening and extending the fractures. When the water is pumped out the proppants remain behind, propping the fractures open. This enables oil or gas to migrate to the borehole. Beyond the reach of the newly introduced fractures, however, the oil and gas remain stuck in place. A second innovation—horizontal drilling—is equally important, as it allows drilling compa- nies to drill horizontally through the thin oil-bearing strata, vastly increasing what petroleum geologists call the "pay zone".

The physical qualities and affordances of shale formations matter throughout the process, reflected in the technologies and labour processes needed to exploit them, and the vast supply chains of materials and energy required to "frack" a well. Indeed, perhaps better than any other technology, the articulated appara- tuses of fracking teach us something about the *evolution* and *concretisation* of technological systems in relation to their associated milieus (Simondon 2016). Imagine a horizontal drilling apparatus laid out in a field like a newly discovered fossil, or the machinery required to frack a well abandoned in the parking lot of a suburban shopping mall. Remove the geological formation from the picture and the technology makes no sense; in its parts, and as an articulated totality, fracking technology *presupposes* the geological formation in its physical form.

But this is not the only way in which the geological formation is an active rather than passive element in social and political life. Owing to the impermeable nature of tight oil, and the specific technologies required to make the oil flow, fracked oil wells have very different spatial and temporal characteristics than conventional oil wells, manifest in what industry insiders call a well's "decline curve". A decline curve measures a well's decline in productivity over time. As noted earlier, conventional oil wells have gentle decline curves and long well lives, as oil or gas migrates to the borehole through permeable rock. Fracked "tight oil" wells, in contrast, have steep decline curves, with declines in production as high as 70–90% during the first year of operation. When a shale formation is fracked, all the oil comes out at once, and after the initial surge, production drops precipitously, since everything beyond the reach of the introduced fractures remains locked in place. For small independent oil producers like those that dominated the North Dakota oilfields in the first years of the boom, this presented a daunting financial challenge, since most had to maintain cash flow through sustained production in order to service debt they had acquired to scale up production at the beginning of the boom. The steep decline curve, an effect of the formation and the technical assemblage developed to exploit it, required that more and more wells had to be drilled in rapid succession in order to keep production levels stable, what those in the industry describe as a "drilling treadmill". In North Dakota, the imperative to speed up production was reflected in worker fatality rates exponentially higher than conventional oil fields elsewhere in the United States. For reasons that were both economic *and* geologic, in North Dakota, *tight* oil became *fast* oil, with consequences written on the bodies of migrant oil workers. A geological formation isn't merely a static or passive resource from which oil is extracted, its unique affordances and dynamics *add something* to the stories that political economy tells (Braun forthcoming).

In short, geology matters. Indeed, fracking reminds us of a larger ontological and existential point, namely that human life is *geological* from the get go. The point is not just that humans are geological actors, an "intemperate force" within earth history. For as Kathryn Yusoff (2013) reminds us, what constitutes "humanity" at any given historical moment is composed within and an effect of "inhuman" forces. The social is composed *through* the geologic, not on top of it. If humans are today a geologic force that in becoming so risks its own extinction and the mass extinction of many others, this is only because "we" have *reanimated* an earlier extinction event: "in unearthing one fossil layer we create another contemporary fossil stratum that has our name on it" (Yusoff 2013:784). Indeed, geologic life is not only true of fossil capital, but true of *all* modes of life on earth, human and otherwise, for the present form of globalising capital is only one of many different ways in which life can be organised, lived and experienced on our rocky planet. For Yusoff, geologising the *anthropos*, rather than anthropo-morphising geology, is critical for imagining and enacting other forms of geological life. "Only when this work is done", she explains,

> does it become possible to make a counterintuitive move, and turn against the 'gifts' of fossil fuels and against the human that is its inheritance (the geopolitical subject of late capitalism), and into other energetic relations that redirect, reimagine, and aestheticise the forces of geopower in equally sensible ways. (2013:791)

Yusoff helps us understand what Elizabeth Grosz (2008) calls "geopower", the potentiality of matter that human life capitalises upon and adds to, but can never separate itself from. But in her careful attention to fossilisation—those previous and still to come—she arguably reinstates a resolutely European and Eurocentric subject at the centre of geological life. Here again, fracking can help us out. Across the western plains of the United States, the expansion of the oil industry is predicated not just on the exploitation of an earlier extinction event, but on the violent appropriation of land, a historical violence that is effaced equally by big oil, the state, and Anthropocene discourse. To understand fracking, we must also understand the processes in settler colonialism by which native land was and still is transformed into *property*, a recursive process of dispossession "by which new proprietary relations are generated, but under structural conditions that demand their simultaneous negation" (Nichols 2017:12; see also Deloria Jr. 1996). In North Dakota, fracking occurs on land that was recognised as Mandan, Hidatsa, and Arikara territory in the Fort Laramie Treaty of 1851. Since 1851, the Mandan, Hidatsa, and Arikara peoples (MHA Nation) have suffered multiple displacements and deprivations, including arbitrary reductions in their treaty lands, the genocidal violence of residential schools, uprooting from fertile agricultural lands along the Missouri River to make way for the Garrison Dam, part of the Pick-Sloan projects designed to protect settler society from annual floods, and alienation of additional land on the Ft. Berthold Reservation through "allotment", a process by which almost a third of the land on the reserve was turned over to white settlers. Today, massive oil wealth is extracted from the traditional territories of the MHA Nation, both inside and beyond the boundaries of Ft. Berthold, by national and multi-national oil companies and the settler state. In turn, the Dakota Access pipe-line—owned by oil and gas companies headquartered elsewhere in the US and Canada—transports oil from western North Dakota through the traditional territo-ries of the Oceti Šakowiŋ Oyate (known by some at the Great Sioux Nation) and under the Missouri River, a source of water and livelihood for residents of the Standing Rock reservation.

For many climate activists fracking is significant for "locking in" dependency on fossil fuels, rushing humanity headlong into a coming catastrophe. But fracking in North Dakota—and elsewhere in the US and Canada—also reminds us that the oil industry in North America is built on a history of state violence and Indigenous displacement, and that for much of humanity, the catastrophe does not lie in the future but occurred in the past and continues on in the present. If, as Christina Sharpe (2016) brilliantly outlines, to be black is to live "in the wake" of the earlier catastrophes of the Middle Passage and slavery, then for Indigenous peoples in the oil-rich lands of the great plains of the United States and Canada, to be Indigenous is to live amidst what Kyle Whyte (2017) describes as a radical and violent form of enforced environmental change. While global elites may tremble in the face of the *coming* catastrophe, Indigenous people, Whyte explains, "already inhabit what our ancestors would have understood as a dystopian future" (2017:207). Fracking reminds us that fossil capital is inherently a racial formation and not just a carbon-intensive one, and that the *anthropos* of the Anthropocene is anything but a universal humanity. If fracking reanimates the geological past and throws local and global communities into potentially dystopic environmental futures, it builds upon and perpetuates other pasts—including those of settler violence and Indigenous dispossession—and assumes without question, although not without resistance, the universality of "settler" time (Rifkin 2017).

Celebrated within (white) nationalist discourse as the path to US energy independence, fracking is perhaps more accurately understood as a technical and environmental manifestation of globalising capitalism's relentless demand for more energy. But it is also much more. As a geo-technical apparatus, it reminds us that in its search for energy, capital must somehow capture the inhuman forces of the earth, a process that introduces diverse actors and unexpected spatio-temporalities to cycles of capitalist accumulation, with consequences for workers and environments alike. And, as the example of North Dakota reveals, fracking teaches us that fossil capital is not just a particular mode of geologic life, but a fundamentally racialised one, deeply entangled with the racial politics of settler societies, in which the reanimation of ancient fossils benefits few at the expense of many others.

References

Braun B (forthcoming) Taking earth forces seriously. In K Lutsky, O Saloojee and E Scott (eds) *Viscosity: Mobilizing Materialities*. Minneapolis: University of Minnesota Press

Deloria Jr. V (1996 [1988]) *Custer Died For Your Sins*. Norman: University of Oklahoma Press

Grosz E (2008) *Chaos, Territory, Art: Deleuze and the Framing of the Earth*. New York: Columbia University Press

Malm A (2016) *Fossil Capital: The Rise of Steam Power and the Roots of Global Warming*. London: Verso

Moore J W (2015) *Capitalism in the Web of Life: Ecology and the Accumulation of Capital*. London: Verso

Nichols R (2017) Theft is property! The recursive logic of dispossession. *Political Theory* 26 (2):3–28

Rifkin M (2017) *Beyond Settler Time: Temporal Sovereignty and Indigenous Self-Determination*. Durham: Duke University Press

Sharpe C (2016) *In the Wake: On Blackness and Being*. Durham: Duke University Press

Simondon G (2016 [1958]) *On the Mode of Existence of Technical Objects*. Minneapolis: Univocal

Whyte K P (2017) Our ancestors' dystopia now: Indigenous conservation and the Anthropocene. In U K Heise, J Christensen and M Niemann (eds) *The Routledge Companion to the Environmental Humanities* (pp 206–215). New York: Routledge

Yusoff K (2013) Geologic life: prehistory, climate, futures in the Anthropocene. *Environment and Planning D: Society and Space* 31(5):779–795

Fragments

Colin McFarlane

Department of Geography, Durham University, Durham, UK;
colin.mcfarlane@durham.ac.uk

What has an exploded garden shed got to do with the tradition of radical geography? Cornelia Parker's 1991 sculpture, *Cold Dark Matter: An Exploded View*, like much of Parker's work, deals with objects that seem familiar and that somehow persist, but which have been transformed and fragmented, their identities—and possibilities—altered (see Figure 1). As John Scanlan (2005:100, 119) has put it, "fragments of matter, and indeed, matter transformed—crushed, melted, sketched, or recovered—appear throughout ... [her] work", resurrecting seemingly "dead" matter into "beguiling new forms". Parker's work captures something of the destructive power of the world, of the force of fragmentation, yet also speaks to how fragments might persist.

Parker's exploding garden shed provides an unlikely provocation for radical geography, and gets to the heart of what I want to argue in this reflection on fragments. For all its power, Parker's sculpture remains frozen, static in space. The fragments may have been, in Scanlon's rather theological description, "resurrected", but they are nonetheless left hanging there, with seemingly little capacity

Figure 1: Cornelia Parker, *Cold Dark Matter: An Exploded View*, 1991 (source: https:// commons.wikimedia.org/wiki/File:Cold_Dark_Matter_.jpg; CC BY-SA 4.0)

Keywords in Radical Geography: Antipode at 50, First Edition. Edited by the *Antipode* Editorial Collective.
© 2019 The Authors/Antipode Foundation Ltd. Published 2019 by John Wiley & Sons Ltd.

to act. What, though, if the fragments could move, or be put to different kinds of work? What if they were more than just the spatial products of fragmentation, and instead generative elements in different kinds of spatial relations?

Like Parker's sculpture, space has been shown to be fundamentally fragmented. "Fragmentation" is a keyword for thinking space: the production, division, and creative destruction of space, from the neighbourhood to the globe. A central task of radical geography has been to elucidate the causes, forms, and alternatives to spatial fragmentation.

Critical geographic thought has inherited a rich set of theoretical resources for thinking the fragmentation of space. In *The Production of Space*, for instance, Henri Lefebvre (1991:342) influentially argued that space should be understood as "homogenous yet at the same time broken up into fragments". Capitalism is conceived as actively requiring the fragmentation of space in order to sustain itself, for instance in the geographical (dis)placing of labour, or in the targeting of specific spaces for accumulation and speculation (and see Merrifield 2014). Along with a host of other key voices in the radical spatial tradition (e.g. Harvey 1973; Massey 1984; Smith 1996), Lefebvre argued that if we are concerned with under-standing how space is produced, then we need to attend to how it is divided, carved-up, controlled, and commodified. It is a position that has held a com-manding influence over how geographers think space.

Less visible though, in the geographical tradition, have been the fragments themselves. We might be tempted to say that this is well and good—radical spa-tial thinking has always sought, as Lefebvre (1991:92) once put it, to "rip aside appearances", to focus analytical and political attention on the processes and not the products. The imperative here is to in a sense look for the sources of the explosions and resolutions, for instance in the creative destruction of cycles of capitalist transformation. And yet, as Lefebvre also knew well (Buckley and Strauss 2016), the products—the fragments of space—are themselves sites through which to develop radical geographies.

The task of exploring the radical geographies of fragments matters, not least because as the world becomes increasingly urban, growing numbers of people inherit or arrive to what Ananya Roy (2016:206) has called "a geography of shards and fragments". Life is shaped and reshaped in the ways that people piece together a world "out of fragments" (Simone and Pieterse 2017:71). And when we attend to how fragments are put to work by all manner of people and to all kinds of ends, we see that there are sometimes political possibilities attached to fragments (e.g. McFarlane 2018; Tronzo 2009). There are lots of ways in which we might think of fragments, and in what follows I highlight just two that connect closely to the radical geographical tradition: *material fragments* and *knowledge fragments*.

First, in addition to the spatial fragments that Lefebvre and others address, there are material fragments. In critical research on urban space, for example, where the term fragmentation is most prominent, we see this history in a rich diversity of work ranging from that on gated enclaves, gentrification and sociospatial polarisation, to debates around splintering urbanism or urban conflict (e.g. Caldeira 2000; Harvey 1973; Smith 1996). Steve Graham and Simon

Marvin's landmark book, *Splintering Urbanism* (2001), for example, examined the "splintering" of public space and provisions in the context of urban infrastructure. They demonstrated how neoliberalism, and in particular the relations between privatisation, liberalisation, and the application of new technologies, shaped a globalising process of "unbundling" infrastructure. This process led to the collapse of what Graham and Marvin called the "modernist infrastructural ideal"—standardised, monopolised and integrated infrastructures for all—in the process intensifying inequalities across an increasingly fragmented urban space.

When we think about material fragments, it is worth reflecting on the difference between "fragment" and "splinter". If the latter has proven useful for thinking the relations between fragmentation and urban inequalities, the splinter itself in this debate has largely remained a *product* of capitalist transformation. Its potential conceptual and political agency, then, falls from analytical view, or surfaces only in narrow parameters. There is a longer history with the term fragment, in contrast, only some of which I discuss below, that emphasises the generative relations and political possibilities enacted with fragments.

Material fragments can take on new lives. As research on, for example, waste economies and recycling has shown, fragments can be remade in all kinds of unpredictable new contexts (e.g. Gregson et al. 2010; McFarlane and Silver 2017). The changing relations that fragments are drawn into points to the double-status of fragments as nouns and verbs. As a *noun*, the fragment is a material form in the landscape, a broken or otherwise inadequate thing, a visible marker of poverty and inequality. As a *verb*, it is a material process, both in the sense that it is produced historically through fragmentation, and in the sense that it can be pulled into different relations, forms of work, and angles of vision. Sure, the fragments sometimes just hang around, much like Parker's sculpture, but oftentimes they are animated and active in all kinds of spatial relations.

Second, there are knowledge fragments. Again, there are long critical intellectual traditions here. We might think, for example, of Walter Benjamin's literary form of connecting fragments of text and assembling them in montage, most notably of course in *The Arcades Project* (2003; see also Buck-Morss 1989; Pred 2000). The spatialities that Benjamin conjured—whether they were of Paris, Berlin, Naples, Marseilles or Moscow—often registered through a set of disjointed textual fragments, thrown into juxtapositions that gave a sense of the fragmented nature of modern urban life and city transformation. Benjamin's montage of fragments interrogated urban space as both coherent and fragmented, singular and multiple, permanent and fleeting, predictable and nonlinear, and ordered and disruptive. Benjamin developed a kind of pedagogy of fragments, a vast experiment with fragments of knowledge aimed at disrupting what is known and jolting new ways of seeing and acting. Knowledge fragments here are generative: a kind of anti-narrative montage, the city and the modern thought simultaneously through different angles of vision and possibility.

A central question running through Benjamin's work—and here he pre-empts what we now think of as postcolonial critique—is whether and how knowledge fragments on the margins of modern life might provoke new ways of thinking space or possibility. We cannot know the answer to such a question in advance:

the potential force of the knowledge fragment, in both Benjamin and in a post-colonial rendering, is to ask the question again and again, and to look for where the question pushes us as we conduct critical geographical research. Knowledge fragments cannot simply be placed with the garden shed, exploding or otherwise, because sometimes they challenge the very nature of the pre-given "whole".

The knowledge fragment plays an important role, for example, in subaltern studies research, an area of work that has had vital contributions for radical geographic thought (e.g. Jazeel and Legg forthcoming; McFarlane 2018; Robinson 2006; Roy 2011). Fragments of knowledge are fundamental to the subaltern studies project because they present clues to other histories and to new forms of conceptualisation, often hinted at in archival research but speaking to a different way of conceiving categories like insurgency, consciousness, politics, class, even history itself. At stake in attending to knowledge fragments here is not empirical variation alone, but new "vantage points". Partha Chatterjee, for example, uses the idea of "fragment" in *The Nation and its Fragments* (1993) as a short-hand for forms of difference and resistance in Bengal that cannot be adequately understood within mainstream representations of nationalism and the modern (and, on the fragments of Indian society, see Gyanendra Pandey 2006).

A knowledge fragment is marked out as such in these accounts in two broad ways. First, because of its position to or within a wider set of political, social and cultural power–knowledge relations. Fragments emerge in the making of dominant national cultures and political economies, and can therefore shift over time depending on how they are framed and shaped. Second, the knowledge fragment can be a form of generative expression, whether as a piece of drama or a collection of poems or a political protest, or some other form, that presents clues to a different way of understanding history, or the nation, or community, or citizenship.

For Dipesh Chakrabarty (2002:274), it is important analytically and politically to stay with knowledge fragments, rather than return to an analytical "whole". As Pandey (2006:296) argues, the fragment here is not just "a 'bit'—the dictionary's 'piece broken off'—of a preconstituted whole", but a potential site for provoking and generating new knowledge and politics. Chakrabarty (2002:275) uses "fragment" as an orientation toward the sociospatial world as beyond any single representational system. Knowledge fragments demand recognition that the world is more than simply "plural", but is "so plural as to be impossible of description in any one system of representation"—instead, the challenge is to "learn from the subaltern"—or, as we might put it here, to learn from the fragment analytically and politically:

> To go to the subaltern in order to learn to be radically "fragmentary" and "episodic" is to move away from the monomania of the imagination that operates within the gesture that the knowing, judging, willing subject always already knows what is good for everybody, ahead of any investigation.

For example, Swati Chattopadhyay (2012:251–252), in her reconceptualisation of infrastructure in relation to contemporary urbanism in India, argues that subaltern practices exist on the "edges of visibility", beyond definition and representation and in excess of authority. These practices can take ordinary spaces such as streets, neighbourhoods, walls or vacant spaces and turn them, temporarily, into "spatial fragments" that belong "neither to the everyday nor to the exceptional ... [they are] created out of a series of conjunctures, of bodies and objects, movements and views, noise and warmth, walls and roads, events and memories" (2012:119). For Chattopadhyay, these spatial fragments are prompts to a reconceptualisation of infrastructure and a challenge to how urban theorists might "unlearn the city" (2012:252). They may also act as knowledge fragments that serve as resources for rethinking and remaking space and the political.

Set against this backdrop, knowledge fragments become lures to different kinds of archives about space. If Benjamin performed a pedagogy of fragments, postcolonial scholarship radically disrupts what we take to be the archive for that pedagogy. It forces attention on pedagogies of writing, imaging, making, talking, seeing, walking, telling, hearing, relating, and so on, from all manners of lives and spaces (Mbembe and Nuttall 2004). The challenge of the fragment here is to compel us to stop and ask: what kind of knowledges are we listening to or seeing as we formulate a conception and politics of space? Edgar Pieterse (2011) has argued that some of the catalysts of these kinds of archives might include ordinary spaces like the street, the informal settlement, the waste dump, the taxi rank, the mosque and church. Similarly, writing about wastepickers in municipal garbage grounds in India, Vinay Gidwani (2013:774) has argued that "the primary intellectual and political task of the postcolonial scholar as archivist" is to derive ways of thinking from the "marginalized, remaindered, and stigmatized". Knowledge fragments, in this reading, connect deeply to the wider postcolonial affirmation of thinking space and its possibilities from and with multiple spaces and cultures of knowledge production.

Whether as a material or as a form of knowledge, the fragment has been a companion to the history of critical geographic thought, sometimes quietened or ignored, and at other times a forceful provocation for thinking space and its political possibilities. There is a long and fascinating tradition of thinking with fragments—from classics, art history and archaeology to philosophy, social theory and urban theory—and tracing those histories and what they offer critical geographic thought would be an exciting project. I have said nothing here, for example, about the use of fragments in strands of critical theory (e.g. Theodor Adorno's [1973] writings on fragments), or postmodern experiments with fragment thinking (e.g. Baudrillard 2006), or in art practices (e.g. Tronzo 2009).

There is also a set of questions around how we might conceive and politically put to work the relations between fragments (material or knowledge) and different conceptions of *wholes* to make sense of contemporary space. Here, Parker's sculpture is provocative in another sense: the sculpture suggests a whole, but a whole that can barely be grasped, a whole that changes its identity because of fragmentation in space and which is caught between one form and another, indeed seemingly refusing to take form. The whole—the complete, the fulfilled, the totality—lingers around the image and form of the fragment, and the relationship between the fragment and the totality or whole is fundamental to the politics of space, whether the space of the home, city, nation, globe, or other spatial demarcations and productions. Charting, politicising and re-expressing distinct fragment–whole relations is, then, a useful conceptual and political ground for geographic thought.

References

Adorno T (1973 [1966]) *Negative Dialectics* (trans E B Ashton). London: Routledge

Baudrillard J (2006 [1995]) *Fragments: Cool Memories III, 1990–1995* (trans E Agar). New York: Verso

Benjamin W (2003 [1939]) *The Arcades Project* (trans H Eiland and K McLaughlin). Cambridge: Harvard University Press

Buckley M and Strauss K (2016) With, against, and beyond Lefebvre: Planetary urbanization and epistemic plurality. *Environment and Planning D: Society and Space* 34(4):617–636

Buck-Morss S (1991) *The Dialectics of Seeing: Walter Benjamin and the Arcades Project.* Cambridge: MIT Press

Caldeira T (2000) *City of Walls: Crime, Segregation, and Citizenship in Sao Paulo.* Berkeley: University of California Press

Chakrabarty D (2002) *Habitations of Modernity: Essays in the Wake of the Subaltern Studies.* Chicago: University of Chicago Press

Chatterjee P (1993) *The Nation and its Fragments: Colonial and Postcolonial Histories.* Princeton: Princeton University Press

Chattopadhyay S (2012) *Unlearning the City: Infrastructure in a New Optical Field.* Minneapolis: University of Minnesota Press

Gidwani V (2013) Six theses on waste, value, and commons. *Social and Cultural Geography* 14(7):773–783

Graham S and Marvin S (2001) *Splintering Urbanism: Networked Infrastructures, Technological Mobilities, and the Urban Condition.* Oxford: Blackwell

Gregson N, Crang M, Ahamed F, Akhtar N and Ferdous R (2010) Following things of rubbish value: End-of-life ships, "chock-chocky" furniture, and the Bangladeshi middle class consumer. *Geoforum* 41:846–854

Harvey D (1973) *Social Justice in the City.* Baltimore: Johns Hopkins University Press

Jazeel T and Legg S (forthcoming) Introducing subaltern geographies. In T Jazeel and S Legg (eds) *Subaltern Geographies: Subaltern Studies, Space, and the Geographical Imagination.* Athens: University of Georgia Press

Lefebvre H (1991 [1974]) *The Production of Space* (trans D Nicholson-Smith). Oxford: Blackwell

Massey D (1984) *Spatial Divisions of Labour.* London: Macmillan

Mbembe A and Nuttall S (2004) Writing the world from an African metropolis. *Public Culture* 16(3):347–372

McFarlane C (2018) Fragment urbanism: Politics at the margins of the city. *Environment and Planning D: Society and Space* https://doi.org/10.1177/0263775818777496

McFarlane C and Silver J (2017) The poolitical city: "Seeing sanitation" and making the urban political in Cape Town. *Antipode* 49(1):125–148

Merrifield A (2014) *The New Urban Question.* London: Pluto

Pandey G (2006) *Routine Violence: Nations, Fragments, Histories.* Stanford: Stanford University Press

Pieterse E (2011) Rethinking African urbanism from the slum. *LSE Cities* November. http://lsecities.net/media/objects/articles/rethinking-african-urbanism-from-the-slum/en-gb/ (last accessed 11 July 2016)

Pred A (2000) *Even in Sweden: Racisms, Racialized Spaces, and the Popular Geographical Imagination.* Berkeley: University of California Press

Robinson J (2006) *Ordinary Cities: Between Modernity and Development.* London: Routledge

Roy A (2011) Slumdog cities: Rethinking subaltern urbanism. *International Journal of Urban and Regional Research* 35:223–238

Roy A (2016) Who's afraid of postcolonial theory? *International Journal of Urban and Regional Research* 40(1):200–209

Scanlan J (2005) *On Garbage*. London: Reaktion

Simone A and Pieterse E (2017) *New Urban Worlds: Inhabiting Dissonant Times*. Cambridge: Polity Press

Smith N (1996) *The New Urban Frontier: Gentrification and the Revanchist City*. London: Routledge

Tronzo W (ed) (2009) *The Fragment: An Incomplete History*. Los Angeles: Getty Research Institute

Garrison Communities

Beverley Mullings

Department of Geography and Planning, Queen's University, Kingston, ON, Canada;
mullings@queensu.ca

Defined as geographically discrete, fortressed urban areas marked by poverty, gang violence, political manipulation and confrontational relationships with law enforcement institutions, garrison communities are spatial expressions of both state power and community survival in Jamaica. Garrison communities entered the popular Jamaican lexicon in the 1980s as growing attention began to be paid to an emerging political culture where politicians and local gangs colluded to exercise control over poor urban communities. Through practices like election rigging, extortion, and regulated access to scarce benefits like municipal contracts, housing or utilities, garrison communities became sites of extreme spatial control whose borders are maintained primarily through violence.

The term "garrison community" is attributed to political scientist Carl Stone (1986), who first identified patterns of voting within local government elections in West Kingston. Stone pointed to what he described as *garrison constituencies*—constituencies where election outcomes could be influenced by the presence of communities that voted uniformly—sometimes as high as 100%, in favour of a single political party. Discernible as early as 1974, Stone defined garrison communities as militarised strongholds based on political tradition, cultural values, beliefs, myths and socialisation that were maintained through a framework of violence and patron–client relations. Stone's early writings focused on the patterns of political control made possible by these forms of community voting behaviour, but by the 1990s scholars began to focus on the impact of garrison communities on security and social life on the island.

The earliest garrison communities emerged within the large-scale government housing schemes built during the 1960s and 1970s on two sites of informal "squatter" settlement in West Kingston—an area encompassing the capital city's historic downtown, the harbour to the south, and Spanish Town Road to the north. The first place of settlement for rural migrants to the city during the 1930s and 1940s, neighbourhoods like Denham Town, Trench Town, Hannah Town and Matthews Lane, had become dense networks of overcrowded tenements and informal settlements by the 1960s, overrun with diseases like tuberculosis and typhoid and associated with a growing criminogenic sector. For example, Tivoli Gardens, one of the best-known garrison communities, was built between 1963 and 1965 on the site of an informal settlement known as "Back-O-Wall" that was demolished as part of a programme of urban renewal. Like many other "slum" communities, Back-O-Wall was a place of livelihood, self-organisation and politics where residents built homes from discarded and retrieved materials, created their

Keywords in Radical Geography: Antipode at 50, First Edition. Edited by the *Antipode* Editorial Collective.
© 2019 The Authors/Antipode Foundation Ltd. Published 2019 by John Wiley & Sons Ltd.

own systems of public sanitation, generated livelihoods within the city's informal and underground economies, and secured the land they occupied by building stockades that could not be easily breached by the police. Although the social and environmental conditions in Back-O-Wall were not significantly worse than those of the surrounding tenement neighbourhoods housing the capital's poorest populations, it increasingly came to be seen as a potential threat to the system of racial capitalism that operated at that time in Jamaica and across the Americas as a whole.

As early as the 1930s, key political and social sectors viewed Back-O-Wall's large and outspoken Rastafarian community as a threat to Jamaica's racial order because of its trenchant racial critique of Euro-American imperialism, its embrace of blackness and Africa, and by the 1960s, its contribution to a radical conscious-ness rooted in the grammar of black liberation. Rastafari were feared as a poten-tially subversive group who could not be easily pacified or brought under the control of either of the country's two main political parties—the Jamaica Labour Party (JLP) or the People's National Party (PNP). As a group, they were therefore treated with suspicion and hostility by much of Jamaica's aspirant middle classes, who feared their capacity to stir up the ethno-racial discord that simmered just under the surface of everyday social interaction (Smith et al. 1967). This sense of threat intensified after 1960, when a raid of the headquarters of the Reverend Claudius Henry, an elder prominent in efforts to repatriate members of the com-munity to Africa, revealed letters addressed to Fidel Castro requesting support in overthrowing the "oppressors" in Jamaica (see Paul 2013). Almost immediately, the Rastafari community was dubbed a threat to national security and thus a dis-posable population that could be attacked or killed with impunity. The eventual demolition of Back-O-Wall in the name of urban renewal and the creation of government housing schemes like Tivoli Gardens marked an entrenchment of the practice of exclusion, containment and pacification that has always been a characteristic of urban governance in Jamaica (Clarke 1975).

The construction of large government housing schemes in West Kingston in the 1960s was expected to create a new type of citizen. Yet, by the 1970s these housing schemes and the tenement communities around them were rapidly becoming citadels divided across party lines and governed by armed para-military groups and the self-styled local area enforcers, or Dons, who ruled them. The challenge to maintain total community control intensified, however, for both Dons and local politicians during the 1980s as the Jamaican economy succumbed to the scourge of rising sovereign debt, declining growth and the increasingly anxious scrutiny of the government of the United States. As the battle for the island's political future intensified, so too did competition among politicians of all persuasions for the support of strong Dons who could assure electoral success. During this period, the most successful Dons were those who were able to estab-lish a new spatial fix through involvement in the transhipment and trafficking of drugs, largely crack cocaine, from South America. Participation in the international drug trade significantly increased the incomes earned by Dons, but it also increased levels of violence across the country as a whole. For example, powerful Dons like Christopher "Dudus" Coke, head of the notorious Shower Posse (otherwise known as "Presidential Click") and local enforcer for Tivoli Gardens throughout this period, derived much of his power from the considerable wealth that he amassed from the trafficking of crack cocaine. Coke, who was extradited in 2010 to the United States and subsequently sentenced to 23 years in federal prison on charges of racketeering related to drug trafficking and assault, was believed to be worth approximately US$3 million at the time of his arrest.

The power of Dons like "Dudus" Coke to maintain virtual immunity from prosecution (National Committee on Political Tribalism 1997) deepened in the 1990s as Jamaica came under the disciplinary governance of the Washington Consensus institutions. The mandate to restore market fundamentals required deep cuts to state spending and the privatisation of most national assets. The emphasis on restoring "market fundamentals", however, ignored the effects of neoliberalisation on the social institutions at work in the making of everyday urban life. Women bore the brunt of these early policies because they were disproportionately affected by the loss of employment when the public sector was downsized; the loss of public access to health and education as public expenditure in these sectors declined; and the rising cost of living that ensued as state subsidies on basic living necessities were removed (Mullings 2009). The effects of these policies for Jamaica's poorest urban residents were dire. For example, in 1990 the proportion of Jamaica's urban population living in slum conditions was approximately 30%, but by 2005 this percentage had doubled to 60.5%,[1] a figure also reflecting the devastating effects of structural adjustment on investment in the social and physical infrastructure in West Kingston.

The withdrawal of the state from investments in the social welfare of the poorest communities over the 30-year period from the 1980s to the 2010s contributed to the proliferation of Garrison communities and the growing power of Dons, who increasingly filled the welfare gap left by the state. As Levy and Chevannes (2001) observe, local Dons within garrison communities are respected for providing community members with many of the resources—housing, electricity, school fees, lunch money, and employment opportunities—required for everyday social reproduction. This was evident in the public support that women in Tivoli Gardens displayed in May 2010 when the Jamaican police and military launched Operation "Garden Parish", a raid aimed at taking "Dudus" Coke into custody for subsequent extradition to face trial in the United States. Armed with placards declaring "Leave Dudus Alone" and "Jesus Die for Us, We will Die for Dudus", the prominent role played by women, many of whom vowed to protect Dudus even with their lives, spoke to the contradictory role of Dons in securing the welfare of residents within the garrison communities they control. While garrison communities offer residents access to economic and physical security, they also function as spaces of great insecurity for residents who transgress the rules decreed by a Don. For women, the contradictory nature of garrison security/insecurity is heightened by the fact that their bodies often serve as boundary markers between rival garrisons and sexual objects available to Dons should they so wish. In keeping with feminist theorists who argue that discourses of nationalism often seek to maintain the integrity and honour of the nation through the sexual regulation of women, garrison communities rely on governance practices that severely limit women's agency and capacity to be full citizens.

The growing participation of Dons in the international drug trade significantly transformed the spatiality of garrison communities during the 1990s and the new millennium. Garrisons can now be found in secondary cities like Montego Bay, Portmore and Spanish Town, and many function as nodes in the transnational networks of gangs involved in that trade. Powerful gangs like the "Shower Posse" once controlled by "Dudus" Coke now operate within diaspora communities in the United States, the United Kingdom and Canada, where high levels of unemployment, poverty, racism and social exclusion continue to draw young people into gangs. But the high levels of brutality and violence associated with drug- and gun-related activities offer few opportunities for autonomy as they consign young people who work for Dons to "Live like a 100 watt light bulb—bright every day and then you just blow" (Gayle 2017).

The impunity with which Dons within Garrison communities once operated changed significantly after the 2010 Tivoli incursion. The standoff between the community and the armed forces resulted in a declaration of a state of emergency and the death of 76 residents. The raid ushered in a reassertion of state authority via heavy policing and strong anti-gang legislation (Government of Jamaica 2014). While this approach initially lowered the homicide rate, it failed to sustain the order it sought to create. And like other Central American countries that implemented similar "mano dura", or heavy-handed, approaches, this policing method continues to raise significant human rights concerns.

In the wake of the Tivoli incursion, there have been a growing number of conversations between different interest groups and government representatives regarding the sustainability of garrison communities. Both government and nongovernmental organisations have developed initiatives aimed at reducing the high levels of youth violence and homicide in garrison communities. For example, one such community organisation named S-Corner has developed a number of initiatives aimed at respect-building and the creation of social enterprise initiatives, in association with a number of non-criminal youth groups, popularly referred to as "corner crews" or "defence crews" (Levy 2012). Similarly, the Peace Management Initiative (PMI), a violence interruption programme, has partnered with local community groups as well as academic and faith-based groups to provide mediation and counselling to young people at risk of being drawn into violent relationships. This initiative has been successful in creating a space for young people to respond to and diffuse community violence. Noting the growing number of young people who are beginning to question governance structures reliant on the dominance of Dons, Horace Levy (2009) suggests that more sustainable modes of community life rooted in the possibilities of social enterprise and processes of healing and reconciliation are slowly emerging. Working with the PMI, one community was even able to broker a truce—albeit short-lived—between warring factions of the dominant gang in the area. The extent to which these initiatives can successfully generate opportunities for more liveable lives remains to be seen, as significant structural issues at the heart of garrison communities such as poor employment opportunities and weak levels of investment in physical and social infrastructure still remain. They suggest, however, a growing interest among youth and garrison community members broadly in more emancipatory modes of existence and a willingness to seek alternative modes of community survival. Though not expressed in the language of black liberation as was the case in the 1960s, these youth-led dialogues draw attention to the need for profound structural change at scales that go beyond the confines of the garrison.

Endnote

[1] See UN Data: http://data.un.org/Data.aspx?q=slum&d=SDGs&f=series%3aEN_LND_SLUM (last accessed 7 August 2018).

References

Clarke C (1975) *Kingston, Jamaica: Urban Development and Social Change, 1692–1962.* Berkeley: University of California Press

Gayle H (2017) Light on violence: Here's why boys join gangs. *The Gleaner* 27 January. http://jamaica-gleaner.com/article/news/20170127/light-violence-heres-why-boys-join-gangs (last accessed 7 May 2018)

Government of Jamaica (2014) "Criminal Justice (Suppression of Criminal Organizations) Act, 2014." https://japarliament.gov.jm/attachments/341_The%20Disruption%20and%20Suppression%20of%20criminal%20organizations.pdf (last accessed 7 August 2018)

Levy H (2009) *Killing Streets and Community Revival.* Kingston: Arawak Press

Levy H (2012) "Youth Violence and Organized Crime in Jamaica: Causes and Counter-Measures—An Examination of the Linkages and Disconnections." Final Technical Report, Institute of Criminal Justice and Security, University of the West Indies, Kingston, Jamaica / International Development Research Centre, Ottawa, ON, Canada

Levy H and Chevannes B (2001) *They Cry Respect: Urban Violence and Poverty in Jamaica.* Kingston: Department of Sociology and Social Work, UWI Mona

Mullings B (2009) Neoliberalization, social reproduction, and the limits to labour in Jamaica. *Singapore Journal of Tropical Geography* 30(2):174–188

National Committee on Political Tribalism (1997) "Report of the National Committee on Political Tribalism." Office of the Prime Minister P. J. Patterson, Kingston, Jamaica

Paul A (2013) Our man in Mona: A conversation between Robert A. Hill and Annie Paul. *Active Voice.* https://anniepaul.net/our-man-in-mona-an-interview-by-robert-a-hill-with-annie-paul/ (last accessed 4 September 2018)

Smith M G, Augier R and Nettleford R (1967) The Rastafari movement in Kingston, Jamaica. *Caribbean Quarterly* 13(3):3–29

Stone C (1986) *Class, State and Democracy.* New York: Praeger

Geopoetics

Sarah de Leeuw

Geography and the Northern Medical Program, University of Northern British Columbia, Prince George, BC, Canada;
sarah.deleeuw@unbc.ca

Eric Magrane

Department of Geography, New Mexico State University, Las Cruces, NM, USA;
magrane@nmsu.edu

If You Write Poetry / Go to the ground. (Forman 1997)

A Proposal, A Proposition

Geopoetics is a slippery-flippery concept, one that jumps about and slyly avoids getting captured in definitive ways. In its avoidance of definition, its inability to be neatly penned, cleanly reproduced, or patented for profitable circulation, geopoetics is a bit like a loch-monster: many have something to say about it but it constantly evades capture. The elusive somewhat tricky and sneaky nature of geopoetics is, we believe, exactly the nature of the concept's productive potential, the source of its potential radicalism and exactly why radical critical geographers might want to embrace it. To love a geopoetic proposal is to embrace the idea that every poem, written or read, is an opportunity to reorder or refresh the world. This is a radical proposition, an ethical one. What world, what earth, is to be made? What world, what earth, is to be reproduced? What relationships are to be privileged and honoured? Who is making the made world, and according to whose form and representation? For whom is the world being (re)imagined? Who benefits and, conversely, who does not? Let's re-line the world, calls out a geopoetics.

Geopoetics invites all of these considerations, and more. Geopoetics also demands that geographers and poets take responsibility upon accepting the open-ended invitation. Auden (1979) wrote, "poetry makes nothing happen". We don't buy that. And we invite radical geographers, critical social theorists, and poets to up-end the ostensible nothingness of poetry alongside us. Don't be shy. Don't worry about "not being a poet". Acknowledge, instead, the radical possibilities of geopoetics. Perhaps geopoetics resists being captured and subsumed into political economy, perhaps geopoetics slows a reader/viewer/listener down and presents a different conception of time. Perhaps geopoetics refuses nationalism and colonial borders, lineating words on new territory. Perhaps geopoetics speaks to the human gut and the billions of microbial organisms living therein. Perhaps,

Keywords in Radical Geography: Antipode at 50, First Edition. Edited by the *Antipode* Editorial Collective.
© 2019 The Authors/Antipode Foundation Ltd. Published 2019 by John Wiley & Sons Ltd.

as poet and queer sex-work activist Amber Dawn (2013) has suggested, poetics (in geography) can save lives. Perhaps geopoetics is buried deep in the ground, is made of stone and time and wind and water, perhaps geopoetics is nothing new at all—really, it is 4.5 give-or-take billion years old (when using a geologic calendar, which is just one kind of calendar).

We call for a geopoetics that is resistant to individuals and systems that are fucking up the planet and consolidating power on the backs of others,

that aerates the soil and lets the worms in,

that looks square on at the wreckages of the current moment and sees them as outliers rather than the "natural order" of things,

that is playful and attentive to both contemporary poetics and critical geographic thought,

that expands the page beyond the page,

that expands the means in which geography is practiced,

that takes the craft of poetry seriously but does not impose one school over another, does not police,

that takes a knee on the sidelines, in the margins, in solidarity,

that is difficult and dangerous, that is not didactic, that is awake,

that is troubling and unsettling, that slows,

that is self-aware,

that resists empire, a poem is not an empire.

Background

Part of critical geography's work is pushing against dominant forms of power, is always making and understanding the world in radically new ways (Harvey 2006). These new ways would, ideally, conceptualise and produce space and place anew, in part by recognising, respecting, and celebrating differences through radical and activist-oriented practices and perspectives. Feminist, queer, environmental, racialised, class-conscious, anti-racist, anti-colonial, Marxist, anarchist, Indigenous, and differently abled geographers (to name a few) are deeply vested in this work, as scholars, activists, disruptors, students, artists, educators, agitators, researchers and writers. Geography's recent creative (re)turn is certainly situated within this longstanding (although perhaps not fully realised) disciplinary history of radicality and criticality. Still, there remain a few degrees of separation between, on the one hand, the work of radical/critical geographers and, on the other hand, geographers engaged in creative geography (de Leeuw and Hawkins 2017; Marston and de Leeuw 2013).

We propose bringing creative and radical/critical geographies into closer proximity with each other.

Our proposal is anchored in geopoetics, a practice-based concept receiving increased attention by geographers (Magrane 2015). Our proposal to *radicalise geopoetics* contemplates how geographers interested in radicalising earth-writing (geo-graphing, the etymological root of geography) might critically engage geopoetics as reading, writing, and performative practices. We offer the important caveat that for us, "earth" always denotes much more than soils and waters, or the physical geographies of a planet—we draw from literary interpretations (and the occasional more-than-human geographer) of "earth" as also and fundamentally connoting the human body: as in "earth to earth, ashes to ashes". Earthwriting, or geo-graphing and thus geopoetics, is consequently also very much an embodied endeavour, one replete with the messy and the fleshy (to draw on Cindi Katz [2000]) and one always open for radicalising, especially by critical human geographers.

And With That, How?

Don't be frightened. Which is not to say fear is not valid, born perhaps from cruel circumstances possibly imposed. But the earth is calling. The ground is calling. Your mouth holds words, a world of words, you can form worlds. Never overlook the worlds of others. And sometimes know the making of your world can wait (wait wait wait) as you give time and space for the worlds of others.

Read. Read. Read poets like Layli Long Soldier and Maggie Nelson, read poets like Marlene Nourbese Philip, Solmaz Sharif, and Billy-Ray Belcourt, read poets like Camille Dungy and Brenda Iijima, read poets like Megan Kaminski and Linda Russo, read poets like Sherwin Bitsui and Craig Santos Perez, read poets like Ross Gay and Claudia Rankine. They are offering you worlds, their words are worlds anew. With this in mind, read and teach and write anew—with a poetic sensibility.

Work with poets. Work with the natural world—in ways that recognise the world as a "biocultural" (Kristeva et al. 2018) configuration. Take "text" as a site for unparsing, unpacking, lineating and orating differently. Advocate engagements beyond text that adheres to historical expectations in the discipline—speak text, sing text, be silent in text.

Try singing. Try being quiet. Try listening. Try crying. Try loving. Try giving up. Try sinking your hands in soil in rot in wet in gut. Remember if you're on hallowed ground to be careful and respectful. Remember all ground is hallowed. Much of it is not yours. Be careful. Be alive.

Be alive with your unease. If life feels easy, you have overlooked something. Find that something. Because there are those who do not have an easy life, and that is not by choice. Enter unfamiliar grounds, shakingly.

Moan and call out. You too are breakable in this earth.

End Without Ending: To the Ground, With Love (a Decomposition)

The call for contributions in the first issue of *Antipode* (Stea 1969:1) began with an epigraph from Walt Whitman (1867a):

Listen! I will be honest with you,
I do not offer the old smooth prizes, but offer rough new prizes …

So it seems poetry has been forefront to a radical geography a la *Antipode* from the onset. Whitman, who wrote from fields teeming with the bodies of US 19th century civil war dead, whose long (dare we say queer?) lines were both ecstatic and seething with corpses (Whitman 1867b):

O how can it be that the ground itself does not sicken?
How can you be alive you growths of spring? …

Now I am terrified at the Earth, it is that calm and patient,
It grows such sweet things out of such corruptions …

May we grow sweet things. May we all grow, may we never stop transforming.

Think of geopoetics as compost (Magrane 2015; Rasula 2002): and here we will conclude with a (decomposition) of *Antipode*'s first piece, the aforementioned call, titled "Positions, Purposes, Pragmatics: A Journal of Radical Geography".

The original reads as a mini-manifesto, relevant today though dated in its antiquated patriarchal language (e.g. "A society which measures man's [sic] worth in terms of volume of publications accumulated is no less sick than one which measures his [sic] worth in terms of dollars amassed" [Stea 1969:1]) and too limited/bordered within "the American University". Stea's (1969) language was oblivious to its masculine colonial normativity. Its bounded elitism. And yet. Relineated. Radically reconfigured. Composted and sung out anew …

The following short piece is composed from fragments of language directly from Stea's piece (and, finally, from the epigraph by Whitman):

alter cloistered
halls, seek

sentiments
perpetuating themselves

    damn articles!
    the academic ostrich!

society which measures
values it pretends

to transcend, if
you share

(a)new
ordering goal

a more viable
    patterning

substituting
rough new prizes.

References

Auden W H (1979) In memory of WB Yeats. *Collected Poems* 247–249

Dawn A (2013) *How Poetry Saved My Life: A Hustler's Memoir*. Vancouver: Arsenal Pulp Press

de Leeuw S and Hawkins H (2017) Critical geographies and geography's creative re/turn: Poetics and practices for new disciplinary spaces. *Gender, Place, and Culture* 24(3):303–324

Forman R (1997) If you write poetry. In id. *Renaissance*. Boston: Beacon

Harvey D (2006) The geographies of critical geography. *Transactions of the Institute of British Geographers* 31(4):409–412

Katz C (2001) Vagabond capitalism and the necessity of social reproduction. *Antipode* 33 (4):709–728

Kristeva J, Moro M R, Ødemark J and Engebretsen E (2018) Cultural crossings of care: An appeal to the medical humanities. *Medical Humanities* 44(1):55–58

Magrane E (2015) Situating geopoetics. *GeoHumanities* 1(1):86–102

Marston S A and de Leeuw S (2013) Creativity and geography: Toward a politicized intervention. *Geographical Review* 103(2):iii–xxvi

Rasula J (2002) *This Compost: Ecological Imperatives in American Poetry*. Athens: University of Georgia Press

Stea D (1969) Positions, purposes, pragmatics: A journal of radical geography. *Antipode* 1 (1):1–2

Whitman W (1867a) Song of the open road. In id. *Leaves of Grass*. New York: Fowler & Wells

Whitman W (1867b) This compost. In id. *Leaves of Grass*. New York: Fowler & Wells

Illegality

Lise Nelson

Department of Geography, Pennsylvania State University, University Park, PA, USA;
lknelson@psu.edu

The concept of illegality explored here is rooted in the discursive figure of the "illegal alien", the product of material and ideological processes situated at the intersection of state territoriality, nationalism, capital accumulation, and racism. This figure is part of a larger constellation constituted by the binary of legality and illegality, one that Heyman and Smart (1999) argue is fundamental to liberal democratic state power, and I would add racialised state power. By constructing the state as guarantor of the law within this binary, the sovereignty of the state is legitimised and a space is created to wage war on "illegal" practices, objects, and subjects. Rather than attempt to capture the complex and multifaceted terrain of illegality in this broader frame, this entry focuses on immigration policy and politics, racial formations, and political economies of neoliberal globalisation, which together produce the category of the illegal/undocumented/unauthorised migrant as a ubiquitous form of exclusion in our contemporary world. Although my initial discussion below focuses on the US context, it is clearly a global phenomenon—from Paraguayans working in Argentina, to Laotians in Thailand, to Senegalese in France. To the extent that much of the conceptual literature on migration and illegality in English draws from the US context, it is important to expand our horizons geographically and recognise the concept of illegality includes a range of legal statuses that in distinct ways function across time and space to contain or completely restrict the civic and social rights of people critical to the functioning of the global economy.

To frame it succinctly, the term illegality in the context of migration is used to critique the confluence of geopolitical, racial, and economic processes that discursively construct specific and racially marked bodies as threatening/unauthorised/outside the law. As Nicholas De Genova (2005:8) argues in the context of the United States:

> Migrant "illegality" proves to be a decisive feature of the distinctive racialisation of Mexicans in the United States. [It is critical to see] ... "illegality" as lived through a palpable sense of deportability whereby some are deported in order that most may remain (undeported) as workers. In other words, "illegality" provides an apparatus for producing and sustaining the vulnerability and tractability of Mexican migrants as labor.

As a concept, illegality critically interrogates a "state of being", one produced by immigration policy, political economic relations and racial formations. It is a state of being predicated on the assumption that low-wage, racialised immigrants are outside the law and inherently transgressive and criminal. De Genova and others conceptualise illegality by drawing on Gorgio Agamben's (1998) notion of *homo sacer*, a figure that embodies the bare or depoliticised life distinguished from the pinnacle of "modern" political subjectivity—the citizen. Of particular importance to geographers is the role of state territoriality in this process. Joseph Nevins (2001:147–148), also considering the US–Mexico border, cogently argues that the spatial dimensions of illegality are fundamental to its discursive power:

> For many, there is no reason to debate policies that aim to stop "illegal" immigration, something that is simply wrong ... This is because the "illegal" is someone who is officially out of place—in a space where he [sic] does not belong ... Territoriality helps to obfuscate social relations between controlled and controller by ascribing these relations to territory, and thus away from human agency.

If this spatial transgression is perceived in both material and discursive terms as located at the geographic borders of the nation, a number of scholars have been attentive to how illegality is refracted, experienced and resisted within local institutions, in the context of everyday life, and in relation to geographies of immigrant detention and deportation (Coleman 2007; Harrison and Lloyd 2012; Martin 2012; Nelson and Hiemstra 2008).

Radical geographers invoke the concept of illegality critically, rejecting the normalised denial of fundamental civic and human rights based on presumptions of spatial transgression, racial hierarchy, class position, criminality, etc. Critical scholarship on illegality rests on an ethical position articulated forcefully and repeatedly by advocates and activists in this arena that no human being is illegal. Those who invoke illegality in and beyond scholarship in radical geography recognise the concept as a discursive and material process that functions to naturalise the operation of power and to oppress groups of people.

Beyond its explicitly critical and normative stance, the scholarship on illegality also extends longstanding discussions in radical geography that consider the political economy of capital accumulation in relation to terrains of identity and discursive forms of power. Geographical inquiry on illegality thus extends Marxist considerations and issues raised by feminist, postcolonial and critical race geographers. The central role of capitalism to the constitution, operation, and entrenchment of illegality is clear. In the context of neoliberal globalisation, businesses, labour contractors and state actors across a number of regions have deepened and made more systematic their efforts to recruit low-wage workers from outside their own national borders to work in low-paid, insecure, usually arduous, and often exploitative conditions. These processes of recruitment coincide with processes of displacement and dispossession in sending regions, ones also traceable to the imposition of neoliberal restructuring and globalisation. Displaced and dispossessed workers arrive by a variety of means that range from state-sanctioned, nominally legal recruitment and employment, such as the *kefala* system in the Middle East or guestworker programs in Canada, to ostensibly unsanctioned but large-scale recruitment and employment of "illegal" migrants in the United States and Australia. What unites these various strategies is that all function to recruit, retain and discipline non-native and usually racially marked workers for a range of economic sectors (agriculture, construction, and low-wage service work in particular), while excluding them from long-term settlement, citizenship or social belonging. This exclusion is enacted legally through temporary permits that track people and require their return or structurally through keeping children in sending regions to compel parents not to settle in destination regions. In all cases, however, it is produced through myriad forms of social-civic-legal hierarchies tied to racism, xenophobia, and the threat of violence or deportation. Of course, contemporary geographies of illegality are not historically unprecedented. Histories of slavery and indentured servitude, as well as histories of post-emancipation racial hierarchies that stripped de facto citizenship rights of communities of colour, demonstrate the long history of social/civic/racialised exclusion sitting right next to, even constituting, the economy.

These historical antecedents notwithstanding, neoliberal globalisation has accelerated cross-border labour flows through the deepening and linking of dispossession and recruitment. Furthermore, neoliberalism has created a particular and vexing contradiction between politics, narratives, and practices of *liberalising* some global economic flows (capital, goods and services) and simultaneous efforts *securing* national borders to the flow of labour and to the embodied presence of racialised difference within the presumed homogeneity of the national community (Andreas 2001). It is within these contradictions that narratives about "the illegals" as existential threats—whether that be "false" refugees in Europe, Latinx immigrants in the United States, Nigerians in China, or Guatemalans in Mexico—have become central to neoliberal governmentality (Hiemstra 2010) and to resurgent nationalist populisms across the globe (Domenech 2011; Van Ramshorst 2018). Radical and critical geographers must continue their efforts to undermine these taken-for-granted categories and narratives, and mobilise empirical evidence to support the work of social organisations and movements challenging the ideological and material architectures of illegality across time and space.

References

Agamben G (1998) *Homo Sacer: Sovereign Power and Bare Life.* Stanford: Stanford University Press

Andreas P (2001) *Border Games: Policing the US–Mexico Divide.* Ithaca: Cornell University Press

Coleman M (2007) Immigration geopolitics beyond the Mexico–US border. *Antipode* 39 (1):54–76

De Genova N (2005) *Working the Boundaries: Race, Space, and "Illegality" in Mexican Chicago.* Durham: Duke University Press

Domenech E E (2011) Crónica de una "amenaza" anunciada. Inmigración e "ilegalidad": visiones de Estado en la Argentina contemporánea. In B Feldman-Bianco, L Rivera Sánchez, C Stefoni and M I Villa Martínez (eds) *La Construcción Social del Sujeto Migrante en América Latina: Prácticas, Representaciones y Categorías* (pp 31–78). Quito: FLASCO

Harrison J L and Lloyd S E (2012) Illegality at work: Deportability and the productive new era of immigration enforcement. *Antipode* 44(2):365–385

Heyman J and Smart A (1999) States and illegal practices: An overview. In J Heyman (ed) *States and Illegal Practices* (pp 1–24). Oxford: Berg

Hiemstra N (2010) Immigrant "illegality" as neoliberal governmentality in Leadville, Colorado. *Antipode* 42(1):74–102

Martin L (2012) "Catch and remove": Detention, deterrence, and discipline in US nonciti-
   zen family detention practice. *Geopolitics* 17(2):312–334
Nelson L and Hiemstra N (2008) Latino immigrants and the renegotiation of place and
   belonging in small town America. *Social and Cultural Geography* 9(3):319–342
Nevins J (2001) *Operation Gatekeeper: The Rise of the "Illegal Alien" and the Remaking of the
   US–Mexico Boundary.* New York: Routledge
Van Ramshorst J P (2018) Anti-immigrant sentiment, rising populism, and the Oaxacan
   Trump. *Journal of Latin American Geography* 17(1):253–256

Imagination

Amanda Thomas[iD]

*School of Geography, Environment and Earth Sciences, Victoria University of Wellington,
Wellington, Aotearoa, New Zealand;
amanda.thomas@vuw.ac.nz*

I would have found it hard to live through the last few years without retaining the belief that the world could be very different. The task is to find a way of expressing that ambition in theory and practice both ... we need to be able to think of a future in which surprisingly radical kinds of democratic and inclusionary change *can* take place. That surely is the central, if as yet unfulfilled, message of 1989.   (Smith 1991:413)

Around the time Neil Smith was writing of his need for hope, Francis Fukuyama was declaring the end of history, Margaret Thatcher had proclaimed there is no alternative to neoliberalism, and the possibilities for radical change seemed to fade from view. Twenty years on, Smith (2009) lamented the damage that had been done to the political imagination in the intervening years. Now, dystopian visions of the future abound through popular culture and academic literature, particularly in relation to the environment. The future is understood to be already determined. Any uncertainty and all risks must be measured, pre-empted, and securitised. Little space is left for hopeful, creative and exciting futures.

Yet, geographers like Smith and J.K. Gibson-Graham (Gibson-Graham and Roelvink 2009) have insisted that it's more important than ever to imagine how our worlds might be different, and more just. The task of imagining means exploring difference, indeterminism, and creativity. For David Harvey (2000) geographers need to develop utopias so we have a point to work towards, and an idea of what could be built. This is fundamentally about creating and sustaining hope about future possibilities. Cultivating hope, Ben Anderson (2006) argues, is an ethical obligation for geographers.

Imagination about better futures exists in a spectrum between two poles. At one end, it builds on what we know and is conceivable within our existing worlds. This includes action towards "everyday" utopias, the way mundane practices demonstrate the possibility for something better (Cooper 2013). For instance, already existing practices of care, communal exercises in redistributing food or advocating for welfare recipients, may be formative for dreaming about ideal places (Williams 2017). Work on community economies, everyday activism and autonomous geographies has shed light on these already existing relations and amplified them (Diprose 2017; Larner 2014).

At the other end of the spectrum, imagination remakes things entirely and goes beyond what we currently know. This involves interrogating the existing categories that constrain where our imagination might go, and attempting to

Keywords in Radical Geography: Antipode at 50, First Edition. Edited by the *Antipode* Editorial Collective.
©2019 The Authors/Antipode Foundation Ltd. Published 2019 by John Wiley & Sons Ltd.

transgress them. This is the kind of imagination where social relations are entirely remade and the world is fantastically different. Across this spectrum, the act of imagining utopias has radical potential; dominant narratives are challenged and other possibilities and desires are unleashed (Pinder 2002).

However, with any project of imagining utopias, there is the question of who is doing the imagining and to what effect. Edward Said (1978) introduced the idea of a "geographic imaginary", the way the Orient was constructed by the West. These constructions imagined boundaries, which then became real; imagined an exotic other that became the binary for Europeans to define themselves against; and constructed an other that needed and justified the "civilising" force of colonisation.

In the South Pacific, Aotearoa New Zealand was an imagined utopia for parts of British society. Central to the utopian narrative about the new colony in the 1800s was the promise of a classless society, where every man could make his own fortune and a better social order would prevail. Sargent (2001:2) argues Aotearoa "New Zealand has been described constantly in terms straight from the utopian lexicon, and authors from the earliest [European] voyages ... to the present-day have described New Zealand as a current or at least potential eutopia or good place". Aotearoa New Zealand, Sargent contends, "appears to have developed a stronger utopian tradition than any other country" (2001:14). This utopian promise, never quite realised, is central to Pākehā (white New Zealand) national imaginaries and present in many tropes about white identity. However, Pākehā national imaginaries are also formed through the active forgetting of how we came to be in this place (Bell 2006). Namely, a white utopian imagination was violently imposed on Indigenous Māori, as land was stolen and sovereignty ignored.

Now, as work towards decolonisation is more and more visible in Aotearoa New Zealand (although deeply contested and resisted), there is a need for new utopias that reimagine the relationships between Indigenous and non-Indigenous people. But these must be utopias that are specific to place. As Moana Jackson, renowned Māori scholar and lawyer, argues, we need entirely different political forms that are rooted to this context. He states:

> It seems to me that democracy has to come from the land of the people it serves. So a democracy that truly belongs to this country has to come from this place, it can't come from London or somewhere else. It needs to reflect the relationships with Papatūānuku (Earth Mother)[1] and those who are party to the Treaty [of Waitangi][2] ....
> I think the Treaty asks us to *envision something different*, not just in structural terms but in philosophical, and if you like, ideological terms. (Jackson 2017:39, emphasis added)

For Moana Jackson, working with other Māori, this utopian project has taken the form of Matike Mai, an ambitious attempt to explore constitutional transformation from a Māori perspective (Matike Mai 2016). After 252 meetings around the country, a report was developed that suggested a range of new political forms that would be underpinned by Māori sovereignty, and create room for everyone that arrived after Māori.

Another example of place-based imaginaries draws on a vibrant history of utopian urbanism. The Imagining Decolonised Cities project, based in Porirua, Aotearoa New Zealand, invited the public to design cities that are equitable places for all whānau (families), and are underpinned by Māori values and identity.[3] What would a decolonised city smell, taste, look, sound, feel like? Submissions—which could be in any form—had to be rooted in the values of a particular iwi (tribe) while being forward looking and utopian, and discuss how they were decolonising. The wildly different entries reflected something like Chakrabarty's (2000) imagination (see also Closs Stephens 2011), which is not about the inclusion of different perspectives but the existence of multiple ontologies (Howitt and Suchet-Pearson 2006).

This is one of the core challenges of utopian thinking. Given the insights from postcolonial theory, it's important that practices of radical imagination are not about one utopic vision, nor state-led utopian projects that tilt into authoritarianism (Smith 1991; Wilson and Bayón 2018). Rather radical geographies are increasingly exploring utopianism as a process, rather than a fixed and static image of the future. Utopia as a process means cultivating multiple imaginations of good, just lives that engage with the specifics of place.

However, the Imagining Decolonised Cities project also highlighted the complexity of focusing on imagination when communities in the here and now don't have justice. This project continues to negotiate the tension between a need for hopeful, forward-looking cities rooted in Māoriness (Kiddle 2018), with the reality that the "geographies of hope promised by postcolonial scholarship are in constant tension with persistent geographies of marginalization, disadvantage and desperation" (Coombes et al. 2013:694). This is a productive tension; Anderson argues for the kind of hope that "does not … make peace with existing world", including past and persistent injustice (2006:705, drawing on Pieper 1967). Understood in this way, hope means refusing a belief that everything will be alright, and that we are on a path of linear progress. Instead, it is about balancing critique with possibility—things can be better but this takes labour.

Important research has sought to achieve this balance by examining everyday activism, and has drawn attention to the important ways change is generated through small acts. This work of creating utopias in the present, I argue, needs to be paired with feral dreams about distant, radically just futures. These radical reimaginations, in questioning what is possible and thinking the impossible, might shift our horizons and hold the potential for new worlds to form. Imagination as a theoretical task and a methodology in radical geography holds the promise of creativity instead of technocratic management, and hope rather than securitisation and crisis management.

Endnotes

[1] Also earth, and wife of Rangi-Nui, Sky Father. All living things originate from them together.
[2] The Treaty of Waitangi, first signed in 1840, is a treaty between the British Crown and more than 500 Māori chiefs. There are certain principles, including partnership, that flow from the Treaty.
[3] See http://www.idcities.co.nz (last accessed 16 August 2018).

References

Anderson B (2006) Transcending without transcendence: Utopianism and an ethos of hope. *Antipode* 38(4):691–710

Bell A (2006) Bifurcation or entanglement? Settler identity and biculturalism in Aotearoa New Zealand. *Continuum* 20(2):253–268

Chakrabarty D (2000) *Provincializing Europe*. Princeton: Princeton University Press

Closs Stephens A (2011) Beyond imaginative geographies? Critique, co-optation, and imagination in the aftermath of the War on Terror. *Environment and Planning D: Society and Space* 29(2):254–267

Coombes B, Johnson J T and Howitt R (2013) Indigenous geographies II: The aspiration spaces in postcolonial politics — reconciliation, belonging and social provision. *Progress in Human Geography* 37(5):691–700

Cooper D (2013) *Everyday Utopias: The Conceptual Life of Promising Spaces*. Durham: Duke University Press

Diprose G (2017) Radical equality, care, and labour in a community economy. *Gender, Place, and Culture* 24(6):834–850

Gibson-Graham J K and Roelvink G (2009) An economic ethics for the Anthropocene. *Antipode* 41(s1):320–346

Harvey D (2000) *Spaces of Hope*. Edinburgh: Edinburgh University Press

Howitt R and Suchet-Pearson S (2006) Rethinking the building blocks: Ontological pluralism and the idea of "management". *Geografiska Annaler: Series B, Human Geography* 88 (3):323–335

Jackson M (2017) "We have come too far not to go further": Interview with Moana Jackson by Dylan Taylor and Amanda Thomas. *Counterfutures* 4:27–51

Kiddle R (2018) Māori placemaking. In R Kiddle, P Stewart and K O'Brien (eds) *Our Voices: Indigeneity and Architecture* (pp 46–61). New York: ORO Editions

Larner W (2014) The limits of post-politics: Rethinking radical social enterprise. In J Wilson and E Swyngedouw (eds) *The Post-Political and Its Discontent: Spaces of Depoliticisation, Spectres of Radical Politics* (pp 189–207). Edinburgh: Edinburgh University Press

Matike Mai (2016) "He Whakaaro Here Whakaumu Mō Aotearoa: The Report of Matike Mai Aotearoa—the Independent Working Group on Constitutional Transformation." http://www.converge.org.nz/pma/MatikeMaiAotearoaReport.pdf (last accessed 16 August 2018)

Pinder D (2002) In defence of utopian urbanism: Imagining cities after the "end of utopia". *Geografiska Annaler: Series B, Human Geography* 84(3/4):229–241

Pieper J (1967) *Hope and History. Translated by D Kipp*. San Francisco: Ignatius Press

Said E (1978) *Orientalism*. London: Routledge and Kegan Paul

Sargent L T (2001) Utopianism and the creation of New Zealand national identity. *Utopian Studies* 12(1):1–18

Smith N (1991) What's left? A lot's left. *Antipode* 23(4):406–418

Smith N (2009) The revolutionary imperative. *Antipode* 41(s1):50–65

Williams M J (2017) Care-full justice in the city. *Antipode* 49(3):821–839

Wilson J and Bayón M (2018) Potemkin revolution: Utopian jungle cities of 21st century socialism. *Antipode* 50(1):233–254

Knowledges

Kate Derickson

*Department of Geography, Environment and Society, University of Minnesota,
Minneapolis, MN, USA;
kdericks@umn.edu*

What is at stake for radical geography, politically and epistemologically, when "knowledge" becomes "knowledges"? The plural form of the word emerges out of an intellectual and political tradition that is concerned with the power-laden, geographically and socially limited nature of any claim "to know". To refer to knowledges in their plurality is to refuse what Donna Haraway (1991) has called the "god trick"—to proceed as though it is possible to see (and know) the social formation it its totality. Moreover, it is to recognise that claims to know are always already conditioned by their own relationship to the social formation. These are not merely academic distinctions, but rather reverberate throughout the project of radical scholarship by centring the question of what radical scholarship can do, what it is for, and ultimately, what radical scholarship *is* by emphasising the power dynamics at work in the act of making knowledges.

In articulating an approach she calls "minor theory", Cindi Katz (1996) reflects on the way that one's own absence from theoretical frameworks can be a strong motivator for questioning how knowledge gets made and the work that theory does. Thinking from these contradictions can yield new insights that cannot be contained by the explanatory power of a critique of capitalism. Moreover, attention to the violence of that drive to containment, to account for, to explain, yields a different attitude or idea about what knowledge can do. Alongside and in conversation with postmodern and post-structural thought more generally, radical and critical geographers have problematised who can know what, what is at stake in knowing, and what makes plural, situated knowledges rigorous. These interventions created much angst in the some radical geographical traditions. For example, what social relations provide windows into the systemic workings of the social formation if not capitalism? And what do we lose politically when we "let go" of these categories, or the ability to do objective analysis? If we no longer make strong claims to having the "truth"?

These questions have been raised in relation to thinking about difference other than class, with varying degrees of fidelity to the Marxist intellectual project and associated ontologies, epistemologies and methodologies. Such analyses have highlighted the political stakes and limitations of knowledge production through attention to the genealogies and effects of categories of analysis (i.e. "woman" and "Third World women"; Mohanty 1988), their presumed ontologies (i.e. essentialist, anti-essentialist; Gibson-Graham 1996), and their relationship to alterity and negation through the deployment of binaries (i.e. woman as not man; Haraway 1991). Crucially much of this scholarship and activism has

Keywords in Radical Geography: Antipode at 50, First Edition. Edited by the *Antipode* Editorial Collective.
© 2019 The Authors/Antipode Foundation Ltd. Published 2019 by John Wiley & Sons Ltd.

illustrated how attention to these sites can tell us more not only about how capitalism works, but how resistance to capitalism might and *ought* to work.

In radical geography these lines of inquiry are often attributed to the related interventions of feminist science and technology studies (see for example Haraway 1991; Harding 1986) and post-structuralist theories of power (see for example Butler 1990; Foucault 1978). In this brief piece, however, I want to highlight the linkages between the ideas at work in the politics of knowledge production and the social movements out of which they emerged. Naming transformational moments in political thought is always a dicey endeavour, but we can mark, for the sake of space in this piece, two complementary trajectories of interventions in left political thought that contributed to shifts in understandings of the relationship between knowledge production and anti-capitalist, anti-racist, liberatory futures.

The Combahee River Collective emerged in part in response to the failure of second wave feminism and some elements of the Black radical tradition to account for the lived experiences of Black women (Taylor 2017). Writing and thinking together, the scholars who made up the Collective made two particularly lasting contributions to how we think about the relationships between knowledges and politics, including the concepts of "interlocking oppression" and "identity politics" (Taylor 2017:4). The concept of interlocking oppression laid the foundation for the idea of intersectionality as "the idea that multiple oppressions reinforce each other to create new categories of suffering" (ibid.). For the Collective, identity politics meant that one's identity and lived experience could be a source of political radicalisation, and could become the grounds of confronting one's oppression. Like the feminist principle that the "personal is political", identity politics as articulated by the Collective has become misunderstood and abused as an individualistic insistence on the centrality of identity, when the intention was to insist that resisting forms of oppression in one's own life that stem from social location and systemic social relations is itself political. Taken together these formulations challenge the foundations of Marxist politics and Marxist knowing; if oppressions are multiple and articulate with one another to create new formulations, and there is an intimate relationship between the ability to know about the forms those oppressions take and our own embodied lived experience, then not everyone can make knowledge about *all* the forms that oppression and marginalisation take. And one's lived experience matters for what kinds of knowledge one can make. The Collective did not understand itself to be anti-Marxist, but rather extending its analytical power.

A second, related intervention has roots in Gayle Rubin's (2006) influential piece from the mid 1970s that proposes the sex/gender system as an analytical framework. This brought heterosexual sexuality and binary gender into relief as systems as "products of human activity which are changeable through political struggle" (Haraway 1991:137). This framework introduced the sex/gender system as an axis of oppression, shaped understandings of gender as socially constructed, and sexuality as a site of politics. Rubin's analysis was shaped not only by intellectual debates, but also in the radical politics of Ann Arbor, where she was involved with radical lesbian and feminist consciousness-raising groups (Rubin and Miller 1998).

If lived experience invites an understanding of capitalism as actually capitalist patriarchy, then knowing about class alone is insufficient for politics. Nancy Hartsock (1983) posited the feminist standpoint approach, which built on and extended Marxist ways of knowing. Instead of knowing from the lives of the working class as the most objective places location from which to know about capitalism, Hartsock argued that capitalist patriarchy necessitated knowing from the lives of women. This analysis would be the grounds for a feminist materialist standpoint, or an engaged position from which to objectively know about the real relations of domination. But as Haraway (1991) pointed out, this investment in the category "women" and its definition through "objectification, exchange and appropriation", on which analyses of gender often depended in major bodies of feminist theory by white women, made the limits of the category and politics of "global sisterhood" which emerged from it hard for many white feminists to grasp (see also Mohanty 1988). For her part, Haraway (1991) posits the figure of the cyborg as an epistemology—one that has no recourse to nostalgia and innocence but is aware of history; born out of militarism and capitalist patriarchy but with no fidelity to its origins, and a figure that refuses the possibility of what she calls the "god trick" or fantasy that a knower can get outside the social formation to know in objective ways. Haraway and others writing in conversation with her (see for example Harding 1986) argued that partial, situated knowledges were the only kinds of knowledges possible, and in so doing, raised political questions about the work of knowledges that masquerade as otherwise.

These debates and conversations came to bear on the discipline of geography both in terms of how we think about the categories of class, race, and gender, and how we understand the act and process of creating knowledges as itself political and sites of politics. Julie Graham (1988) traced the confrontation between Marxism and postmodernism at the 1987 Annual Meeting of the Association of American Geographers, arguing that the issue at hand for postmodern critiques of Marxism was essentialism, or how to understand the ontological nature of categories of class, production and accumulation and whether and how they organise the social formation in a real way. Graham argued that Marxism need not make an essentialist move, and along with her collaborator Kathy Gibson, came to make an influential contribution to Marxist geography by arguing for anti-essentialist approaches to Marxism that hold class constant as a strategy of analysis but reject the related move of asserting class processes have more causal power, diverging decisively from critical realist epistemologies that underlie most Marxist thought in geography (Gibson-Graham 1996, 2006; see also Sayer 2000). This intervention was influential for feminist geographies and some economic geographies, but it was also strongly contested by those who considered this political and intellectual move to be giving up Marxist geography's "main strength" (Peet 1992:122).

More recently, drawing on a long tradition of Black feminist and radical thought, the sub-field of Black geographies has challenged radical geographical knowledges, inviting the discipline to consider the relationship between the categories it deploys, the questions it asks, and the work those knowledges do in the discipline and beyond. Contributions to this tradition illustrate the ways in which work that seeks to understand the dispossession of Black people can be perversely attached to Black suffering, and produce the Black subject as always already abject, overlooking and minimising the way Black people create a "black sense of place" (McKittrick 2006) that is not defined by or contained by white supremacy. Woods (2002) asks whether critical geography can amount to anything more than the work of "coroners", diagnosing how oppression and dispossession happened. These interventions cannot be neatly tied to a genealogy of the discipline in this short space, but we can understand them as part of a broader provocation to consider the way that the knowledges we make, even as they are about dispossession, inequality, oppression and accumulation, are not *necessarily* contesting it.

In her recent book, Lisa Lowe suggests that we ought to be attentive to what she calls the "politics of our lack of knowledge" (2015:39) or the things that we do not know or choose to forget. Hers is a broader critique of the cultural and economic formation, but the provocation can be turned back on radical and critical knowledge projects specifically to create a new vantage point on the stakes of making radical knowledges. If the god trick is a fantasy, as Haraway insists, then the anti-racist, anti-capitalist project requires the creation of knowledges that travel via what Katz (2001) calls "countertopographies", or can be "worlded" to use Ananya Roy and Aiwha Ong's (2011) term. These knowledges ought not only count and chart forms of dispossession (see McKittrick 2014), but rather take seriously, learn from, respond to and mobilise utopian visions of otherwise being. This entails a commitment to radical knowledge as plural knowledges, the work and political value of which is more than just what we come to know, but what we call into being in the process of knowing.

References
Butler J (1990) *Gender Trouble*. New York: Routledge
Foucault M (1978) *The History of Sexuality: An Introduction*, Vol. 1. New York: Vintage
Gibson-Graham J K (1996) *The End of Capitalism (As We Knew It): A Feminist Critique of Political Economy*. Minneapolis: University of Minnesota Press
Gibson-Graham J K (2006) *A Postcapitalist Politics*. Minneapolis: University of Minnesota Press
Graham J (1988) Post-modernism and Marxism. *Antipode* 20(1):60–66
Haraway D J (1991) *Simians, Cyborgs, and Women: The Reinvention of Nature*. New York: Routledge
Harding S (1986) *The Science Question in Feminism*. Ithaca: Cornell University Press
Hartsock N C (1983) The feminist standpoint: Developing the ground for a specifically feminist historical materialism. In S Harding and M B Hintikka (eds) *Discovering Reality: Feminist Perspectives on Epistemology, Metaphysics, Methodology, and Philosophy of Science* (pp 283–310). Dordrecht: Kluwer
Katz C (1996) Towards minor theory. *Environment and Planning D: Society and Space* 14(4):487–499
Katz C (2001) On the grounds of globalization: A topography for feminist political engagement. *Signs* 26(4):1213–1234
Lowe L (2015) *The Intimacies of Four Continents*. Durham: Duke University Press

McKittrick K (2006) *Demonic Grounds: Black Women and the Cartographies of Struggle.* Minneapolis: University of Minnesota Press

McKittrick K (2014) Mathematics black life. *The Black Scholar* 44(2):16–28

Mohanty C T (1988) Under Western eyes: Feminist scholarship and colonial discourses. *Feminist Review* 30:61–88

Peet R (1992) Some critical questions for anti-essentialism. *Antipode* 24(2):113–130

Roy A and Ong A (eds) (2011) *Worlding Cities: Asian Experiments and the Art of Being Global.* Oxford: Wiley

Rubin G (2006 [1975]) The traffic in women: Notes on the "political economy" of sex. In E Lewin (ed) *Feminist Anthropology: A Reader* (pp 87–106). Oxford: Wiley

Rubin G and Miller K (1998) Revisioning Ann Arbor's radical past: An interview with Gayle Rubin. *Michigan Feminist Studies.* https://quod.lib.umich.edu/cgi/t/text/text-idx?cc=mfsfront;c=mfs;c=mfsfront;idno=ark5583.0012.006;g=mfsg;rgn=main;view=text;xc=1 (last accessed 17 September 2018)

Sayer A (2000) *Realism and Social Science.* London: Sage

Taylor K-Y (2017) *This Is How We Get Free: Black Feminism and the Combahee River Collective.* Chicago: Haymarket

Woods C (2002) Life after death. *The Professional Geographer* 54(1):62–66

Love

Oli Mould

Department of Geography, Royal Holloway, University of London, Egham, UK;
oli.mould@rhul.ac.uk

The Ancient Greeks had four words for "love"—*eros*, *philia*, *storge*, and *agape*—each one implying a quite radically different meaning of that most visceral and fundamental of human emotions, love.

The first, *eros*, is perhaps best known to Anglophone speakers. Eros was the Greek god of love (the Roman equivalent being Cupid), and has been depicted throughout antiquity as a winged cherub (one of the most famous statues of Eros adorns the Shaftesbury memorial fountain in central London, opposite the media-saturated spectacle of Piccadilly Circus). But the word *eros* stood for a passionate, sexualised love (hence the term "erotic"). This is a love that one individual has for another, driven by biological and physiological desires, but also a compassionate love of another individual, a long-term partner (albeit for Lacan, this is love merely for the self but projected onto another). In both cases though it is an individualised love, devoid of any connectivity external to that desire of self-sustenance.

This kind of love is a politically and socially indifferent force, but only fairly recently so. The "free love" movement that accompanied the hippie subculture in the 1960s was seen as revolutionary. It rejected societal technologies of love, such as marriage, commitment, family life, and the rearing of children and monogamous relationships. It was a reaction to the perceived institutional, state-led, and class-based shackles that directed sexual acts to being part of a familial reproductive system, something to produce more workers for the capitalist system.

However, with the rise of postmodern capitalism and the appropriation of critical movements (a la Boltanski and Chiapello 2005), such sexual freedom was reduced to a commodified desire. Horvat (2016) has succinctly argued that the May '68 uprisings in Paris, based as they were on a desire for freer love between students, was a time in which the force of love *could* be a revolutionary force for real change in society. However, as we now know, this period saw the complete commodification of love by the new agile, flexible forms of capitalist appropriation (see Mould 2018). Now, in the digitally mediated age of the 21st century, the radicality of free love has been appropriated by a notion of promiscuity. Sexual freedom has lost its revolutionary capacities, and instead become absorbed into the capitalist dialectic. On the one hand, promiscuity, particularly in women, is utilised as an accusatory gesture by mainstream conservative values. It has even been used to excuse rape and physical abuse. On the other, the onset of dating websites, and location-based dating apps such as Tinder and Grindr, have commodified the encounter, taken away the "fall" into love, and turned into a

Keywords in Radical Geography: Antipode at 50, First Edition. Edited by the *Antipode* Editorial Collective.
© 2019 The Authors/Antipode Foundation Ltd. Published 2019 by John Wiley & Sons Ltd.

"swipe" into love (Horvat 2016). So the erotic love that the Greeks spoke of has been an attribute of capitalist accumulation and its masculine bias for many decades.

The second version of love that the Greeks articulated was *philia*, meaning friendship. We see this mostly as a suffix to denote a love for something in particular. For example, a Francophile being someone who loves all things French, technophilia being the love of technology, or of particular note to geographers, topophilia expressing the love of place (Tuan 1990). The Greeks though originally used the term to express a deep friendship, a "sisterly or brotherly love". Such a love was more than familial though as it was free from "natural" occurrences. You can't choose your family, but you can your friends. In this sense, *philia* expresses a companionship devoid of erotic love; a love that is in many ways "unnatural" (in that we don't need friends to reproduce as a species). Such a love is appreciative, reciprocal, and productive, and is utilised to achieve a shared goal; in many ways it is a political, class-based love.

Said (1983) has eluded to a similar ideal as "filiation", which he cites as a precritical form of kinship, in contrast to *affiliation* which is to compensate for, and is critical of, the blind spots that can occur from a more "naturalised" filiation. Yet, Jaques Derrida's (2005) famous essay *The Politics of Friendship* denotes that friendship (or more accurately, fraternity) is a critical component of a more radical form of politics and democracy. To think fraternity beyond biological ties is to acknowledge someone as equal, another self. He invokes an Aristotelian notion of friendship that is a fundamental part of democracy. Indeed politics is only possible once a "canonical" friendship is forged, one that invokes a shared origin that has "constancy beyond discourses", including biological fraternity (i.e. being "naturalised" friends via similar biological characteristics such as having the same skin colour, for example). Derrida, contra Said, stresses that a *philia* or filiation that is not limited by biological discourses is able to maintain a critique to the friend/enemy logic, and thereby maintain the possibility of transcending that binary and move beyond wars and enemies. So fraternity extends beyond bloodlines and shared DNA; it is space of possibility. It is little wonder the slogan of the French Revolutionaries was "liberty, equality, and fraternity".

However, like the free love movement, such a fraternity that transcends biological links and seeks a common interest can have destructive powers when utilised intensively. One of the critiques of neoliberal capitalism is it is highly nepotistic (Merrifield 2013); it thrives through the imposition of "cliques". This is most readily recognisable in the accusation of political elites often being chosen from elitist networks. For example, nine of the last 12 British prime ministers studied at Oxford University. And one only has to take a quick glance at who Donald Trump hired into prominent positions in his administration after he entered the White House in 2016. Rex Tillerson, the ex-CEO of ExxonMobil, became Secretary of State; Steven Mnuchin, a former banker for Goldman Sachs, became Secretary of the Treasury; and Wilbur Ross, a former multi-million dollar investor, became Secretary of Commerce. These were all people with whom Trump had done business with in the past. Indeed Trump is the latest in a long line of American Presidents who fills his administration with personal contacts, and there are many different examples that could be used (both Republican and Democrat presidencies). Suffice to say the term "revolving door" is used liberally to imply the reciprocal networks of power that span government, lobbyists, business, and senior figures in public institutions, and allow for wealth and benefit to flow at the expense of those outside them. These cliques are a prime example of how the political will of (a particular narrow definition of) fraternities can be utilised to centralise resources, distort democracy, and perpetuate injustice. Furthermore, there is a gendered bias to all of this; fraternity is of course a masculine term, and the corridors of power in many of the world's national governments are overwhelmingly white men keen to exclude anyone who challenges that power imbalance.

Related to *philia* (but not entirely the same), the third definition of love that the Greeks articulated was that of *storge*, loosely meaning a familial love, specifically between a parent and a child. It refers to a love that is abundant, but brought about "naturally". A parent's love for their child is as much a product of biological and neuro-chemical responses as it is of familiarity. Of course, this need not always be via blood lines as foster parents, adopted guardians, and inter-generational carers also develop *storge* in response to prolonged intimacy that replicates the more normalised nuclear family setting. Invariably, the socio-cultural setting of the family will directly influence the practice of *storge*, with the variance in cultural and national "versions" of parental behaviour an obvious barometer (one only has to think of the stereotypes of different nationalities depicted in popular culture).

Other articulates of *storge* pertain to it as the "friendship" love; a love between two people that starts as friendship but then develops into a deeper, perhaps even romantic love (and then back again). Often devoid of sexual intimacy yet nevertheless practiced between the traditional "couple", such a love is dependent upon the passing of time. It is often characterised by couples that remain good friends even after the "break up" of their romantic or erotic relationship. But like the erotic versions of love, *storge* has also become another axiom of consumerist-soaked late capitalism. Friendships that are deep, detailed, discursive, and lovingly curated over time are being replaced by a more nebulous and vacuous notion of virtualised friendship. Facebook is the main culprit of course with its very vernacular, but the development of our friendship circles becomes nothing more than an exercise in data gathering and a constant performance of what an idealised friendship should entail. Other social media and instant group-communication platforms have come to dominate the interface of friendship and in so doing stripped communication of the visceral, non-verbal, affective, place-based encounter. "Friendship" can now be measured in how quickly you respond to each other on WhatsApp or how you react to a "friend's" post on Facebook. And despite being called social media, the rate of loneliness and mental health epidemics, particularly in the young, are at record levels, meaning it is anything but social.

Then there is *agape*, which, etymologically at least, pertains to the love of humankind for God, and God for humankind. But such a love need not be limited to religious discourses; it is a love that is not so easily co-opted by soft hegemonic powers because it denotes an *unconditional* love. *Agape* thrives when love is given without any semblance of reward; hence it is completely antithetical to contemporary socio-cultural renderings of a conditional or self-sustaining love. A love that is selfless and ubiquitously equitable is the exact opposite of late capitalism that thrives on selfishness, conditionality, and inequality. It is a love that has no conditions. As Horvat (2016) has described, it is a revolutionary love.

Jaques Ellul, a prominent Christianarchist, argued vociferously for a love that was more than generous giving, sympathy, or pity. His argument was for a radical love forgoing solid comforting ground for the uncertainty of poverty, marginality, and unstable places. Ellul collided the teaching of Jesus and Marx, and argued that the love shown in the Biblical Gospels is a subversive force, one that rejects the very-human articulations of morality, law, or ethics. Indeed he argues: "Love, which cannot be regulated, categorized, or analysed into principles or commandments, takes the place of law. The relationship with others is not one of duty but of love" (Ellul 1986:71). So Ellul was following a lineage of radical Marxist and anarchist thinkers (notably Leo Tolstoy) who used the radicality of Jesus' teachings on love as a means to articulate it as a subversive *practice* rather than an emotion, or affective register. Ellul was heavily involved in the French resistance to Nazi occupation during WWII, sheltering Jews and providing them with false papers to escape encampment. He also set up clubs for juvenile delinquents, resisted large-scale developments that would cause environmental damage, and was supportive (by means of his training as a lawyer) to conscientious objectors based on his commitment to non-violence as a form of civil disobedience (Greenman et al. 2012).

Because *agape* is a rendition of love that is completely selfless, based in praxis and compelling people to forgo comfort in order to help others, it is in complete opposition to how society is structured today. Neoliberal versions of capitalism are founded on self-interest, the marketisation of everything, and the deeper imprinting of codification into every realm of life (see Brown 2015). *Agape* is active in resisting this totally. By forgoing the self in favour of the Other, by resisting the line that we must constantly "grow" and "accrue", and forcing a deeply personal, intimate, socialised, and *de*codified relationship, *agape* is antithetical to contemporary models of social organisation under capitalism.

Examples of such agapetic love exist all around us, within the fissures of neoliberal capitalism and in the gaps between the observed parts of our world (see Hardt and Negri 2004). Mass mobilisations of people to help construct and maintain refugee camps in defiance of the State, acts of communal solidarity in the face of an ultra-violent religious fundamentalism (from both the East and West canons of mass religion), personal sacrifices to support the unknown neighbour in need, public workers forgoing personal safety to retrieve immigrants and working people from burning buildings made precarious by neoliberal urban governance; there are countless stories of an active, unconditional love that goes unrewarded. But such love energises the Other; it fuels the agency of the marginalised and oppressed (albeit perhaps temporarily) and aids in the formulation of that agency as a revolutionary force.

Love, as Morrison et al. (2013) have rightly pointed out, is so often a taboo subject within geography. To talk of love is somehow to talk outside the confines of social scientific scholarship, to delve into the very irrational realm of emotion, feelings, and even that most caustic of phrases, "self-help" (which itself is nonsense given how radical interpretations of love go against the entire notion of the self in the first instance). It is very *uncritical*. But they also argue that love "is relational and deeply political" (Morrison et al. 2013:506). Love as *agape* is political precisely because it cuts across (and is indifferent to) bodies, cultures, races, religions, disciplinary boundaries, discourse, language, affect, and thought. It is a love that is reckless in praxis but constructive in deed, one that reconstitutes the public sphere from the appropriative clutches of privatisation (Hardt and Negri 2004; Nash 2011). It takes its shape in radical geographies and broader movements; notably feminism (see Moore and Casper 2014), but also within queer studies, decolonisation, and anti-racist struggle precisely because it shuns the comforting ground that comes with hegemonic power. This refusal to pin the theorisation of love to any discipline or hierarchical mode of thought renders the benefits of this radical love allusive to the central powerful majority, and provides an ethical resource for those at the exploited margins.

To theorise love in this way, to see how it bleeds from one register of social marginalisation (e.g. sexism) to another (e.g. racism) has the power to highlight the differences in how those oppressive registers work, but also puts into practice those modes of resistance which work across space and place—in other words, how love can be among other things, *geographical*. As Nash (2011:20) has eloquently argued when theorising love within black feminism, a "visionary love-politics effectively and hopefully uses a refrain like 'where is the love?' and transforms it … into a political call for transcending the self and transforming the public sphere".

In a world characterised by increasing hate, one in which the languages of genocide have crept back into the mainstream by a political-media industrial complex too obsessed with holding onto power, talking about this radical praxis of love—one that actively and unconditionally loves your neighbour (wherever she may be)—is more critical than ever.

References

Boltanski L and Chiapello E (2005) *The New Spirit of Capitalism*. London: Verso

Brown W (2015) *Undoing the Demos: Neoliberalism's Stealth Revolution*. Cambridge: MIT Press

Derrida J (2005) *The Politics of Friendship*. London: Verso

Ellul J (1986) *The Subversion of Christianity*. Eugene: Wipf and Stock

Greenman J, Schuchardt R and Toly N (2012) *Understanding Jacques Ellul*. Eugene: Wipf and Stock

Hardt M and Negri A (2004) *Multitude: War and Democracy in the Age of Empire*. New York: Penguin

Horvat S (2016) *The Radicality of Love*. New York: Wiley

Merrifield A (2013) *The Politics of the Encounter: Urban Theory and Protest Under Planetary Urbanization*. Athens: University of Georgia Press

Moore D and Casper M J (2014) Love in the time of racism. *Ada: A Journal of Gender, New Media, and Technology* 5. https://adanewmedia.org/2014/07/issue5-moorecasper/ (last accessed 19 September 2018)

Morrison C, Johnston L and Longhurst R (2013) Critical geographies of love as spatial, relational, and political. *Progress in Human Geography* 37(4):505–521

Mould O (2018) *Against Creativity*. London: Verso

Nash J C (2011) Practicing love: Black feminism, love-politics, and post-intersectionality. *Meridians: Feminism, Race, Transnationalism* 11(2):1–24

Said E (1983) *The World, the Text, and the Critic*. Cambridge: Harvard University Press

Tuan Y F (1990) *Topophilia: A Study of Environmental Perceptions, Attitudes, and Values*. New York: Columbia University Press

Margin

Sophie Hadfield-Hill

*School of Geography, Earth and Environmental Sciences, University of Birmingham,
Birmingham UK;
s.a.hadfield-hill@bham.ac.uk*

The margin, how it is lived, experienced, navigated, politicised, framed, used, theorised and represented forms the basis of much geographical scholarship on difference, age, gender, race, sexuality and more. How we understand "the margin" or "marginality" within radical geography is crucial to pushing the boundaries of social justice. This piece identifies three aspects of the margin which are important considerations in radical scholarship: (1) the shifting place of the margin; (2) its temporalities; and (3) the place of the body.

What we define as the margin, how it is viewed, experienced and conceptualised shifts across time and space. Dominant minority world characterisations tend to refer to those people and infrastructures who are perceived to be at the edge of society—spatially, socially, economically, culturally and politically unable to fully participate in civic life. Urban geographers, since the early 1900s have depicted marginality to identify with those at the "fringes of the modern industrial city—alienated, dissolute, forever aspiring to the goals that could never be achieved" (Goldstein 2008:202). Economic geographers meanwhile have followed the circulation of capital and identified territories on the margins of economic development (Williams et al. 2017); associated with this are the informal labourers who operate at the margins of the capitalist economy. Development geographers find themselves unpacking what life is like at the margins, the power relations that have led to such marginalisation and everyday moments of conflict and agency which occur on the margins. Despite the use of the term margin and marginality within different spheres of geography, how it is conceptualised has received scant attention which has led to the recent call for further attention to differing marginalities, across space and time (Andrucki and Dickinson 2015; Williams et al. 2017). Recently, radical geographers have begun to deconstruct the bipolarity of the connotations which are so frequently associated with the core and the margin, where the core or centre holds, governs and distributes the power and those at the edge, at the margin, are left out and left over.

Very often geographical accounts of the margin focus on a specific place. Margins are spatial, they have a physical geography; they are habitually demarcated by lines, boundaries and infrastructures. The margin is often referred to as that place over there, or in between, perhaps abandoned, illegally occupied, overgrown and messy. The anchor for much geographical imagination of the margin is rooted in a place which is identified as having less structure and order. Griffiths (2015:22) explains that these places may "become populated by traces, the residues of transitory occupations, transgressions, gatherings and accumulations that happen out of sight, an under-consciousness of the city". Margins, also known as edgelands or wasteland (Gandy 2013), are populated by people, weeds and wastes—those left behind, on the edge of the progression to modernity, the edge of civilisation (Said 1978) and carry "the image, and stigma, of their marginality" (Shields 1991:3). In her personal reflections, bell hooks (1989) describes the place of the train track in her community, a physical reminder of being on the margin. She describes her everyday encounters with the margin and movement towards the centre, and back again at the end of the day. She explains that:

> across those tracks was a world we could work as maids, as janitors, as prostitutes, as long as it was in a service capacity ... we could enter that world but we could not live there ... we always had to return to the margin, to cross the tracks, to shacks and abandoned houses on the edge of town. (hooks 1989:20)

Much of the framing of marginality within geography is grounded in widely taken for granted assumptions and theorisations of core and periphery. Urban planning theory—in the formulation of cities and the framing of the world-order demarcating centre and periphery countries in the distribution of power—have influenced much of this thinking about what constitutes the centre and the margin. These models, associated politics and indeed the *place* of the centre and the periphery (and by association, the margin) have until recently been static (Andrucki and Dickinson 2015). Indeed, Andrucki and Dickinson (2015:204) argue that "there is a persistent tendency to understand centres and margins as priori spatial categories fixed, for example, by previous rounds of capital accumulation". Recent shifts in global politics, south–south cooperation and the intensified debates about western hegemony have prompted a reconsideration of what constitutes the centre and the margin (Andrucki and Dickinson 2015). In this vein, radical geographers need to extend their thinking of margin and marginality to consider the "intersecting 'axes' of centrality" (Andrucki and Dickinson 2015:205), being aware of the multiplicity of marginality and the possibility for bodies to be at once central and marginal. Indeed, building on the example from bell hooks, she explains that her own marginal position was constructed through the wider discourse of who and what was at the centre and the margin but also an appreciation that those on the margin are vital in the everyday goings-on of the centre, and indeed the whole. hooks (1989:20) also argues that it is important to acknowledge the potential of the margin to be a site of radical resistance, a space where people don't want to "give up or surrender ... but rather as a site one stays in ... it offers to one the possibility of radical perspective from which to see and create, to imagine alternatives, new worlds".

Radical geographers can further push, extend and open up conceptualisations of the margin, to recognise the multiple marginalities that intersect in everyday life, across spatial terrains and through time. We should be open to multiple, shifting, embodied, intersecting experiences of marginality—not only of those who experience marginalisation on the basis of their skin colour, religion, social status, sexuality or their neighbourhood, but to open up our reference points to the margin, to account for other feelings of marginality which people encounter on an everyday basis, moments which may be fleeting, temporary or deeply engrained; experiences which shape people's past, present and future.

One of the challenges is to open up our conceptualisation of who and what is marginal. In this vein, Williams et al. unpack everyday experiences of Muslim, white-collar professionals navigating India's new service economy. They argue that "Muslim professionals are simultaneously central to India's liberalising growth dynamic, and marginal, both within narratives of economic success as well as lived realities of these new Indian workplaces" (2017:1280), thus prompting new thinking about the everyday realities of social, political and cultural marginalisation. Similarly, in analysing interviews from two recent ESRC-funded research projects on children, young people and their families' experiences of urban transformation in India (Hadfield-Hill and Zara forthcoming) and everyday articulations of the food, water, energy nexus in Brazil (Kraftl et al. 2018), it is evident that experiences of marginality shift across time and space; these are multiple, within and between bodies. Consider, for example, a participant who speaks of the anxiety she feels when sitting selling fruit on the street, in fear of being moved on by the security guards; here she is "out of place" in an area ostensibly designed and planned as a space for the mobile middle class. Her marginal position is embodied, as she walks to the end of the tarmacked road, to the mud track which is left un-tarmacked, left-over, un-cared for, on the margin (akin to hooks' experience of the railway line). This is a commonly articulated story from those who are perceived to be at the margins. But what about the participant who feels she is marginalised within her own home, second place to her husband's other wife? What about the fee-paying student whose family are fortunate enough to send him to a prestigious higher education college, how are his feelings of marginality encountered when he finds himself in an isolated "prison of mountains", in a space where he feels cut off, isolated, marginalised? How are these experiences of marginalisation exacerbated through and across the seasons, for example when the monsoon arrives? If we are open to there being multiple marginalities which intersect across space and time, we would be sympathetic to these diverse experiences.

In other research on the food–water–energy nexus, young people were asked to create a visualisation of their relationship with and experience of the nexus (Kraftl et al. 2018). These nexus visualisations revealed a series of precarities which intersect with expressions of marginality. Articulations of marginality came, not solely from their socio-economic status or place-based geography, but were spoken about in terms of a range of nexus threats which impacted on their everyday experiences. Nexus threats ranged from extreme weather events, political turmoil and familial threats to nexus relations (i.e. drugs, money, work, ill-health). This research prompts us to continue to search for innovative ways to uncover marginal experiences, methodologically and theoretically.

It might also be useful to think about the overlaps between age, transition and marginality. What might this add to our understanding of the lifecourse? What might marginality mean when we consider those moments in the lifecourse which don't follow the normalised hegemonic predictable pathway—when life takes a different, uncertain turn? Consider, for example, the current UK housing crisis and those young people who are marginalised by societal discourses about home ownership, or perhaps those who experience infertility, occupying a marginal status, socially and spatially. It is common for young people to be referred to as "on the margin", or occupying a liminal space between childhood and adulthood, waiting for the next life stage which opens up possibilities and options. Yet throughout life, at particular moments, marginality is a lived reality for many; this complicates our understanding of the margin and its spatial and temporal intricacies. Given that much geographical scholarship refers to the margin as being rooted in place, the edge of a city, the periphery, across the railway track, down there, how do we account for mobile marginalities? Being on the edge, on the side lines can happen on the move; as we navigate through space, boarding a bus or walking down the street, being open to the shifting time/space of marginality is important. It is critical to acknowledge the ways in which experiences of the margin can be mobile and manifest in multiple spaces at multiple times.

This piece has primarily focused on the experiences of people at the margins, but what of the other infrastructures, natures and non-human others that are perpetually on the margins of human life? Radical geographers should unpack what marginality might mean for the diversity of non-human life which is routinely exploited and entangled with human bodies and infrastructures. Indeed, Gandy (2013:1305) recognises that wastelands and left-over marginal spaces are increasingly recognised as "part of the ecological infrastructure of the city"—for diverse roles such as biodiversity and flood management. Opening up our understanding of marginality to appreciate those non-human others, whose diverse bodies may be on the margins would further complexify our narration of marginality.

Radical geographers need to be open to other interpretations of the margin, broader than a bipolar demarcation between core and periphery. First, we need to be aware of the shifting place of the margin, how the margin is so much more than a physical demarcation of space, a border, a line, a crossing—margins move with us, margins are mobile. Second, there needs to be an awareness of its multiple temporalities, marginality is influenced by moments of crisis, broader sociopolitical factors and ruptures in personal trajectories. Marginality also stays with us, it becomes part of who we are, part of our identities, our past, present and future. Finally, a more critical look at the body is needed, indeed diverse bodies, in their experience of being on the margin, how bodies resist, adapt and transverse the multiple scales of marginality. Radical geographers should be sensitive to multiple margins, operating at different scales, at different times, in different human and non-human bodies, intersecting across space and time.

Acknowledgments
This work was supported by the Economic and Social Research Council: *New Urbanisms in India* [ES/K00932X/2] and *(Re)Connect the Nexus: Young Brazilians' experiences of and learning about food-water-energy* [ES/N013190/1].

References

Andrucki M J and Dickinson J (2015) Rethinking centres and margins in geography: Bodies, life course, and the performance of transnational space. *Annals of the Association of American Geographers* 105(1):203–218

Gandy M (2013) Marginalia: Aesthetics, ecology, and urban wastelands. *Annals of the Association of American Geographers* 103(6):1301–1316

Goldstein D M (2008) Dancing on the margins: Transforming urban marginality through popular performance. *City and Society* 9(1):201–215

Griffiths R (2015) "Re-Imagining the Margins: The Art of the Urban Fringe." Unpublished PhD thesis, Royal Holloway, University of London

Hadfield-Hill S and Zara C (forthcoming) Complicating childhood–nature relations: Negotiated, spiritual and destructive encounters. *Geoforum.*

hooks b (1989) Choosing the margin as a space of radical openness. *Framework: The Journal of Cinema and Media* 36:15–23

Kraftl P, Balastieri J, Campos A, Coles B, Hadfield-Hill S, Horton J, Soares P, Vilanova M, Walker C and Zara C (2018) (Re)thinking (re)connection: Young people, "natures" and the water-energy-food nexus in São Paulo State, Brazil. *Transactions of the Institute of British Geographers*, accepted.

Said E W (1978) *Orientalism.* London, UK: Routledge & Kegan Paul

Shields R (1991) *Places on the Margin: Alternative Geographies of Modernity.* London: Routledge

Williams P, James A, McConnell F and Vira B (2017) Working at the margins? Muslim middle class professionals in India and the limits of "labour agency". *Environment and Planning A* 49(6):1266–1285

Mental Health

Linda Peake

The City Institute, York University, Toronto, ON, Canada;
lpeake@yorku.ca

Beverley Mullings

Department of Geography and Planning, Queen's University, Kingston, ON, Canada;
mullings@queensu.ca

Addressing issues of mental health became virtually unavoidable in the Anglo-American academy during this last decade, particularly in relation to students. Scholars and university administrators started to pay attention to a range of experiences understood as losing varying degrees of control of one's emotions, one's feelings, one's thoughts, one's perceptions of reality, and one's sense of oneself. In seeking to avoid language that represents these experiences solely in terms of a disorder or a problem, we employ here the terminology of people with experience of mental and emotional distress as a way to discuss issues related to mental health broadly defined, in ways that focus on an individual's experiences rather than on the terms used to categorise them.

Conversations about mental health have begun to proliferate among graduate students and other scholars contingently employed in the academy. Particularly among those in precarious employment positions, there is often a visceral sense of anxiety, and frequently, diagnoses of depression and medication. The increasing visibility of graduate students experiencing mental and emotional distress is indicative of how the ground upon which "normal" is defined has been shifting rapidly beneath our feet. As milder forms of depression and anxiety have become almost benign in the cultural consciousness (Lunau 2012), what is considered to be the "new normal" is a form of academic subjectivity that is prone to loss—to anxiety, depression, and breaking down—perhaps stated most eloquently by the queer cultural theorist Ann Cvetkovitch (2012:18–19):

> Academia breeds particular forms of panic and anxiety leading to what gets called depression – the fear that you have nothing to say, or that you can't say what you want to say, or that you have something to say but it's not important enough or smart enough. In this particular enclave of the professional managerial class, there is an epidemic of anxiety-induced depression that is widely acknowledged informally but not always shared publicly or seen as worthy of investigation ... Why is a position of relative privilege, the pursuit of creative thinking and teaching, lived as though it were impossible? ... My aim is to take seriously the forms of unhappiness and hopelessness produced even by these relatively privileged and specialised projects and ambitions.

For racialised scholars the heightened levels of anxiety and panic that are fast becoming a normalised part of academic life are compounded by

Keywords in Radical Geography: Antipode at 50, First Edition. Edited by the *Antipode* Editorial Collective.
© 2019 The Authors/Antipode Foundation Ltd. Published 2019 by John Wiley & Sons Ltd.

the marginalising relationships that many experience in the academy (Joshi-McCutcheon-Sweet 2015). From the weak presence of Black scholars in tenured faculty positions, to the proliferation of circulating discourses that cast the histories and knowledge systems of racialised populations as inferior, the dominant role of whiteness in the production of knowledge throughout the academy produces micro-aggressions that can heighten mental distress. As Felly Nkweto Simmonds (1999:52) notes:

> In the final analysis, I might be an academic but what I carry is an embodied self that is at odds with expectations of who an academic is. I can be invited and/or dismissed as the token (Black, woman, "Third World"), and can be expected or presumed to be taking one or more of these positions in how I teach/what I teach.

Like many in other disciplines, geographers are beginning to seriously question how to balance the existential ledger, namely, that despite the (often middle-class) "privilege" of being in the academy, many can also feel despairingly hopeless. What does this say not just about mental health but also about the university? What does this tell us about the relationship between knowledge production, feelings and politics in these neoliberal times? In this entry we put anxiety, the affective lived experience of precarity, and other distressing states of mental being onto the agenda of radical geographers in order to address the conditions under which we produce knowledge, how we create value, and the new normal's implications not only for our workplaces and our professional institutions, but for politics as well.

The relationship between the social and the physiological across different states of mental health is one that continues to be contested among scholars. While Cvetkovitch (2012), who views depression as an embodiment of feelings and an important part of what it means to be human, acknowledges that she fails to address biology in her work, others, such as Fitzgerald and Callard (2015:19), argue that "it is increasingly difficult for the social sciences to maintain a potent hold on the expansive category of 'human life' while remaining indifferent to the complex neurogenetic textures of human capability". For as scholars like Bentall (2016) demonstrate, experiences of mental and emotional distress affect brain structure in ways that suggest that the mind cannot be separated from the body. Working within and across the social and human sciences in relation to the mind, body, and embodiment is an area that radical geographers are only beginning to explore and for which there are more questions than answers.[1] What uncertain modes of knowledge production would this require? At least it would require a willingness to attend to what Fitzgerald and Callard (2015:19) call an "entanglement of researchers, instruments, writing practices, discourses, observations, archives, bodies, topologies" that speak to the "tangled biological and sociocultural processes of human life".

While there is a range of ontological lenses through which we can think about mental and emotional distress (from biomedical models to those that fix on the cultural realm), an understanding based on the (bio)psychosocial dimension of peoples' lives asserts that people "lose control" as a result of a complex mix of socio-economic and (bio)psychological circumstances. In terms of external conditions, such as social problems or past personal stresses (for example, poverty, racism, abuse, social isolation), we would add to this list how engaging with the organisation of the neoliberal academy affects mental health. Suggesting a relationship between the dominance of the rationality of the market in the academy and increasing numbers of people experiencing mental and emotional distress may generate a degree of scepticism among those who see no connection between these two issues. Yet, there are few aspects of university life that have not been affected by the spending cuts and re-orientation of the university to make it more commercial in orientation, business-like in its knowledge practices and corporate in its self-presentation (Polster and Newson 2015).

Various (symptom-based) studies of higher education in Canada, the USA, and the UK are indicative of a quantitative increase in the number of students experiencing symptoms of distress (see Peake and Mullings 2016). While these studies alert us to the crisis of mental health on university campuses, they overwhelmingly focus on the impact of individual students' experiences of, for example, lack of sleep, debt, or economic uncertainty. Few studies have explored the relationship between these increasing numbers and the changing structural environments within which knowledge is produced. This is a significant gap in the literature that exists on the debilitating impacts of the neoliberalisation of the academy on knowledge production as a whole (The SIGJ2 Writing Collective 2012), and to which geographers have made significant contributions.

The graduate years can be mentally taxing. There is a growing awareness that in the neoliberal university, the pressure to excel and achieve high grades is leading to "destructive perfectionism" (University of Pennsylvania 2015). Graduate students face considerable levels of stress as they juggle the short-term, often hourly-paid teaching contracts and non-academic jobs needed to pay their bills with their need to develop their CVs in order to pursue a permanent academic post or other career. The conditions of work—solitary and unstructured (in the case of non-science bench PhDs), competitive, uncertain, and financially stressful, with time pressures and no certainty of a job—can generate mental and emotional distress. After graduation, often weighed down by student debt and the pressures of intergenerational inequality, many now have to string together short-term contracts, themselves often hard to secure, only too aware of the dearth of tenure-track academic jobs, postdoctoral posts and the current reality of the financial and geographical instability of a career in the academy. Studies that examine how the corporatisation of higher education affects mental health are only just starting to appear in geography (see Horton and Tucker 2014). One study of geography graduates in the US reveals their co-optation into a neoliberal ethos of individuality, competition and heightened pressures of productivity and performance, resulting in feelings of inadequacy, guilt, and isolation (Hawkins et al. 2014), and indicative of a survival-oriented campus culture.

If part of the crisis of mental and emotional distress across university campuses is the very culture and composition of the academy itself, what are the prospects for changing it? We offer some thoughts about what a different kind of academic space could look like and the challenges this involves.

One starting point is how personal experience can be used to challenge what counts, and what counts as knowledge, through investigating what it feels like for people experiencing mental and emotional distress to work in the academy. What does the landscape of everyday feelings in academia look like? While there are a number of first person accounts of experiences of anxiety, depression and breaking down written by public mental health advocates, activists, artists and writers, there are very few by geographers in the academy (but see England 2016). Yet, such accounts can be used as a basis for creative thought, forming a cultural archive from which to generate new ways of thinking about agency, potentially to generate the affective foundation of hope that is necessary for political action but certainly to use depression and anxiety as ways of addressing questions about everyday life and politics through grounded theory and the systemic nature of these experiences. Perhaps the best known investigation of how the new normal might be employed as a resource for political action rather than as its antithesis, is the Public Feelings project associated with Lauren Berlant.[2] The project is a way to think of resistance, what some call "weak activism", in which the version of the human does not depend on wholeness and completeness. Similarly the Institute for Precarious Consciousness (2017) addresses how "failure" to succeed (on the academy's terms) and subsequent paralysis can be transformed into a sense of injustice and a "move towards self-expression, and a reactivation of resistance". Calling attention to "how bodies feel internally in relation with material social space" (Joshi-McCutcheon-Sweet 2015:300), geographers are also beginning to demonstrate the oppressive and distressing impact of racial microaggressions within geography departments on scholars of colour (Joshi-McCutcheon-Sweet 2015; Mahtani 2014; Tolia-Kelly 2017). By focusing on everyday scenes of microaggression these scholars demand a space to recognise (without pathologising) the mental, physical and emotional strain—"racial battle fatigue"—generated in the mundane knowledge production practices in geography.

In an effort to address "failure", the pace of productivity and oppressive practices within geography, critical conversations have begun among geographers who are members of the Great Lakes Feminist Geography Collective in North America on the possibilities of building a slow scholarship movement in geography (Mountz et al. 2015). This latter tactic is part of a broader feminist ethics of care within the academy, an ethics grounded in voice and relationships. The creation of spaces of care and sociability is fundamental to exploring how we might embed a culture of self-care into academic spaces, including strategies to achieve life–work balance. While discussions about such strategies are gaining pace, the silences in the academy suggest much more work is needed until we are comfortable with mainstreaming the de-stigmatisation of mental and emotional distress.

Over the last five years a small but steadily growing transnational community of geographers has deliberately been taking conversations about mental health out of the back spaces of academia and into its professional spaces (e.g. departmental meetings, disciplinary conference spaces and academic journals). Since 2014, activities in these spaces have included the initiation of a listserv, Mental Health and the Academy,[3] Canadian Association of Geographers (CAG) and

American Association of Geographers (AAG) conference sessions and publications (see, for example, *The Canadian Geographer / Le Géographe canadien* 2016). Given that disciplinary professional associations are lagging behind universities in adopting initiatives, commissioning reports and looking at practices that recognise the diversity of mental states among scholars there has been a direct engagement with the AAG through the establishment, in 2014, of the AAG Task Force on the Status of Mental Health. This is advocating for new responses to the challenges that are shaping knowledge production and mental health in the academy, committed to providing environments where all may thrive.[4] Advocacy on behalf of the new normal demands nothing less.

Endnotes

[1] Such an approach already has a long history in Black studies that geographers have started to draw upon (see, for example, McKittrick 2016).
[2] See http://bcrw.barnard.edu/videos/public-feelings-salon-with-lauren-berlant/ (last accessed 21 August 2018).
[3] MHGEOG-L@LISTS.QUEENSU.CA
[4] The Taskforce delivered its report to the AAG in June 2018. To request a copy please contact us at lpeake@yorku.ca or mullings@queensu.ca. Geographers in the RGS-IBG have also started to discuss the formation of a similar group.

References

Bentall R (2016) Mental illness is a result of misery, yet still we stigmatise it. *The Guardian* 26 February. https://www.theguardian.com/commentisfree/2016/feb/26/mental-illness-misery-childhood-traumas (last accessed 21 August 2018)
Cvetkovitch A (2012) *Depression: A Public Feeling.* Durham: Duke University Press
England M (2016) Being open in academia: A personal narrative of mental illness and disclosure. *The Canadian Geographer / Le Géographe canadien* 60(2):226–231
Fitzgerald D and Callard F (2015) Social science and neuroscience beyond interdisciplinarity: Experimental entanglements. *Theory, Culture, and Society* 32(1):3–32
Hawkins R, Manzi M and Ojeda D (2014) Lives in the making: Power, academia, and the everyday. *ACME* 13(2):328–351
Horton J and Tucker F (2014) Disabilities in academic workplaces: Experiences of human and physical geographers. *Transactions of the Institute of British Geographers* 39(1):76–89
Institute for Precarious Consciousness (2017) We are all very anxious. *Workshop for Intercommunal Study* 9 May. http://intercommunalworkshop.org/we-are-all-anxious/ (last accessed 21 August 2018)
Joshi-McCutcheon-Sweet S P E (2015) Visceral geographies of Whiteness and invisible micro-aggressions. *ACME* 14(1):298–323
Lunau K (2012) The mental health crisis on campus. *Maclean's* 5 September. https://www.macleans.ca/education/uniandcollege/the-mental-health-crisis-on-campus/ (last accessed 21 August 2018)
Mahtani M (2014) Toxic geographies: Absences in critical race thought and practice in social and cultural geography. *Social and Cultural Geography* 15(4):359–367
McKittrick K (2016) Rebellion/invention/groove. *Small Axe* 20(1):79–91
Mountz A, Bonds A, Mansfield B, Loyd J, Hyndman J, Walton-Roberts M, Basu R, Whitson R, Hawkins R, Hamilton T and Curran W (2015) For slow scholarship: A feminist politics of resistance through collaborative action in the neoliberal university. *ACME: An International E-Journal for Critical Geographies* 14(4):1235–1259

Peake L and Mullings B (2016) Critical reflections on mental and emotional distress in the academy. *ACME* 15(2):253–284

Polster C and Newson J (2015) A Penny For *Your Thoughts: How Corporatization Devalues Teaching, Research, and Public Service in Canada's Universities*. Ottawa: Canadian Centre for Policy Alternatives

Simmonds F N (1999) My body, myself: How does a black woman do sociology? In J Price and M Shildrick (eds) *Feminist Theory and the Body: A Reader* (pp 50–63). New York: Routledge

The Canadian Geographer / Le Géographe canadien (2016) Special issue: Cultivating an ethic of wellness in Geography. 60(2):159–281

The SIGJ2 Writing Collective (2012) What can we do? The challenge of being new academics in neoliberal universities. *Antipode* 44(4):1055–1058

Tolia-Kelly D P (2017) A day in the life of a Geographer: "Lone", black, female. *Area* 49 (3):324–328

University of Pennsylvania (2015) Report of the Task Force on Student Psychological Health and Welfare. *University of Pennsylvania Almanac* 61(23). https://almanac.upenn.edu/arc hive/volumes/v61/n23/contents.html (last accessed 21 August 2018)

Mercury

Becky Mansfield

Department of Geography, Ohio State University, Columbus, OH, USA;
mansfield.32@osu.edu

Mercury is an excellent emblem of emerging thinking about the nature of life. It might seem paradoxical that a *chemical*—a heavy metal at that—is an emblem of *life*. Yet this turn to the chemical is one of the key features of this understanding of life—an understanding in which knowledge about contemporary hazards is breaking down presumed boundaries between humans and nature, bodies and environments, and even the living and non-living (Romero et al. 2017).

For example, knowledge of the action of chemicals draws from and contributes to changing ideas about the malleability of the body. Against ideas about genetic determinism, knowledge of how chemicals act even at the genetic level has contributed to a much larger shift toward "biosocial", "postgenomic" models of the body (Mansfield 2018). In this regard, lead may be more familiar than mercury as a public health and environmental justice concern, especially given the prominence of the Flint water crisis over the past few years. Like lead, mercury is a neurotoxin to which people are exposed differentially, with greatest effects on the very young even when exposed at very low levels. Both mercury and lead are about the openness of the body to socio-environmental influences.

At the same time, knowledge about the action of chemicals also draws from and contributes to changing ideas about the malleability of the earth system. Against ideas about pristine nature, knowledge of the action of chemicals that people mobilise into the atmosphere has contributed to a much larger shift toward "socionatural" models of the planet, most recently captured in ideas about the "Anthropocene" (Mansfield 2018). Carbon may steal the spotlight in this regard, as a contributor to global climate change and all its cascading effects. Mercury, too, is an earth-system altering product of fossil-fuel driven industrial capitalism. Both mercury and carbon are about the openness of the planet to unequal anthropogenic influences.

In other words, mercury is an excellent emblem of increasingly popular ideas about both socionatural planetary change and biosocial bodily change. Even more, I will show that knowledge about mercury *materialises connections between* global environmental change and toxicity, between planetary and bodily processes. To make the case in this short essay, I tell a common story of mercury's effects, sources, movements, exposures, and governance, drawing primarily from a series of reports produced by the UN since 2002, called the Global Mercury Assessments (AMAP/UNEP 2015; UNEP 2002, 2008, 2013). Along the way, I draw out the ways mercury appears to be natural and social, geological and biological, elemental and global, moving and persistent, and a threat and an opportunity. I

Keywords in Radical Geography: Antipode at 50, First Edition. Edited by the *Antipode* Editorial Collective.
© 2019 The Authors/Antipode Foundation Ltd. Published 2019 by John Wiley & Sons Ltd.

tell the story of mercury as a mercurial story: a story of change that is changing our stories of the world.

Mercury as Geo and as Bio

Mercury is a constituent element of the earth; it exists in its elemental form and as numerous compounds both inorganic and organic (UNEP 2002:28).

A variety of chemical and *biological processes can change mercury from one form into another* (UNEP 2002:33). Especially important is that aquatic bacteria turn elemental mercury into the organic compound methylmercury. Although all forms of mercury are hazardous, methylmercury is particularly harmful to health and development.

Mercury crosses the blood–brain "barrier" and is a neurotoxin (UNEP 2002:38). Starting in the 1950s in the Japanese city Minamata, people and animals experienced a range of unexplained and serious sensory symptoms, loss of muscular control, and even coma and death. It turned out that Chisso Corporation's methylmercury-laden industrial effluent caused this "Minamata disease". It was by puzzling out this case that toxicologists began to understand the range of mercury's actions.

Mercury affects neurodevelopment even at very low doses (UNEP 2002:38). One of the surprising findings from Minamata was that doses of methylmercury that did not affect adults did profoundly affect individuals whose mothers were exposed when pregnant. Studies in the decades since show that mercury affects fetal neuro-development, and that even very low doses produce subtle yet long-lasting effects in behaviour and intellectual development.

Thus, knowledge of mercury challenged two dictums of environmental health: that the placenta is a barrier between the fetus and the world, and that even hazardous substances are safe at low doses. Challenging these ideas about thresholds separating body from environment, knowledge about mercury was an early spur to contemporary ideas about permeability and malleability.

Volcanos, Gold, and Coal

Most mercury is sequestered in the earth's crust, brought to the surface by geologic processes such as volcanos. Yet only about 10% of annual emissions of mercury are due to these processes (UNEP 2013:i).

People mine mercury to use intentionally in numerous products and processes, including in thermometers and lightbulbs, in dental fillings and skin-lightening creams, to make chlorine and lye and to extract gold (UNEP 2013:6–7).

People release mercury into air and water; about 30% of annual emissions are due to human activity in the present (UNEP 2013:6–7). Releases occur during manufacture, use, and disposal associated with the intentional uses just mentioned; they are also the by-products of other activities such as cement production and burning fossil fuels.

Over a third of annual mercury emissions today result from artisanal and small-scale gold mining (UNEP 2013:ii). This gold mining not only releases mercury to water and air, but is an occupational hazard that directly exposes workers: 10–15 million mostly poor people in parts of Latin America, Asia, and Africa (AMAP/ UNEP 2015:2; UNEP 2008:13).

About one quarter of annual mercury emissions result from burning fossil fuels, especially in coal-fired power plants (UNEP 2013:ii). Most of these emissions are from East and South Asia, North America, and Europe (AMAP/UNEP 2015:10).

Movement = Persistence

Once in the environment, mercury moves and is a "global pollutant". While some mercury lands near its source, atmospheric processes transport elemental mercury globally and then deposition spreads mercury across the earth, oceans, and polar regions (AMAP/UNEP 2015:1; UNEP 2008:1).

Once deposited, mercury continues to move. Some is sequestered long term in sediments, but much of it enters global cycles, where it continually changes form and moves within and between air, water, soil, and living things (UNEP 2008:31). Mercury can also be re-emitted to the atmosphere from all these compartments: the UN estimates that 60% of annual emissions are actually re-emissions, mostly of anthropogenic mercury (UNEP 2013:i).

Anthropogenic environmental change affects cycling and re-emissions (UNEP 2008:29, 2013:iv). For example, climate change can transform mercury sinks into sources, as warming temperatures melt permafrost across the Arctic and release the trapped mercury.

There are many temporal lags (UNEP 2013:iii, 8). Mercury released years, decades, even centuries ago cycles and is remobilised today. Mercury in the surface ocean waters is double what it was pre-industrially, and mercury in some species of Arctic marine mammals is 12 times what it was. Today's releases add to those of the past.

Dispossession by Methylmercury

The behaviour of mercury in aquatic ecosystems makes it a key cause of concern (UNEP 2013:24, 26). Aquatic bacteria methylate mercury into its more toxic form. Aquatic organisms easily absorb this methylated mercury and store it in their tissue. Concentrations of methylmercury increase up the food chain—and aquatic food chains are especially long. Large fish, marine mammals, and the people who eat them can have extremely high concentrations of methylmercury in their bodies.

While fish is the main source of methylmercury exposure for most people, exposure differs by class, race, nationality, and indigeneity (UNEP 2002:iv). For example, fish and marine mammals are a central and culturally significant part of the diet for many indigenous people, including across the Arctic. Fish is a large and economically significant part of the diet of many poor people, for whom "sport" fishing (even in polluted waters) is subsistence fishing.

Indigenous people have been at the forefront of fighting back (Arquette et al. 2002; Harper and Harris 2008). For decades, communities affected by mercury (and other pollutants) have called for and conducted research, developed new models of risk, demanded evidence-based fish consumption advisories, and insisted that advisories are never a solution: only clean-up is a solution.

Regulation and Deregulation as Accumulation Strategies

Global mercury is an opportunity for governance. In 2017 the Minamata Convention on Mercury entered into force to regulate mercury throughout its lifecycle (UNEP 2017). Recognising new knowledge about mercury's anthropogenic sources, fate, and effects, this international law joins the older Basel, Rotterdam, and Stockholm conventions to govern the sources, use, trade, storage, and releases of hazardous chemicals.

Global mercury is also an opportunity for "green" growth. According to the Secretary General of the UN, the Minamata Convention is an opportunity "to accelerate the transition to a fairer, greener economy" (UNEP 2017:3). Malleable nature is neoliberal nature: regulation is an accumulation strategy.

The United States has long been slow to regulate mercury. In 1990, Congress directed the Environmental Protection Agency to develop regulations for mercury; yet it was not until 2014 that the EPA finalised its Mercury and Air Toxics Standards regulating power plants (EPA 2018). Not only is a lawsuit against these regulations still working its way through the courts, but under Trump the EPA is moving to substantially weaken these rules by proposing new methods of costs–benefit analysis that would downplay the benefits to public health and emphasise the costs to power companies (Eilperin and Dennis 2018). Deregulation is an accumulation strategy.

In Sum

Over the past few decades, mercury's story has been one of openness and malleability, of unforeseen hazards that challenge our understanding of the nature of nature. It is not that there are parallel emerging ideas about the body (postgenomics) and the planet (the Anthropocene), both of which mercury happens to represent in different ways. Rather, these ideas about bodily and planetary life intertwine; that they are emerging together is not incidental. Mercury is emblematic because it materialises these connections as it moves, transforms, persists, and acts biologically. Mercury appears as the cross-cutting stuff of the world: resolutely natural and social, molecular and global, body and environment, chemistry and biology, non-life and life.

In this, mercury is also emblematic as it inspires multiple scientific, industrial, infrastructural, and regulatory controversies. That is, these emerging ideas about life are both profound and profoundly controversial—there is excitement but also scepticism and outright denial, as we see in the US today. The view of Trump and Pruitt is quite old-fashioned: the planet and the body are closed and sovereign; chemicals released by today's practices have no real effect; there is nothing to worry about so let's get back to making money.

While it seems imperative to push back on this denialist stance, the lesson is that *both* denial *and* acceptance of new ways of thinking can be mobilised to support ongoing capital accumulation and new forms of exploitation, inequality, and injustice. Mercury is part of the long story of earth and part of the (post)colonial, late-capitalist story of unequal and differential environmental change as both crisis and opportunity. We need to ensure that the "opportunity" is not to remake the socionatural world into another money-making scheme in which harm of some is in service to "life" in "general". We cannot assume that non-dualist modes of life will be more just, we must push to make it so.

References

AMAP/UNEP (2015) *Global Mercury Modelling: Update of Modelling Results in the Global Mercury Assessment 2013*. Oslo: Arctic Monitoring and Assessment Programme / Geneva: United Nations Environment Programme Chemicals Branch

Arquette M, Cole M, Cook K, LaFrance B, Peters M, Ransom J, Sargent E, Smoke V and Stairs A (2002) Holistic risk-based environmental decision making: A native perspective. *Environmental Health Perspectives* 110(s2):259–264

Eilperin J and Dennis B (2018) In rollback of mercury rule, Trump could revamp how government values human health. *Washington Post*. Washington, DC. https://www.washington post.com/energy-environment/2018/10/01/rollback-mercury-rule-trump-could-revamp-how-government-values-human-health/?utm_term=.d177ba420bfd (last accessed 8 October 2018)

EPA (2018) "Mercury and Air Toxics Standards (MATS)." United States Environmental Protection Agency. https://www.epa.gov/mats (last accessed 20 August 2018)

Harper B L and Harris S G (2008) A possible approach for setting a mercury risk-based action level based on tribal fish ingestion rates. *Environmental Research* 107(1):60–68

Mansfield B (2018) A new biopolitics of environmental health: Permeable bodies and the Anthropocene. In T Marsden (ed) *The Sage Handbook of Nature* (pp 216–234). London: Sage

Romero A M, Guthman J, Galt R E, Huber M, Mansfield B and Sawyer S (2017) Chemical geographies. *GeoHumanities* 3(1):158–177

UNEP (2002) *Global Mercury Assessment*. Geneva: United Nations Environment Programme Chemicals Branch

UNEP (2008) *The Global Atmospheric Mercury Assessment: Sources, Emissions and Transport*. Geneva: United Nations Environment Programme Chemicals Branch

UNEP (2013) *Global Mercury Assessment 2013: Sources, Emissions, Releases and Environmental Transport*. Geneva: United Nations Environment Programme Chemicals Branch

UNEP (2017) *Minamata Convention on Mercury: Text and Annexes*. Nairobi: United Nations Environment Programme

Monument

Kanishka Goonewardena

Department of Geography and Planning, University of Toronto, Toronto, ON, Canada;
kanishka.goonewardena@utoronto.ca

The condensed representations of space, time and social relations that we call monuments tend to make their presence—or absence—most palpable at moments of radical political upheaval. The fate of many statues of Marx, Engels and Lenin in countries of "actually existing socialism" after 1989 is an obvious case in point; as is that of a few Confederate statues pulled down in Southern US states especially last fall, over 150 years after the Civil War. Monuments surely have proven to have lives beyond the desires of their creators, and even beyond those of their destroyers. In the prevailing political climate of the US, there is no guarantee that we will not see one more memorial to some postmodern Robert E. Lee; while it is also telling that a statue of Engels abandoned to a post-Cold War dustbin of history in the Ukrainian village of Mala Preshchepina has recently found its redemptive passage to a dignified site in Manchester, where Engels fell in love with Mary Burns and the working class, as movingly documented in Berlin-based, British-born artist Phil Collins's 2017 film "Ceremony: The Return of Friedrich Engels". Monuments like these stand in the spatio-temporal force-field of political struggles, like lightning rods and transmission towers channelling their electric energy.

We love them (Engels); we hate them (Robert E. Lee). But can we be for or against monuments as such? Yes and no, writes Henri Lefebvre (2003:21–22) in *The Urban Revolution,* in his signature dialectical style. We are against the monument because it is "essentially repressive" as the "seat of an institution" of dominant power. "Any space that is organized around the monument", he says, "is colonized and oppressed". This is especially true of "great monuments ... raised to glorify conquerors and the powerful". Lefebvre even claims that the "misfortune of architecture is that it wanted to construct monuments". Yet the monument in his view is also "the only conceivable or imaginable site of collective (social) life". Although it "controls people", it "does so to bring them together". As such, "great monuments were transfunctional (cathedrals) and even transcultural (tombs)", radiating vital "ethical and aesthetic power". The monument differs from buildings as the festival does from everyday life or products from *oeuvres*: whereas the urban both shapes and is shaped by social life in the course of everyday routines, "monuments project onto the land a conception of the world". It is a conception not merely of the actually existing world, but of a world that was or is possible. In this sense, monuments "embody a sense of being elsewhere" and "have always been u-topic".

"For millennia", Lefebvre (1991:220) claims in *The Production of Space,* "*monumentality* took in all the aspects of *spatiality* ... the perceived, the lived and the

Keywords in Radical Geography: Antipode at 50, First Edition. Edited by the *Antipode* Editorial Collective.
© 2019 The Authors/Antipode Foundation Ltd. Published 2019 by John Wiley & Sons Ltd.

conceived". While the perceived and lived aspects of monuments pertain to their ability to affect and be affected by social practice, the conceived moment corresponds to their concentrated capacity for *representation*. This representational aspect is rarely more compelling and instructive than in monuments to what Benedict Anderson (1991) theorises as the modern "imagined community" of the nation. Here the monument—a tomb of an unknown soldier, a parliament building, a public space—ranks alongside the museum, the census and the map as a device capable of fashioning out of an agglomeration of strangers the magnetic consciousness of a nation: a human community "always conceived as a deep, horizontal comradeship", notwithstanding social divisions and other differences. "Ultimately", Anderson avers, "it is this fraternity that makes it possible, over the past two centuries, for so many millions of people, not so much to kill, as willingly to die for" for abstracted collections—and partitions—of humanity (1991:7). This curated resolve to die if not kill for the nation too is telegraphed by the imaginative power of national monuments, by virtue of their capacity to intimate nothing less than the secular transcendence of death. In Anderson's words, they bind "the dead and the yet unborn", transform "fatality into continuity", and so represent the nation as immortal (1991:11).

Yet more formal features of the monument may be discerned from its service to the seemingly eternal entity called the nation. Precisely in this role of representing the nation, monuments display their power to link memory, identity and utopia in the cosmological manner characteristic of the great world religions, by symbolically coming to terms with a fundamental question on which modern thought has no sure handle. It is the "man-in-the-cosmos" question of "death and immortality", according to Anderson, about which "neither Marxism nor Liberalism are much concerned" (1991:10)—barring exceptions like Italian communist Sebastiano Timpanaro, who reflects in *On Materialism* (1970) on the baleful weight of nature on human life in the finitude of history (see Anderson 2001). "The cultural significance of such monuments becomes even clearer", Anderson sardonically suggests, "if one tries to imagine, say, a Tomb of the Unknown Marxist or a cenotaph for fallen Liberals" (1991:10). The apparent absurdity of such propositions stems for Anderson from "the great weakness of all evolutionary/progressive styles of thought, not excluding Marxism", which address matters of life and death only "with impatient silence" (1991:10). But monuments to the nation know how to reckon with them: "if nation-states are widely conceded to be 'new' and 'historical', the nations to which they give political expression"—as represented in their monuments—"always loom out of an immemorial past, and, still more important, glide into a limitless future" (1991:11–12). In its secularisation and modernisation of the cosmic temporality of religious consciousness, we also witness "the magic of nationalism", i.e. its ability "to turn chance into destiny". So "with [Régis] Debray we might say, 'Yes, it is quite accidental that I am born French; but after all, France is eternal'" (1991:12).

This "magic" reveals the fundamental appeal of the monument and its essentially *ideological* nature. It is hard to overemphasise this ideological essence of the monument, if only because Anderson himself is keen on insisting that nationalism is categorically distinct from "ideologies" such as Liberalism and Marxism, but functionally analogous to the consciousness of religions, given its proven power to deal with radical existential questions like who we are and what our proper place is in the grand scheme of things. Yet if ideology is understood not simply as the pre-scribed program of this or that political outlook, but more like what Antonio Gramsci termed "conception of the world" or what anthropologists studying so-called primi-tive societies used to call "cosmology", then representations of the nation in monu-ments and by other means would seem to be nothing if not ideological. Indeed, if "ideology", as Louis Althusser (2014:181) famously theorised it, "represents individ-uals' imaginary relations to their real conditions of existence", then one looks in vain for a *concept* that is more appropriate to Anderson's "imaginary community" and its monumental representations. For Althusser's explication of how ideology "interpel-lates" individuals as "subjects", by means of "representations" of their "imaginary relations" with "real conditions", raises the question (2014:188): what do monu-ments do in the contested political field *but* "interpellate individuals as subjects"?

The radically contentious nature of such ideologies—evidently not Althusser's strongest point—also explains why monuments have been so passionately con-tested, both politically and aesthetically. It is not only a matter of what to represent, but how to represent it—as is strikingly evident in the saga of the memorial built in 1926 by the German Communist Party (KPD) to Rosa Luxemburg and Karl Lieb-knecht, its leaders murdered on 15 January 1919 by proto-Nazi Freikorps under bizarre instructions issued by the Social Democratic Party (SPD). Its design was for-tuitous. Ludwig Mies van der Rohe later recalled how he came to be entrusted with this monument, thanks to a chance encounter with KPD cofounder Eduard Fuchs:

> One of the first houses I built was for Hugo Perls in Berlin. Mr. Perls sold his house in the early 20s to a Mr. Edward Fuchs ... After discussing his house problems Mr. Fuchs then said he wanted to show us something ... It was a huge stone monument with Doric columns and medallions of Luxemburg and Liebknecht. When I saw it I started to laugh and I told him it would be a fine monument for a banker.[1]

Mies proposed instead a revolutionary design for the martyred revolutionaries of the November Revolution: rather than bourgeois statues and classical columns, an abstract-expressionist bare brick wall boldly composed of austere cubic forms (see Figure 1). Having been deeply moved by this masterpiece of modernism as an architecture student in Sri Lanka in the 1980s, I went looking for it on my first visit to Berlin in the summer of 2000. Rather than a sight of the monument, however, what I got was a lesson in history and politics, after stumbling upon the following inscription in the fascinating Friedrichsfelde cemetery:

> On this foundation stood the revolutionary monument to Karl Liebknecht, Rosa Luxemburg and many other fighters of the German worker's movement

> 1926 built by the German Communist Party, according to plans by Ludwig Mies van der Rohe

> 1935 destroyed by fascists[2]

It is symptomatic of "de-nazified" post-war Germany and Europe that this monu-ment was never rebuilt in spite of several spirited campaigns by left groups, quite

Figure 1: Monument to Rosa Luxemburg and Karl Liebknecht by Ludwig Mies van der Rohe, 1926 (CC BY-SA 3.0; available at https://de.wikipedia.org/wiki/Re volutionsdenkmal)

unlike the Vendôme Column in Paris, which was restored just three years after being ceremoniously toppled by the Communards on 16 May 1871, and still stands as the tallest sign of *revanchism* à la Neil Smith, within a 40-minute walk from the famous Sacré-Coeur Basilica, the similarly reactionary symbolism of which is addressed in one of David Harvey's (1979) finest essays. Gustave Courbet had legendarily declared that this phallic erection of Vendôme topped off by a statue of Napoleon was "devoid of all artistic value", seven months before the Commune decreed that it be demolished. On the other side of the barricades, anti-Communard poet Catulle Mendès penned the anguish of his class at the impending fate of this emblem of Napoleonic imperialism:

> Don't think that demolishing the Vendôme Column is just toppling over a bronze column with an emperor's statue on top; it's unearthing your fathers to slap the flesh-less cheeks of their skeletons and to say to them: You were wrong to be brave, to be proud, to be grand! You were wrong to conquer cities, to win battles. You were wrong to make the world marvel at the vision of a dazzling France. (Quoted in Ross 2008:5)

Within days, as Kristin Ross aptly notes, Communard Louis Barron saw it differently:

> I saw the Vendôme Column fall ... Immediately a huge cloud of dust rose up, while a quantity of tiny fragments rolled and scattered about, white on one side, gray on the other ... This colossal symbol of the Grand Army—how it was fragile, empty, miser-able. It seemed to have been eaten out from the middle by a multitude of rats, like France itself ... and we were surprised not to see any [rats] run out ... The music

played fanfares, some old greybeard declaimed a speech on the vanity of conquests, the villainy of conquerors, and the fraternity of the people, and we danced in a circle around the debris, and then went off, very content with the little party. (Quoted in Ross 2008:5–8)

Struggles over monuments are ultimately struggles over the world we want, mediated by representational problems pertaining to their aesthetic form and content. But as we move from the ideology of national "imagined communities" towards universal liberation with Communards and kindred world spirits, what kind of monument would do justice to an emancipated humanity? The Soviets had a bit more time than the Communards to think about it, as we see in Vladimir Tatlin's other-worldly Monument to the Third International composed in the revolutionary style of Russian Constructivism: a 400 m tall structure of steel and glass straddling the river Neva flowing through Petrograd, housing various offices of the Comintern in geometric volumes revolving at different speeds, all pointing in a dynamic assemblage of spirals, cylinders and cubicles to one side—the future (see Figure 2). Tatlin's unsurpassed achievement here consists not only in designing a monument to the project of communism, but also in creating a new artistic language for the task—following Marx's insistence in *The 18th Brumaire* that the *socialist* revolution must find its poetry from not the past but the future. Much like the communism of the Communards and Soviets, however, Tatlin's design could not be realised, due to a preponderance of hostile historical-geographical forces.

Figure 2:   Monument to the Third International (Model) by Vladimir Tatlin, 1920 (public domain; available at https://en.wikipedia.org/wiki/Tatlin%27s_Tower)

Yet this is unlikely to be the last such monument intended as not mere artistic reflection but also active ingredient of the desirable future, so long as, as Althusser (1992) once said, "the future lasts a long time".

Endnotes

¹ https://de.wikipedia.org/wiki/Revolutionsdenkmal (last accessed 5 May 2018).
² My translation from the German.

References

Althusser L (1992 [1985]) *L'avenir dure longtemps*. Paris: Stock
Althusser L (2014) *On the Reproduction of Capitalism: Ideology and Ideological State Apparatuses* (trans G M Goshgarian). London: Verso
Anderson B (1991) *Imagined Communities: Reflections on the Origin and Spread of Nationalism* (new edn). London: Verso
Anderson P (2001) On Sebastiano Timpanaro. *London Review of Books* 10 May. https://www.lrb.co.uk/v23/n09/perry-anderson/on-sebastiano-timpanaro (last accessed 5 May 2018)
Harvey D (1979) Monument and myth. *Annals of the Association of American Geographers* 69(3):362–381
Lefebvre H (1991 [1974]) *The Production of Space* (trans D Nicholson-Smith). Oxford: Blackwell
Lefebvre H (2003 [1970]) *The Urban Revolution* (trans R Bonnono). Minneapolis: University of Minnesota Press
Ross K (2008 [1988]) *The Emergence of Social Space: Rimbaud and the Paris Commune*. London: Verso

New Left

David Featherstone

School of Geographical and Earth Sciences, University of Glasgow, Glasgow, UK;
david.featherstone@glasgow.ac.uk

The issues which concern the New Left go almost entirely unmentioned in the geographic journals. (Peet 1969:3)

Richard Peet's contribution to the inaugural issue of *Antipode* both dramatised and contested the systematic exclusion of radical ideas and politics from geographical scholarship. Peet drew attention to a whole set of pivotal concerns such as the struggles against the Vietnam War, the anti-apartheid movement, the Israeli occupation of Palestinian land, "ghetto formation" and the dominance of the military–industrial complex which were not getting sustained attention by geographers. Bearing the force of the tumultuous events of 1968—the year when in Ruth Wilson Gilmore's (2008:24) terms "revolutionaries around the world made as much trouble as possible in as many places as possible"—Peet's essay sought to make these concerns integral to a politicised radical discipline. He argued that the "nascent New Left in geography" could contribute by helping to "design a more equitable society", through struggling for the "achievement of radical change", and organising for effective action to challenge the hold of conservatives within academic geography (Peet 1969:4). Re-reading this bold and imaginative agenda 50 years on invites the question of what kind of relations were shaped between the New Left and radical geographies and what are their legacies and contemporary relevance. This essay seeks to engage with this question and to consider these relations but also to trace the ongoing importance and relevance of New Left political trajectories in the current political conjuncture.

Rethinking the Geographies of the New Left

The New Left has had various histories and geographies, but is particularly associated with two key moments. The first of these was in 1956, when dissident activists left official Communist Parties in huge numbers over the Soviet invasion of Hungary and the revelations of the enormity of the repression under Stalin, and sought to envision a Left politics which rejected both Stalinism and Western-style capitalism. The second moment, the diverse struggles and uprisings which occurred in 1968, was the context for the founding of *Antipode*. These events, as Michael Watts has argued, "were revolutionary not because governments were, or might have been, overthrown but because a defining characteristic of revolution is that it abruptly calls into question existing society and presses into action" (2001:160). They have, however, frequently been told through narratives which

Keywords in Radical Geography: Antipode at 50, First Edition. Edited by the *Antipode* Editorial Collective.
© 2019 The Authors/Antipode Foundation Ltd. Published 2019 by John Wiley & Sons Ltd.

have depoliticised them and positioned them as nothing more than a set of individualised youth protests heralding the birth of a radically individualised society which ultimately paved the way for neoliberal political projects.

In her book *May '68 and Its Afterlives*, Kristin Ross usefully contests such simplistic recuperations of the 1968 movements, arguing that throughout the 1960s in France, "the discourses of anti-capitalism and anti-imperialism were woven together in an intricate mesh" (2002:11). This intervention is important as it suggests the need to think carefully about the geographies of 1968 and the New Left and to offer a nuanced and critical evaluation of their impact. Given that the New Left was amorphous, diverse and contested, tracing its lineages through radical geographical debates in the post-1968 period is not a straightforward task (and is beyond the scope of this short piece). Further, as Alex Vasudevan's (2015:10–11) discussion of the importance of "squatting-based activism" to the West German New Left attests, the influence of these moments was as much on the style and modes of political engagement and practice as about particular intellectual contributions.

The transnational political vision and anti-colonial internationalisms of the New Left impacted on the pages of *Antipode* in various ways, including through emergent discussions between geographers and theorists of underdevelopment. These concerns were developed in two issues published in 1977 on "Socio-Economic Formation" and "Geography and Underdevelopment" respectively, which sought to engage with "problems of the Third World" and challenged the hegemony of western "development" experts by including authors from the global South. Edited by Milton Santos, David Slater and Richard Peet, they were also shaped by the new spaces of engagement with Marxism which challenged the reductive impact of "Stalinism, the centralisation of the communist parties, and the Cold War" on exploring Marxist scholarship and categories (Santos and Peet 1977:1).

While Santos was Brazilian and Slater was from the UK, they had begun to work together while they were both based at the University of Dar es Salaam. This encounter situates *Antipode* as part of an important set of intellectual circulations pivoting around sites in newly independent countries such as Tanzania with its progressive leader Julius Nyerere (Ferretti and Pedrosa 2018:9; see also Sharp forthcoming). These sites brought different critical intellectual trajectories together, though Santos's "original project of giving voice to intellectuals from the South" proved fraught and contested (Ferretti and Pedrosa 2018:9). Slater's two-part intervention on the "Geography and underdevelopment" (1973, 1977) was also strongly influenced by a number of key intellectuals associated with theories of underdevelopment, many of whom were also key political activists. Thus his first essay draws on the work of the Guyanese political theorist, pan-Africanist and radical Walter Rodney, who was part of an "independent Caribbean New Left" and was also based at the University of Dar es Salaam in the early 1970s (Austin 2013:16). Slater drew on his work to challenge so-called "balanced" assessments of the role of European influence in Africa. He mobilised Rodney's work to analyse the "immense disparities" between exports and investment in "terms of the construction of roads and railways, the provision of schools and the creation of social services" (Slater 1973:25).

The milieu of New Left political movements, as Slater's experience suggests, were formative for many radical geographers, albeit in different and contested ways. Thus Doreen Massey (2013:254) recalls the impact of the unequal gendered politics of these movements, arguing that she was initially put off Marxism by dominant readings of Marx which reproduced "very essentialist" ideas about the "sexual division of labour". Massey's reflections suggest some of the uneven intellectual and political encounter(s) between radical geography and the New Left. Indeed dominant articulations of Marxist geography were often quite resistant to the commitment to foregrounding forms of labour and political agency which were integral to the work of key New Left intellectuals such as E.P. Thompson. Such concerns were largely eschewed by key figures in radical geography, notably David Harvey and Neil Smith, whose work arguably proved more adept at tracing capital's geographies than engaging with opposition to it.

In this regard it was not until the emergence of labour geography in the 1990s and early 2000s that such intellectual concerns became more explicitly central to Left geographical scholarship (Herod 2001). This has been part of a more sustained engagement with the diverse spatial dynamics and forms of political engagement and agency. Thus labour geographers such as Andrew Cumbers and colleagues drew on the tradition of writing about working class agency associated with Thompson and Raymond Williams, in conversation with Autonomous Marxist authors, to challenge the ways in which dominant Marxist-inflected accounts of economic restructuring left "little sense of agency for individuals and communities" treating them as "passive victims of deeper underlying processes" (Cumbers et al. 2010:46). As the next section suggests, engaging the different ways in which diverse forms of agency become articulated as part of Left political projects is of renewed political relevance in the current conjuncture.

New Left Political Trajectories and the Current Conjuncture

Shortly after Jeremy Corbyn's election as leader of the UK Labour Party, Doreen Massey noted that his support drew "together many flows", "young and old, long histories and new initiatives", and encompassed "elements of both the labour movement and of new social movements" (2015:9). Massey's comments signal how what might be termed old "New Left" ideas and activists continue to be influential, not least through interventions like the Kilburn Manifesto she co-authored with Stuart Hall and Michael Rustin who were veterans of New Left interventions in the late 1950s and early 1960s (see Hall et al. 2013), but also to articulate with emergent Left formations. Such flows and trajectories are not only a feature of the UK Left. Thus Fabien Escalona (2017) argued that "though their theoretical orientations and the strategic challenges they face", "'new' radical Left parties such as Podemos in Spain and Syriza in Greece" are positioned in relation to earlier Left traditions such as Eurocommunism, which are now largely forgotten.

Such longer New Left trajectories have been influential in relation to attempts to think and practice Left politics in more "intersectional" ways. Thus Laura Pulido

(2006:93) noted that New Left organisations like Students for a Democratic Society in the US were strongly shaped by anti-racism, influenced by Black Power, and were part of challenges to forms of Left politics which ignored or silenced differences other than class. These movements both constructed politicised understandings of difference and reconfigured understandings of class in ways which prefigured what might now be referred to as an "intersectional Left" (Seitz 2018:3). In similar terms, opposition forged during the emergence of neoliberal politics was shaped by diverse solidarities such as the diverse support groups for the 1984–1985 miners' strike in Britain, which were in part influenced by activists with New Left trajectories (Kelliher 2017). Doreen Massey and Hilary Wainwright, who were both active in the support group movement during the strike, argued that such relations and differences were generative and had the potential to reconfigure existing understandings of class politics, a position indelibly connected with their own political engagement with the strike (Massey and Wainwright 1985:168).

Such questions of class and difference, however, continue to be struggled over and contested on the Left. Thus David Roediger, a veteran of Students for a Democratic Society, has intervened in recent debates around race and class in US politics and has challenged "the view of David Harvey and many others that race sits outside the logic of capital" (Roediger 2017:19). He has observed the unhelpful effects of the continued mobilisation of an "iron distinction between antiracist and anti-capitalist" (Roediger 2017:1). In this regard, uneven ways of theorising the articulations between class, race and gender, however, continue to have consequences on the forms of analysis and political engagement shaped by both radical geographers and broader Left(s).

Roediger's engagement with Bernie Sanders' presidential campaign emphasises that these are in no way issues with simply theoretical implications and speaks to the importance of constructing diverse Left constituencies and solidarities. Thus David Seitz has argued that some of the discourses used in the 2011 Wisconsin Uprising against Governor Scott Walker's neoliberal and anti-union reforms were structured by a "racialised heternormativity" which positioned "certain subjects understand themselves to be entitled to an exemption from the worst of neoliberalism's onslaught" (Seitz 2018:11). The events in Wisconsin and Sanders' presidential campaign, however, signal ways in which Left possibilities have been developed in post-crisis conjuncture. One key commonality across different contexts in this regard here is a revitalised engagement from the Left with political parties such as the Democrats in the US and Labour in the UK. These interventions relate to longer histories of engagement such as the New Politics campaign in the early 1970s which was linked to the American New Left and sought to shape "mini-conventions" which "would act as 'a transmission belt between movement politics and party politics'" (Hilton 2017:112).

As this suggests, one of the exciting and potentially transformative aspects of these developments is a rekindling of ideas of Left political parties with roots in and links to social and political movements, which have long been important in different contexts, for example, the relations between social movements and the MAS in Bolivia (see Laing 2012). These linkages are rarely straightforward or uncontested, but can be part of shifting the political terrain to the Left. Thus in conversation with Chantal Mouffe, Íñigo Errejón of Podemos has resisted characterisations of Podemos as a straightforward political voice of the 15-M movement. He argues, however, that the 15-M movement "helped to articulate a collective demand for a broadening of rights and an expansion of democracy— for more universality of rights and more democracy, instead of more restrictions and less democracy" (Errejón and Mouffe 2016:70–71). Further, as the veteran political activist and intellectual Hilary Wainwright has argued in the context of the election of the radical housing activist Ada Colau as mayor in Barcelona, "social movements organising and exercising leverage" can create "a social force that can be coupled with the particular institutional power that can be deployed by occupying positions in a municipal council" (Wainwright 2018:124).

New Left trajectories then continue to have an impact on the ways in which the Left envisions alternatives and modes of organising. As Wainwright argues, the "negative lessons of Greece warn harshly against any separation from the radical social movements" which supported parties like Syriza and "on whose transformative power [figures like] Corbyn … depend to achieve the changes they have promised and for which they were elected" (Wainwright 2018:128; see also Hadjimichalis 2018). Given the emergent antagonisms around the post-crisis conjuncture and the rise of racialised right-wing populisms, finding ways of developing such linkages and strategies is an urgent task that radical geographers have much to contribute to.

References

Austin D (2013) *Fear of a Black Nation: Race, Sex, and Security in Sixties Montreal*. Toronto: Between the Lines
Cumbers A, Helms G and Swanson K (2010) Class, agency, and resistance in the old industrial city. *Antipode* 42(1):46–73
Errejón I and Mouffe C (2016) *Podemos: In the Name of the People*. London: Lawrence and Wishart
Escalona F (2017) The heritage of Eurocommunism in the contemporary radical Left. In L Panitch and G Albo (eds) *Socialist Register 2017: Rethinking Revolution* (pp 102–119). London: Merlin
Ferretti F and Pedrosa B V (2018) Inventing critical development: A Brazilian geographer and his Northern networks. *Transactions of the Institute of British Geographers* https://doi. org/10.1111/tran.12241
Gilmore R W (2008) *Golden Gulag: Prisons, Surplus, Crisis, and Opposition in Globalizing California*. Berkeley: University of California Press
Hadjimichalis C (2018) *Crisis Spaces: Structures, Struggles, and Solidarity in Southern Europe*. London: Routledge
Hall S, Massey D and Rustin M (eds) (2013) *After Neoliberalism? The Kilburn Manifesto*. London: Soundings
Herod A (2001) *Labor Geographies*. London: Guilford
Hilton A (2017) Organised for democracy? Left challenges inside the Democratic Party. In G Albo and L Panitch (eds) *Socialist Register 2018: Rethinking Democracy* (pp 99–129). London: Merlin
Kelliher D (2017) Constructing a culture of solidarity: London and the British coalfield in the long 1979s. *Antipode* 49(1):106–124

Laing A (2012) Beyond the zeitgeist of "post-neoliberal" theory in Latin America: The politics of anti-colonial struggles in Bolivia. *Antipode* 44(4):1051–1054

Massey D (2013) "Stories so far": A conversation with Doreen Massey. In D Featherstone and J Painter (eds) *Spatial Politics: Essays for Doreen Massey* (pp 253–266). Oxford: Wiley-Blackwell

Massey D (2015) Exhilarating times. *Soundings: A Journal of Politics and Culture* 61:4–13

Massey D and Wainwright H (1985) Beyond the coalfields: The work of the miners' support groups. In H Beynon (ed) *Digging Deeper: Issues in the Miners' Strike* (pp 149–168). London: Verso

Peet R (1969) A New Left geography. *Antipode* 1(1):3–5

Pulido L (2006) *Black, Brown, Yellow, and Left: Radical Activism in Los Angeles*. Berkeley: University of California Press

Roediger D (2017) *Class, Race, and Marxism*. London: Verso

Ross K (2002) *May '68 and Its Afterlives*. Chicago: Chicago University Press

Santos M and Peet R (1977) Introduction. *Antipode* 9(1):1–3

Seitz D K (2018) "Protect Wisconsin Families"? Rethinking Left family values in the 2011 Wisconsin Uprising. *Antipode* https://doi.org/10.1111/anti.12409

Sharp J (forthcoming) Practicing subalternity? Nyerere's Tanzania, the Dar School, and postcolonial geopolitical imaginations. In T Jazeel and S Legg (eds) *Subaltern Geographies*. Athens: University of Georgia Press

Slater D (1973) Geography and underdevelopment–1. *Antipode* 5(3):21–32

Slater D (1977) Geography and underdevelopment–Part II. *Antipode* 9(3):1–31

Vasudevan A (2015) *Metropolitan Pre-Occupations: The Spatial Politics of Squatting in Berlin*. Oxford: Wiley Blackwell

Wainwright H (2018) *A New Politics From the Left*. Cambridge: Polity

Watts M (2001) 1968 and all that... *Progress in Human Geography* 25(2):157–188

Offshore

Shaina Potts

Department of Geography, University of California, Los Angeles, Los Angeles, CA, USA;
spotts@geog.ucla.edu

Whether in Hollywood scenes of criminals escaping to the Bahamas with suitcases full of cash, or of genius hackers transferring a corporation's money to the Caymans with the tap of a button, the "offshore" has long been associated with representations of remote islands at the edges of the respectable world. Inside and outside the academy, the offshore has been seen as a space "beyond"—beyond regulation, beyond taxation, beyond national monetary space and policing. It has also often been understood as a marginal space, both in the sense of peripheral and of minor—a kind of capitalist fringe in which the illegal and the illicit reign. Since the 1990s, however, this understanding has been challenged by geographers and others who have investigated the ways in which the offshore is, in fact, central to contemporary capitalism.

In this entry, I define offshore broadly to refer not only to offshore finance, but to the logic through which capitalism carves out and exploits legally distinct and supposedly exceptional spaces. Offshore capitalist spaces (along with offshore leisure and military spaces) have mushroomed in importance since the mid-20th century (Urry 2014). The estimated number of offshore financial centres and tax haven jurisdictions (referred to jointly here as OFCs) numbers between two dozen and 91, depending on the criteria used. Special economic zones (SEZs), bounded subnational territories characterised by low taxes and lax regulations, rose from a handful in the 1950s to over 4,000 today, ranging from garment districts in Bangladesh to China's high-tech Shenzhen. SEZs are fundamental to outsourcing, in which transnational corporations relocate production "offshore", usually from high- to low-wage countries. More specialised offshore practices also abound. Flags of convenience (FOCs), which enable pockets of national sovereignty to travel beyond national borders, have boomed. Ogle (2017) refers to this assemblage of OFCs, SEZs and FOCs as "archipelago capitalism", a term we might also extend to include offshore oil rigs (Appel 2012), and even ongoing efforts to extend mining to "off-world" asteroids (Klinger 2018). All these spaces rely on boundaries that separate and define. They depend on being seen as distinct from and in opposition to the "onshore" spaces where "normal" business occurs.

OFCs in particular are frequently presented as exotic, with images of tropical beaches and palm trees dotting business brochures and how-to guides. At the same time, they are associated with laundering drug money and white-collar fraud. Both framings serve broader logics of disembedding by which the offshore is positioned as physically, economically, socially and legally separate from the onshore world (Appel 2012). In reality, the vast majority of capital in OFCs is there legally, and, as the Panama Papers and Paradise Papers leaks showed, many

Keywords in Radical Geography: Antipode at 50, First Edition. Edited by the *Antipode* Editorial Collective.
© 2019 The Authors/Antipode Foundation Ltd. Published 2019 by John Wiley & Sons Ltd.

well known politicians, businessmen and celebrities routinely make use of these spaces. The offshore is not beyond at all, though it functions precisely by being cast as "elsewhere"—an image bolstered by the secrecy and concealment that accompanies most offshore activity. Only by thinking from the margins, do we begin to see how the offshore is in fact constitutive of global capitalism. The tax breaks and regulatory perks offered in offshore spaces are usually only available to those not *from there.* These spaces function through the reification of dichotomies —onshore/offshore, here/elsewhere, citizen/foreigner. The task, as feminist geographers have taught us, is not to debate where the onshore and offshore truly end, but to move beyond these distinctions altogether, even as we investigate how they function in the world around us.

The offshore is a relational space *par excellence.* It exists only insofar as it both provides a barrier between and mediates connection to the onshore. The primary function of offshore space of all kinds is to provide opportunities for *arbitrage*: the exploitation of difference for profit. Peck (2017), for instance, shows how even as the "outsourcing complex" takes on more varied and more complicated forms, the drive for cost-cutting and cheap labour remains fundamental. Offshore spaces in general are defined by differences that can be exploited, whether in wages, taxes, fees or regulations. What is more, they offer distinct differences, competing as niche spaces in a world where niches are increasingly common. Mostly, they provide *unequal* differences, allowing transnational corporations and investors to use small or impoverished economies. But the efficacy even of powerful offshore spaces like the City of London rests on the production of a distinct regulatory space. The offshore world has its own uneven geographies.

Most offshore geographies are imperial geographies, in history and function. SEZs, along with outsourcing in general, replaced more blatant colonial labour exploitation with a liberal market-based version for the post-colonial era. De-colonisation, ironically, fuelled the rise of OFCs too, as newly sovereign states sought ways to survive harsh economic conditions, just as financial deregulation and technological developments made tax haven operations more accessible (Palan et al. 2010). Today, around half the world's OFCs form a "layered hub-and-spoke array of tax havens, centered on the City of London, which mostly emerged from the ashes of the British empire" (Shaxson 2012:17). Many Caribbean tax havens advertise their British heritage and legal genealogy (Maurer 1997). Other OFCs are in the United States and its current or former dependencies.

Offshore spaces also remain imperialist in their extractive function: channelling wealth from less to more powerful people and places. SEZs are sites for intensified extraction of labour power and cheap materials, while OFCs facilitate the extraction of capital from everywhere "else" through tax avoidance, capital flight and more. The offshore can also be understood as extracting sovereignty itself. The offshore is never beyond or "between" jurisdictions—instead, it is always "another jurisdiction", or a jurisdictional "elsewhere". That is, offshore spaces are state spaces, and capitalists want them precisely in order to "use their sovereignty" (Palan et al. 2010:3). The function of this less than sovereign sovereignty is to produce useful borders. These are primarily legal borders, though they may be reinforced by isolated coastlines or barbed-wire fences. These offshore spaces stand in counterpoint to capital's homogenising tendencies. While the concept of time–space compression emphasises the removal of barriers to movements of goods, people, information and capital, offshore spaces erect and exploit barriers to the movement of (certain) regulations and information and prevent capital from being found or moved before its owners want it to be. The creation of "worldwide" capitalist space is simultaneously about homogenisation and fragmentation, differentiation and hierarchisation (Lefebvre 2009). Whether defining OFCs or SEZs, flags of convenience or offshore oil rigs, offshore borders are central to the operations of major banks and transnational corporations. They shape supply chains, accounting practices and capital flight, and they allow powerful institutions to escape regulations and taxation. "Globalisation" as a force that supposedly destroys borders is dependent precisely on the erection of thousands of new ones.

The offshore is not only defined in relation to the onshore, it also always *is* onshore. It is onshore, first, for the people who live there, whether those benefiting or those enrolled as cheap labour. Identifying centre and margin always depends on the perspective of the viewer. It is onshore, second, in that the most powerful OFCs are located "onshore" in places like London, Delaware and Switzerland. Most importantly, it is onshore in the sense that all offshore spaces exist only because they operate as part of "onshore" strategies. This is fuelled by, but goes beyond, simple price arbitrage. Peck demonstrates that outsourcing does far more than relocate jobs and factories—it reorganises and transforms the social, managerial and technical relations of production, (temporarily) providing an "organisational and a spatial fix" for transnational capital. In the process, it reconfigures geographies of capitalism, while always setting the conditions in which the search for the "next shore" plays out (Peck 2017:210). This analysis of the dynamic socio-spatial processes driving offshoring can be extended to offshore logics more broadly.

Powerful states, too, have long shaped the offshore. The London-based Eurodollar market, the world's largest single offshore financial space, only grew from the 1960s on with support from the United States and Britain. The World Bank has directly supported new SEZs in the global South. Moreover, the lines between onshore and offshore are themselves often blurred. Jurisdictions are not, in practice, exclusive, or always coextensive with political boundaries. In the British Virgin Islands, both Britain and the United States have formal control over some of the country's financial activities and criminal prosecution (Maurer 1995). Haberly and Wójcik (2017:238) argue that offshore activity contributing to the 2008 financial crisis was not outside US regulatory space at all, but rather within "the purview of onshore regulators". In short, the most powerful countries are not incapable of regulating the offshore. Rather they benefit from asserting a sharply delimited legal space in some cases, while extending that space in others.

By the same token, recent efforts to rein in tax havens are not simply signs of beleaguered nation-states struggling to regain control in the face of free-footed capital. At worst, these efforts may be no more than public-relations campaigns aimed at cleaning up criminal tax evasion, while leaving the vast majority of offshore financial activity unchallenged. At best, they expose the internal contradictions of the state as an ensemble of often conflicting agents and interests. Either way, insofar as the offshore allows transnational corporations and elites to escape

taxation, they further shrink the revenues available to nation-states already strain-
ing under neoliberal austerity. The Tax Justice Network estimates that over US
$600 billion in potential taxes is lost globally per year, largely on account of OFCs
(Cobham and Jansky 2015). SEZs, too, attract capital by offering lower taxes and
wages, usually in states already dealing with restricted budgets and elevated poverty levels. By temporarily allowing capital to wrest higher profits, while doing
nothing to resolve the underlying crises of contemporary capitalism, offshore
spaces only contribute to the scale of crises to come.

Historians, political scientists and journalists of the offshore tend to agree, first,
that the offshore is central, not marginal, to globalisation, and second, that it is
closely linked to neoliberal capitalism. Many call for stronger regulations and more
transparency, especially for recovering taxes lost to OFCs—all worthwhile goals.
What radical geographers and fellow travellers in other disciplines have been
especially well positioned to see, however, is how offshore spaces are constitutive
of the always uneven and contradictory development of capitalism. The offshore
is a fundamental manifestation of capital's general reliance on the exploitation of
difference and uneven development, and of the globalising division of labour.
The proliferation of offshore spaces demonstrates that increasing spatial fragmentation is at the heart of the production of worldwide capitalism, and that globalisation itself is built on difference and division.

References

Appel H (2012) Offshore work: Oil, modularity, and the how of capitalism in Equatorial
    Guinea. *American Ethnologist* 39(4):692–709
Cobham A and Jansky P (2015) "Measuring Misalignment: The Location of US Multination-
    als' Economic Activity Versus the Location of their Profits." Working Paper No. 42,
    International Center for Tax and Development. http://www.ictd.ac/publication/measur
    ing-misalignment-the-location-of-us-multinationals-economic-activity-versus-the-location-
    of-their-profits/ (last accessed 18 July 2018)
Haberly D and Wójcik D (2017) Culprits or bystanders? Offshore jurisdictions and the glo-
    bal financial crisis. *Journal of Financial Regulation* 3(2):233–261
Klinger J M (2018) *Rare Earth Frontiers: From Terrestrial Subsoils to Lunar Landscapes.* Ithaca:
    Cornell University Press
Lefebvre H (2009) *State, Space, World: Selected Essays* (eds N Brenner and S Elden). Min-
    neapolis: University of Minnesota Press
Maurer B (1995) Writing law, making a "nation": History, modernity, and paradoxes of
    self-rule in the British Virgin Islands. *Law and Society Review* 29(2):255–286
Maurer B (1997) Creolization redux: The plural society thesis and offshore financial services
    in the British Caribbean. *New West Indian Guide/Nieuwe West-Indische Gids* 71(3):249–
    264
Ogle V (2017) Archipelago capitalism: Tax havens, offshore money, and the state, 1950s–
    1970s. *American Historical Review* 122(5):1431–1458
Palan R, Murphy R and Chavagneux C (2010) *Tax Havens: How Globalization Really Works.*
    Ithaca: Cornell University Press
Peck J (2017) *Offshore: Exploring the Worlds of Global Outsourcing.* Oxford: Oxford University
    Press
Shaxson N (2012) *Treasure Islands: Uncovering the Damage of Offshore Banking and Tax
    Havens.* New York: Palgrave Macmillan
Urry J (2014) *Offshoring.* Cambridge: Polity

Organising

Jane Wills

Centre for Geography, Environment and Society, University of Exeter, Penryn, Cornwall, UK;
J.Wills2@exeter.ac.uk

There is no aspect of life that is untouched by the processes and practices of organising. It underpins the development of human cultures and the associated institutions through which ever greater numbers of human beings have found ways of living together. Organising the means to get something done is vital to securing a successful economy, supporting an effective political regime and nurturing a meaningful cultural life. If our leading businesses, political parties or community organisations fail to organise themselves successfully, keeping up with changes in the wider political-economy and culture, they will wither and die.

Yet while organising underpins the established order of things, it is even more essential if we are to challenge the way things are done. Mounting a campaign that has sufficient strength to take on established ideas, vested interests and socio-cultural inertia, requires great strategic and sustained effort. It is also made doubly difficult because the established order is already so embedded or "normalised" in our social organisations and ways of thinking. Any sort of radical challenge to established ways of thinking and doing things requires some sort of organising activity to build momentum for an alternative. Such organising activity might begin modestly by publishing or promoting some new ideas and then finding kindred spirits with a similar view. It may involve a lot of time talking to people, building relationships, raising money and holding events. If successful, it can generate significant changes in individuals and human relationships, stimulating new ways of thinking, feeling and living. It is impossible to think about the profound changes caused by movements such as those to promote dissenting forms of Christianity, women's equality or the abolition of whaling, without respect for the power of organising in making change happen. Organising underpins all radical change, be it from the Left with the plethora of organisations set up to promote socialist, anarchist and communist ideas, or from the Right, as evident in the organised horrors of fascism. Moreover, although organising activity clearly underpins powerful movements that are driven by ideology and that can have very wide geographical and historical reach, it is also important in explaining change on a much smaller scale. It is evident in the way that some communities have a richer density of organisations, social relationships and collective identity than others. Some places are more organised than others, and this matters to local experience but is also important in relation to subsequent change.

Given their focus on changing the world, radical geographers have paid particular attention to the processes and practices of organising. This work has focused on social movements, taking a geographical perspective to consider the

Keywords in Radical Geography: Antipode at 50, First Edition. Edited by the *Antipode* Editorial Collective.
© 2019 The Authors/Antipode Foundation Ltd. Published 2019 by John Wiley & Sons Ltd.

inter-relationships between place, space and organising in a variety of campaigns (Miller 2000; Nicholls 2007, 2009). There is also a smaller body of geographical scholarship looking at other forms of organising. This includes a strand of research focused on the organising activity associated with political parties, considering the geography of efforts to secure electoral success, uniting people around their shared ambitions for political representation (Scott and Wills 2017). There is also an interest in community organising, and the extent to which place provides a platform for institutionally oriented organising around the common good (DeFilippis et al. 2010; Harney et al. 2016; Wills 2012, 2016). This form of organising uses local institutions, and the relationships between them, as the foundation from which to identify shared interests, mobilise people and foster solidarity for political change. Such work has to focus on the particularities of place but it is often reflective of wider concerns and interests while also being connected to trans-local networks of organising alliances. In the United States there are a number of different networks and models of community organising and these have sparked similar developments in other parts of the world (Orr 2007; Walls 2014).

In summary, scholars have developed a two-pronged approach to understanding the intersection of geography and organising, looking at the importance of: (1) geographical inheritance; and (2) geographical strategy. Both reflect the extent to which geography plays a foundational role in the formulation and prosecution of organising activity; geography shapes what is possible and it also plays a key role in organising success.

Understanding the role of geographical inheritance involves paying attention to the ways that existing social organisation and associated culture shape the development and prosecution of new ideas. In his book *Political Process and the Development of Black Insurgency*, Doug MacAdam (1999) explored the civil rights movement in the American south, highlighting three aspects of this "geo-inheritance" which will shape the way that any movement can grow. The first concerns what he called "readiness" or the willingness and ability of the population to act when opportunities allow. Without existing relationships and networks, it is very difficult to transmit a message, to sustain organising and promote mobilisation. Thus, Oberschall (1973) argues that the degree of existing organisation in any community is key to movement development, sustained activity and success. In his research, MacAdam highlighted the particular role of the churches in the American south in acting as anchors for the civil rights movement. The churches were already organised, they had respected and talented local leaders and their own social networks through which to spread the word and facilitate mobilisation. Successful action then further reinforced existing social relationships and mutual support. As Oberschall puts it, "mobilization does not occur through recruitment of large numbers of isolated and solitary individuals. It occurs as a result of recruiting blocs of people who are already highly organized and participants" (1973:125). The established social infrastructure provided by the church helped to support the organising campaigns, building on existing social relations.

In a similar vein, Walter Nicholls (2008) has highlighted the ways in which some urban communities have become "incubators" for social movement ideas and talent; as the number of local organisations grows, relationships develop, and experience shapes expectations, there is cross-fertilisation and it becomes easier to generate more organising activity. There is something of a virtuous circle as more organising begets more organising.

The existing social infrastructure also plays a role in the development of an "insurgent consciousness" that is also key to organising success. If they are to turn out to support a campaign, people need to feel it is worth their while, and they have some chance of success (Piven and Cloward 1979). This again is much more likely in instances when people already have strong personal and institutional relationships. As Doug MacAdam suggests, "in the absence of strong interpersonal links to others, people are likely to feel powerless to change conditions even if they perceive present conditions as favourable to such efforts" (1999:50). As the wider society changes, and particular events help to shift the opportunity structure in favour of political change, this can ignite a passion for organising that is more likely to find nourishment in some places than others. People are unlikely to turn out to support organising efforts if they feel there is little chance of success, that there is no audience or recep-tivity for their cause, and they have weak incentives to reinforce their engage-ment. As such, it is clear that the existing institutions and repositories of social relationships, shared experiences and culture, and a feeling for shifting political opportunities underpin organising activity. This geo-inheritance varies greatly across space and between different places. If some places have fragile forms of social capital, and/or a weaker "fit" with a particular campaign, they will prove less able to engage. As an example, spatial differentiation proved important in the 1984/85 miners' strike in Britain, and a number of geographers sought to understand why some areas did or didn't support the strike (Rees 1985, 1986; Sunley 1986, 1990).

In addition to the question of inheritance, geography can play an important part in relation to more strategic questions in organising campaigns. Geographical strategy relates to the geographical depth and/or reach of organising campaigns and their use of particular geographical tools such as community building or fos-tering networks, and the extent to which this increases the likelihood of winning campaigns (Leitner et al. 2008). Such strategy will be at least partly shaped by the origin and goals of any campaign. Labour organising necessarily has to start from the particular workplaces where workers have sought to win better terms and conditions of work, so too, the early women's movement organised in peo-ple's homes, fostering solidarity that then underpinned a broader campaign. We have also seen how the civil rights movement used religious spaces for similar purposes. However, in order to secure significant change, it has often proved nec-essary to "upscale" the reach of a campaign to widen networks of solidarity and to "target" the people who are able to act to resolve a demand. This can involve making topological connections across space, linking the experiences of people in the provinces to those in the capital, or those in a peripheral branch plant to the corporate headquarters. It can also involve connecting different worlds within any space. The rise of the living wage movement has provided a powerful demonstra-tion of both aspects of geographical strategy. Organisers have connected the experiences and demands of workers in manufacturing supply chains in the global

South to consumers in the global North, highlighting the issue of labour stan-dards and the need for ethical production and consumption (Hale and Wills 2005). By educating and mobilising consumers, organisers are putting political pressure on the companies at the "top" of supply chains to take responsibility for those at the "bottom" (Merk 2009). The same approach has also been used to connect managers and in-house workers with those employed on-site by sub-contractors in jobs like cleaning, catering and security, to ensure the payment of a living wage (Wills and Linneker 2014).

Geographical strategy will be shaped by the particular dynamics of any cam-paign but it is often necessary (and perhaps increasingly necessary) to "move" the demands of one group from one part of the world, to ensure they are heard by those with the power to act, who often sit elsewhere (Herod and Wright 2008). This has been powerfully demonstrated by labour organising campaigns, past and present, and the way they have imaginatively developed solidarity across different scales, including the workplace, city, region, nation and international scales (Herod 1998, 2010; Waterman and Wills 2001). Such organising is also mediated by the political opportunities that are provided by political institutions that also have their own geography such as local, city and national government, as well as international bodies like the European Commission and Parliament, and the United Nations (Wills 2018). There are related opportunities provided via the social clauses attached to free trade agreements and competition rules established by bodies like the World Trade Organisation. As such, the uneven geography of political institutions creates a patchwork of political opportunities that will shape the geo-strategic decisions made in organising campaigns. There is no "best" scale or geo-strategy that can guarantee success in a campaign in advance, and decisions necessarily evolve during the lifetime of any campaign; however, it is clear that geography plays a critical role in underpinning organising activity and outcomes (Leitner et al. 2008). The twin issues of geographical inheritance and geographical strategy will always be important in understanding and prosecuting organising campaigns for radical change.

References

DeFilippis J, Fisher R and Shragge E (2010) *Contesting Community: The Limits and Potential of Local Organizing*. New Brunswick: Rutgers University Press

Hale A and Wills J (eds) (2005) *Threads of Labour: Garment Industry Supply Chains from the Workers' Perspective*. Oxford: Blackwell

Harney L, McCurry J, Scott J and Wills J (2016) Developing "process pragmatism" to under-pin engaged research in human geography. *Progress in Human Geography* 40(3):316–333

Herod A (ed) (1998) *Organizing the Landscape: Geographical Perspectives on Labor Unionism*. Minneapolis: University of Minnesota Press

Herod A (2010) Implications of just-in-time production for union strategy: Lessons from the 1998 General Motors-United Auto Workers dispute. *Annals of the Association of American Geographers* 90(3):521–547

Herod A and Wright M W (eds) (2008) *Geographies of Power: Placing Scale*. Hoboken: John Wiley and Sons

Leitner H, Sheppard E and Sziarto K (2008) The spatialities of contentious politics. *Transactions of the Institute of British Geographers* 33(2):157–172

MacAdam D (1999) *Political Process and the Development of Black Insurgency, 1930–1970* (2nd edn). Chicago: University of Chicago Press

Merk J (2009) Jumping scale and bridging space in the era of corporate social responsibility: Cross-border labour struggles in the global garment industry. *Third World Quarterly* 30(3):599–615

Miller B (2000) *Geography and Social Movements: Comparing Anti-Nuclear Activism in the Boston Area.* Minneapolis: University of Minnesota Press

Nicholls W J (2007) The geographies of social movements. *Geography Compass* 1(3):607–622

Nicholls W J (2008) The urban question revisited: The importance of cities for social movements. *International Journal of Urban and Regional Research* 32(4):841–859

Nicholls W J (2009) Place, networks, space: Theorising the geographies of social movements. *Transactions of the Institute of British Geographers* 34(1):78–93

Oberschall A (1973) *Social Conflict and Social Movements.* London: Prentice Hall

Orr M (ed) (2007) *Transforming the City: Community Organizing and the Challenge of Political Change.* Kansas City: University Press of Kansas

Piven F F and Cloward R A (1979) *Poor People's Movements: Why They Succeed, How They Fail.* New York: Vintage

Rees G (1985) Regional restructuring, class change, and political action: Preliminary comments on the 1984–1985 miners' strike in South Wales. *Environment and Planning D: Society and Space* 3(4):389–406

Rees G (1986) "Coalfield culture" and the 1984–1985 miners' strike: A reply to Sunley. *Environment and Planning D: Society and Space* 4(4):469–476

Scott J and Wills J (2017) The geography of the political party: Lessons from the British Labour Party's experiment with community organising, 2010 to 2015. *Political Geography* 60:121–131

Sunley P (1986) Regional restructuring, class change, and political action: A comment. *Environment and Planning D: Society and Space* 4(4):465–468

Sunley P (1990) Striking parallels: A comparison of the geographies of the 1926 and 1984/5 miners' strikes. *Environment and Planning D: Society and Space* 8(1):35–52

Walls D (2014) *Community Organizing.* Cambridge: Polity

Waterman P and Wills J (eds) (2001) *Place, Space, and the New Labour Internationalisms.* Oxford: Blackwell

Wills J (2012) The geography of community and political organisation in London. *Political Geography* 31:114–126

Wills J (2016) *Locating Localism: Statecraft, Citizenship, and Democracy.* Bristol: Policy Press

Wills J (2018) The geo-constitution: Understanding the intersection of geography and political institutions. *Progress in Human Geography* https://doi.org/10.1177/0309132518768406

Wills J and Linneker B (2014) In-work poverty and the living wage in the United Kingdom: A geographical perspective. *Transactions of the Institute of British Geographers* 39(2):182–194

Peace

Sara Koopman

School of Peace and Conflict Studies, Kent State University,Kent, OH, USA;
skoopman@kent.edu

"No justice, No peace!" I have chanted at many marches, as we interrupted the "peace" to work against an injustice. Radical geography disturbs the peace of mainstream geography and works against injustice, both in the world and in our discipline. As good organisers, we even have a freedom school, our Highlander as it were. The *Antipode* Institute for the Geographies of Justice (of which I am a proud alum) has been training geographers to work for justice since 2007.

Can we work for justice *and* as part of that for a greater, more inclusive, peace for all? Aren't they intertwined? Don't we want to dream big? But we have been more likely to talk about the spatiality of war and conflict than about how peace with justice (or justice with peace) is spatial, and how we build it. This is reflected in *Antipode*, where there are dramatically more articles with the words war or conflict in the title or abstract, than peace, though of course the articles focused on conflict are anti-war.

Ross (2011) argues that this is, in effect, peace scholarship and that these sorts of counts are facile. Perhaps, but there is something here. It is generally more common, in both activism and activist-academic work, to focus on what we are against than what we are for. Gandhi (1948) called the latter constructive (vs. obstructive) non-violence, and emphasised it especially towards the end of his life. Kropotkin's (1902) focus was also on building the world we want, through mutual aid. Indeed he also lamented that there were extensive descriptions of acts of violence, but little chronicling of the everyday ways "the masses" built peace, even while war was being waged.

Antipode was founded in 1969 at the height of anti-war organising in the US as "a reaction against the Vietnam war, racism and pollution" (Wisner quoted in Peake and Sheppard 2014:309). But as Peake and Sheppard (ibid.) go on to put it, in a phrase that is currently cited on the Antipode Foundation's "About" page, "the journal's pages have been 'bound together by a shared no—rejection of the … status quo—and diverse yeses'".[1] We could weave a stronger strand on peace and how we build it into the web of those diverse yeses.

Perhaps some of the reticence to use the word peace in radical geography has been from a sense that the term is too far gone, too co-opted to mean pacification, as Ross (2011) argues. I disagree. Certainly her call to see the violence inside what is presented as peace is important. But let us not therefore give up on the term, which was never fully taken from us. There is a long history of radical peace organising around the world, and indeed by geographers, who since 1891 have been arguing that "peace should not be the peace of death" (Reclus quoted in

Keywords in Radical Geography: Antipode at 50, First Edition. Edited by the *Antipode* Editorial Collective.
© 2019 The Authors/Antipode Foundation Ltd. Published 2019 by John Wiley & Sons Ltd.

Ferretti 2016:572). Feminist pacifist Jane Addams argued in 1899 that peace was not merely the end of war, but that "positive ideals of peace" are built day to day by "the people" even in the midst of war (Carroll and Fink 2007). But perhaps the problem is not that peace is read as pacification, but that its definition has too often been limited to simply not war.

Inwood and Tyner (2011:453), in their call in *ACME* for a pro-peace agenda in geography and particularly our pedagogy, argue that seeing peace as only a negative peace has "limited larger engagements with questions of justice". Theirs is one of repeated calls for geographers to study peace. This is not meant as another manifesto, but rather a discussion of why and how it is a keyword particularly for radical (and critical) geography. Interestingly in the recent debate about what radical geography really is, sparked by Springer's (2012, 2014) arguments that it was too Marxist heavy and should draw more on anarchism, none of the six responses addressed the issue of peace, despite Springer recognising in his opening piece that anarchism is stereotyped as violent but arguing that it has a tradition of non-violence and works for a greater peace than the supposedly peaceable order of the state.

The various manifestos for more of a focus on peace in geography are mixed in terms of their reference to its importance for radical and critical geography. Pepper and Jenkins (1983) did call on various types of geographers to study peace, including radical geographers. Wisner's (1986) call was in *Antipode* and did not specify that it was key for radical geographers in particular, though it was implicit. He writes about how physicists and anthropologists have spoken out strongly against the misuse of their work by the military, but geographers have not. We are only just now taking up his call to start with an inventory of the connections between geography and "organised killing" with the AAG Special Committee on Geography and the Military formed in 2017 in response to a petition by the Network of Concerned Geographers.[2] Indeed, we have a greater responsibility to work on and for peace because our discipline has so long served and been shaped by war and conquest (Driver 2001). But the AAG did do its first ever special issue of the *Annals* focused on peace, and in the introduction Kobayashi (2009:824) highlighted that the contributions were primarily by critical geographers, i.e. "with a commitment to social change based on fundamental rethinking of political economy".

Megoran (2010) argued specifically for critical geopolitics (a subfield) to go beyond "oppositional critiques" and turn to "possibilities for peace".[3] He followed it up with a call for the study of peace across geography that recognised various ways that it was already happening (Megoran 2011). In an intervention in *Antipode* in 2011, Williams and McConnell focused on critical geography specifically, and suggested that more work was being done on peace than was recognised, in part because of the lack of a critical conception of peace as a broad and ongoing process that varies across space and time and includes other "peace-full" concepts such as justice and solidarity. They called for more explicit critical geographies of peace as a way to foster "academic action that is proactive rather than reactive" (Williams and McConnell 2011:930). Megoran and Dalby (2018) together recently argued again for the particular importance of critical geopolitical perspectives on peace in the Anthropocene, and that in theorising peace we can change the ways that we imagine global space in ways that do not just reflect but can help produce peaceful political realities.

These various calls both reflected and created a growing interest in peace in geography. Mamadouh (2005) pointed out that peace was not in key reference texts and barely in textbooks. I pointed this out again in 2011, when peace was still not in any reference texts, nor in the index of most geography textbooks, even critical ones on geopolitics and political geography.

But this is changing. This was reflected in the last edition of *Geopolitics: An Introductory Reader* (Dittmer and Sharp 2014), which has a robust focus on peace in the index and throughout. There has also been greater inclusion in reference texts, including an Oxford University Press bibliography (Koopman 2016), an entry in a new encyclopaedia of geography (Koopman 2017), and indeed, inclusion in these 50 keywords. In 2017 there was an IGU Thematic Conference, "Geographies for Peace", in La Paz, Bolivia—*Geopolitics* published a virtual issue on "Geopolitics and Peace" for it[4]—and the edited collection *Geografías al servicio de los procesos de paz* (Sandoval Montes and Nuñez Villalba 2017) came out of it (see also the most recent edited collection in English; McConnell et al. 2014).

Though perhaps interest in peace in the discipline was long delayed by a limited understanding of it as not war, recent literature strongly agrees that war/peace is a false binary, as Loyd (2012) argues in her review of the field. Even if defined only as a lack of direct political violence between states, peace is never clearly distinct from war. War is inside peace, shaping everyday political life, institutions, and socio-spatial order (see Cowen and Gilbert [2007] for a review of work on this). But armed conflict has also changed and now has even less of a clear beginning or end in time or space, such that Gregory (2011) calls it an "everywhere war". It also blurs the line between civilian and combatant. Kirsch and Flint (2011) outline the dangers of promoting a false dichotomy between war and peace, and argue that they are mutually constitutive, both materially and rhetorically. Feminist geopolitics usefully puts this entanglement in conversation with other binaries it challenges, such as personal/political, global/intimate, and hot/banal (Christian et al. 2016). This is not an argument for conflation, but to see the ways they are produced as separate and the messiness involved. There are real effects to, say, calling Colombia today post-conflict rather than post-accord.

This makes peace a more complicated idea, but also I hope makes it clearer how it is at the root of social and political change, and thus key for radical activists. But what about radical geographers? Recent literature in geographies of peace widely agrees that peace is a precarious and ongoing spatial process that varies across time, place, and scale (Bregazzi and Jackson 2018; McConnell et al. 2014). The term peacebuilding is perhaps redundant. Peace is always being built, never done. Peace is also plural, both in and across place, and radical geographers are well positioned to build more inclusive peaces by putting them into non-innocent conversations with other struggles (Koopman 2011). Positive peaces can include justice, solidarity, care, wellbeing, dignity—a great many of the yeses that radical geography cares about. I am not proposing that peace is one term to rule them all, but rather a useful weft in our greater weave to build a better world.

In 2012 Loyd called for "deeper conversations between geographers of peace and the work on violence being done by scholars of racism, colonialism, sexuality, and gender" (2012:485) and particularly for geographers of peace to engage with work on actions against racial state violence, such as that by Gilmore (2006) and Woods (1998). Decolonial struggles are struggles for peace, though they may not use that word. I echo her call, and am particularly interested in linking geographies of and for peace with decolonial geographies, which were given new strength as the theme of the RGS in 2017 (Esson et al. 2017; Naylor et al. 2018; Noxolo 2017; Radcliffe 2017; see also the entry on "Decolonial Geographies" in this collection). Geography has long been, and continues to be, used for war and colonisation. This master's tool used without that awareness can unintentionally reinforce the master's house. Unlike Lorde (2007), I believe that the master's tools can at times dismantle that house, and that as radical geographers we have a responsibility to hold this master's tool carefully and rework it as a tool for peace and freedom.

Endnotes
[1] See https://antipodefoundation.org/about-the-journal-and-foundation/a-radical-journal-of-geography/ (last accessed 3 October 2018).
[2] See https://actionnetwork.org/petitions/network-of-concerned-geographers (last accessed 3 October 2018).
[3] Though notably some early critical geopolitics work by Dalby (1991, 1993) focused on peace movements, but it was less of a focus as the subfield developed.
[4] See http://explore.tandfonline.com/content/pgas/fgeo-virtual-special-issue (last accessed 17 September 2018).

References
Bregazzi H and Jackson M (2018) Agonism, critical political geography, and the new geographies of peace. *Progress in Human Geography* 42(1):72–91
Carroll B A and Fink C F (2007) Introduction. In J Addams *Newer Ideals of Peace* (pp xiii–lxxiii). Champaign: University of Illinois Press
Christian J, Dowler L and Cuomo D (2016) Fear, feminist geopolitics, and the hot and banal. *Political Geography* 54:64–72
Cowen D and Gilbert E (2007) The politics of war, citizenship, territory. In D Cowen and E Gilbert (eds) *War, Citizenship, Territory* (pp 1–30). New York: Routledge
Dalby S (1991) Dealignment discourse: Thinking beyond the blocs. *Current Research in Peace and Violence* 13(3):140–155
Dalby S (1993) The "Kiwi disease": Geopolitical discourse in Aotearoa/New Zealand and the South Pacific. *Political Geography* 12(5):437–456
Dittmer J and Sharp J (eds) (2014) *Geopolitics: An Introductory Reader*. New York: Routledge
Driver F (2001) *Geography Militant: Cultures of Exploration and Empire*. Oxford: Wiley-Blackwell
Esson J, Noxolo P, Baxter R, Daley P and Byron M (2017) The 2017 RGS-IBG chair's theme: Decolonising geographical knowledges, or reproducing coloniality? *Area* 49(3):384–388
Ferretti F (2016) Geographies of peace and the teaching of internationalism: Marie-Thérèse Maurette and Paul Dupuy in the Geneva International School (1924–1948). *Transactions of the Institute of British Geographers* 41(4):570–584
Gandhi M K (1948) *Constructive Programme (Its Meaning and Place)*. Ahmedabad: Navajivan. http://gandhiashramsevagram.org/pdf-books/constructive-programme.pdf (last accessed 3 October 2018)

Gilmore R W (2006) *Golden Gulag: Prisons, Surplus, Crisis, and Opposition in Globalizing California.* Berkeley: University of California Press

Gregory D (2011) The everywhere war. *The Geographical Journal* 177(3):238–250

Inwood J and Tyner J (2011) Geography's pro-peace agenda: An unfinished project. *ACME*
10(3):442–457

Kirsch S and Flint C (2011) Reconstruction and the worlds that war makes. In S Kirsch and
C Flint (eds) *Reconstructing Conflict: Integrating War and Post-War Geographies* (pp 3–28).
Burlington: Ashgate

Kobayashi A (2009) Geographies of peace and armed conflict. *Annals of the Association of
American Geographers* 99(5):819–826

Koopman S (2011) Let's take peace to pieces. *Political Geography* 30(4):193–194

Koopman S (2016) Geographies of peace. *Oxford Bibliographies.* http://www.oxfordbibli
ographies.com (last accessed 3 October 2018)

Koopman S (2017) Peace. In D Richardson, N Castree, M F Goodchild, A Kobayashi, W Liu and
R A Marston (eds) *International Encyclopedia of Geography: People, the Earth, Environment,
and Technology* (https://doi.org/10.1002/9781118786352.wbieg1175). Hoboken: Wiley

Kropotkin P (1902) *Mutual Aid: A Factor of Evolution.* Whitefish: Kessinger

Lorde A (2007) *Sister Outsider: Essays and Speeches.* Darlinghurst: Crossing Press

Loyd J M (2012) Geographies of peace and antiviolence. *Geography Compass* 6(8):477–489

Mamadouh V (2005) Geography and war, geographers and peace. In C Flint (ed) *The
Geography of War and Peace: From Death Camps to Diplomats* (pp 26–60). Oxford:
Oxford University Press

McConnell F, Megoran N and Williams P (eds) (2014) *The Geographies of Peace: New
Approaches to Boundaries, Diplomacy, and Conflict Resolution.* London: I.B. Tauris

Megoran N (2010) Towards a Geography of Peace: Pacific Geopolitics and Evangelical
Christian Crusade Apologies. *Transactions of the Institute of British Geographers*
35(3):382–398

Megoran N (2011) War and peace? An agenda for peace research and practice in geography. *Political Geography* 30(4):178–189

Megoran N and Dalby S (2018) Geopolitics and peace: A century of change in the discipline of geography. *Geopolitics* 23(2):251–276

Naylor L, Daigle M, Zaragocin S, Ramírez M M and Gilmartin M (2018) Interventions:
Bringing the decolonial to political geography. *Political Geography* https://doi.org/10.
1016/j.polgeo.2017.11.002

Noxolo P (2017) Decolonising geographical knowledge in a colonised and re-colonising
postcolonial world. *Area* 49(3):317–319

Peake L and Sheppard E (2014) The emergence of radical/critical geography within North
America. *ACME* 13(2):305–327

Pepper D and Jenkins A (1983) A call to arms: Geography and peace studies. *Area* 15
(3):202–208

Radcliffe S A (2017) Decolonising geographical knowledges. *Transactions of the Institute of
British Geographers* 42(3):329–333

Ross A (2011) Geographies of war and the putative peace. *Political Geography* 30
(4):197–199

Sandoval Montes Y and Nuñez Villalba J (eds) (2017) *Geografías al servicio de los procesos
de paz: Análisis global, reflexión y aporte desde el contexto latinoamericano.* La Paz: Plural
Editores

Springer S (2012) Anarchism! What geography still ought to be. *Antipode* 44(5):1605–1624

Springer S (2014) Why a radical geography must be anarchist. *Dialogues in Human Geography* 4(3):249–270

Williams P and McConnell F (2011) Critical Geographies of Peace. *Antipode* 43(4):927–931

Wisner B (1986) Geography: War or peace studies? *Antipode* 18(2):212–217

Woods C A (1998) *Development Arrested: The Blues and Plantation Power in the Mississippi
Delta.* New York: Verso

Political Consciousness

Divya P. Tolia-Kelly

Department of Geography, University of Sussex, Brighton, UK;
d.p.tolia-kelly@sussex.ac.uk

My key phrase was inspired by the Gramscian formulation of "hegemony" and the problematics of *political consciousness* for "radical" and "critical" intellectuals in society. We often find ourselves claiming radical or indeed cutting *critical* ground in the challenge of inequalities, including poverty (Martin 1995), social exclusion (Sibley 1998), patriarchy (McDowell 1986), neoliberalism (Amin and Thrift 2005; Tickell 1995), colonialism (Blunt and McEwan 2003; Driver 2001; Livingstone and Withers 1999; Raghuram et al. 2014) and imperialism (Gregory 2004), or indeed spearheading movements for decolonising the academy (Radcliffe 2017) or accounting for power in networks of racism (McKittrick and Woods 2007; Woods 2002) and cultural prejudice (Kobayashi and Peake 1994). However, as Gramsci's work shows, there is a long history of intellectuals operating in duplicity as "arms" of the state in supporting, promulgating and indeed defending ideas that support the formation and powerful work of cultural hegemony. In celebrating *Antipode*'s 50th anniversary of publication, and a site of radical geographical thought, it is important to examine the ways in which "radical" and "critical" are claimed without necessarily challenging the work of cultural hegemony that academia is also inculcated into (Waterstone 2002). The critical radical edge of geography celebrates, promotes and indeed promises *praxis*, however (despite the introduction of market forces in the form of fees) the cultural economies of Higher Education are increasingly wedded to delivering the reproduction of society with "civic" values and hierarchies that are untouched (Bates 1975). There is a symptomatic double-facedness that requires us to both inhabit civic structures of producing "good citizens" while critiquing state governance as it reproduces spheres of domination, through economic policy, the militarisation of borders, the structures of law, judiciary and policing, injustice, inequality and oppression. *Praxis* ultimately is antithetical to being edified in the economies and cultural communities of academia. To be radical, critical and usurping of the status quo, for many academics is to be situated *outside* the gates and towers of rightful belonging; this is despite being supported by intellectual theory produced from the radical traditions and ethos.

What Gramsci argues is that any intellectual class of a society effectively promulgates and reproduces the values of the elite and as such creates a sense of community solidarity which itself subordinates others. Thus, there is a false promise in that the intellectual class can empower, revolutionise or indeed assist in dismantling regimes of truth, and structures of dominion over others:

Keywords in Radical Geography: Antipode at 50, First Edition. Edited by the *Antipode* Editorial Collective.
© 2019 The Authors/Antipode Foundation Ltd. Published 2019 by John Wiley & Sons Ltd.

... an independent class of intellectuals does not exist, but rather every social group has its own intellectuals. However the intellectuals of the historically progressive class ... exercise such a power of attraction that they end ... by subordinating intellectuals of other social groups and thus create a system of solidarity among all intellectuals. (Bates 1975:353)

In this vein then, the European intellectuals' commitment to racial science and to cultural hierarchies in the 19th century can be seen as examples of western intellectuals subordinating "others". By utilising the power of European intellectual realms and accoutrements of academia, dominion over the very value of other bodies and minds as well as delimiting the boundaries of thought and acceptable ideas to within its own intellectual community. This leads us to reflect then on how exactly is the academic community and the discipline of geography *radical* and *critical* in light of its positioning in relation to dependency on the state, and indeed increased dependence given the "impact"[1] and "prevent"[2] agendas that colour higher education's independence and autonomy in thought and praxis (see Holmwood 2011; Martin 2011). There is also an embedded contradiction in Gramsci's account of left-wing intellectuals and their potential for creating a new world order. There is a rejection of the state as non-representative of majority world rights and values, yet the very sustenance of academia is dependent on the state having use for its role in producing citizens and assisting with its work of compliance and policing challenges to its very function.

What to Do?

Praxis is the element that I wish to focus on in this section. The process of praxis commences with a sense of political awareness beyond an understanding of individual needs and struggles. In the workplace the distinction can be made between individual awareness of others' rights and an understanding of collective positioning and solidarity; praxis is the actions undertaken towards attaining better conditions for all. The consciousness evolves then to demands for emancipation and freedoms for all. Praxis is about action that seeks to shift the conditions under which oppression and limitations are defined, met and secured. Praxis has been discussed as the pragmatic actions that create the circumstances from which we can challenge hegemonic ideas and also recognise them as not being in collective interests. Critical thought without practice is simply armchair theory; and as such, working in the realms of theory does not, and cannot create conditions for counter-hegemony or indeed new infrastructures of freedom. As Woods (2005) has so elegantly shown in the US, in 1967–1968 there is a recognition among Black African Americans of their shared oppression, poverty and a society that does not serve them equally. What emerges from the recognition of collective oppression is a Black political consciousness that seeks to imagine a society free of the constraints of hegemony which serves to reproduce US apartheid: "the conditions and consciousness of the rural African American working class shocked and radicalised both King and Malcolm X in the last years of their lives (Woods

1998:187). It is in these conditions that the Poor People's March of 1968 garnered collective political praxis to include "Native Americans, Chicanos, Puerta Ricans and Whites" (ibid.).

It is clear that *praxis* stems from a desire to win *collective* freedoms. Rather than focus on the success of a vanguard of critical thinkers, the *edges* of the limits to academic praxis can be exposed by looking at inequalities that have survived in perpetuity (see Pilkington 2011). Within academia there are "others" that are subordinated and reduced to lesser, and as such bear costs of exclusion, discrimination and prejudice (Smith 1999). Within the very realms of theoretical radicalism and critical geographical challenge, there are negations of the rights of individuals and groups; these are breaches of laws of race equality, gender equality, and sexuality. The lack of focus on praxis can be measured in the uneven landscape of academia as a site of work itself, in the inequalities borne through representation, non-employment, teaching-only contracts and figures for pay, promotion and retention of those discriminated against colleagues and students (see Blackaby et al. 2005; Broecke and Nicholls 2006; HEA/ECU 2011; Shen 2013; Singh 2011). In this century, the campaigns around "Why isn't my professor black?"[3] chime with reports on BAME student attainment gaps as well as BAME staff appointments, promotion and retention (Alexander and Arday 2015). Overall there is evidence for bias in appointment, recruitment, promotion and inclusion within universities of BAME staff and women. As a result of unconscious or implicit bias, monocultures are created when people recruit in their own image. This is particularly true in senior positions. A number of institutions are introducing training which looks at unconscious bias. Despite the training, the cultures of academia are as much about being seen to be doing something rather than actually effecting change. As Ahmed (2006) has stated, there is a dance that occurs where the most passionate of anti-racists get co-opted into committees and groups that are responsible for writing policy, Athena Swan Charter Mark applications and as such there is containment of momentum, political will and moral imperatives which edify certain individuals, managers or departments, but which fail to create an environment of inclusion, respect and valuing of "other" staff:

> A document that documented the racism of the university became usable as a measure of good performance. Here, having a good race equality policy quickly got translated into being good at race equality. Such a translation works to conceal the very inequalities that the documents were written to reveal. In other words, its very existence is taken as evidence that the institutional environment documented by the document (racism, inequality, injustice) has been overcome; as if by saying that we "do it" means that's no longer what we do. (Ahmed 2006:108)

This negation of action towards righting inequalities is the responsibility of all of us. Praxis is very much about recognising and then doing political work to address the structural inequalities that we are complacent about in our everyday life (see Tolia-Kelly 2017). These exclusions are not accidental, benign omissions but point to the very gaps in political consciousness that are embodied in everyday university work. These active omissions of praxis reduce the potential of lived lives of academic colleagues and students within our presumed *meritocracy* that remains unchallenged in the realms where it matters; in recruitment, promotion, retention and our duty of care towards well-being of marginalised colleagues and students. This includes incorporating consciousness beyond our usual grammars, vocabularies and fields of vision to incorporate *other* embodied ontologies that may shake the foundations of our sometimes parochial habitus (e.g. Woods 2005). In praxis radical and critical research has often created a valuable cultural currency among us, which is rewarded by promotion and recognition. What is clear on reviewing the list of geographical radicals listed at the start of this piece is that the institutions reflected by geography's *radical* geographers remain unreconstructed; "meritocracy" remains unchallenged, despite the statistical evidence against this belief system. The very politics of the work of challenging institutional racism, misogyny, violences and exclusion is not necessarily addressed by those who are the most radical in publications and rhetoric. There are a handful of activist-practitioners that are part of the struggle to reshape the political palette of the university as a workplace, at severe personal cost (Routledge 2012). What is missing is the orientation towards collective praxis, to challenge the dimensions of individual rewards for individual impact (see Fuller and Kitchin 2004). What is needed is a raising of our game. Of conscious political action beyond the page. By paying appropriate attention to our own *positioning* as teachers, researchers, employers, colleagues and activists, the very realms within our collective power should become sites of political praxis, and not only things which can promote our careers and radical brandings in our neoliberal institutional life.

Endnotes

[1] See UK Research and Innovation on "Excellence with impact": https://www.ukri.org/innovation/excellence-with-impact/ (last accessed 27 July 2018).
[2] See Universities UK on "The Prevent Agenda": http://www.safecampuscommunities.ac.uk/the-prevent-agenda (last accessed 27 July 2018).
[3] See the *Times Higher Education* on "Why isn't my professor black?": https://www.timeshighereducation.com/blog/why-my-professor-still-not-black (last accessed 27 July 2018).

References

Ahmed S (2006) The nonperformativity of antiracism. *Meridians* 7(1):104–126
Alexander C E and Arday J (2015) "Aiming Higher: Race, Inequality and Diversity in the Academy." Runnymede.
Amin A and Thrift N (2005) What's left? Just the future. *Antipode* 37(2):220–238
Bates T R (1975) Gramsci and the theory of hegemony. *Journal of the History of Ideas* 36(2):351–366
Blackaby D, Booth A L and Frank J (2005) Outside offers and the gender pay gap: Empirical evidence from the UK academic labour market. *The Economic Journal* 115(501):F81–F107
Blunt A and McEwan C (eds) (2003) *Postcolonial Geographies*. London: Bloomsbury
Broecke S and Nicholls T (2006) "Ethnicity and Degree Attainment." Research Report RW92, Department for Education and Skills
Driver F (2001) *Geography Militant: Cultures of Exploration and Empire*. Oxford: Blackwell
Fuller D and Kitchin R (eds) (2004) *Radical Theory/Critical Praxis: Making a Difference Beyond the Academy?* Victoria: Praxis (e)Press
Gregory D (2004) *The Colonial Present: Afghanistan, Palestine, Iraq*. Malden: Blackwell

HEA/ECU (2011) "Improving the Degree Attainment of Black and Minority Ethnic Students." Higher Education Academy/Equality Challenge Unit

Holmwood J (2011) The impact of "impact" on UK social science. *Methodological Innovations Online* 6(1):13–17

Kobayashi A and Peake L (1994) Unnatural discourse: "Race" and gender in geography. *Gender, Place, and Culture* 1(2):225–243

Livingstone D N and Withers C W (eds) (1999) *Geography and Enlightenment*. Chicago: University of Chicago Press

Martin B R (2011) The Research Excellence Framework and the "impact agenda": Are we creating a Frankenstein monster? *Research Evaluation* 20(3):247–254

Martin R (1995) Income and poverty inequalities across regional Britain: The north–south divide lingers on. In C Philo (ed) *Off the Map: The Social Geography of Poverty in the UK* (pp 23–44). London: Child Poverty Action Group

McDowell L (1986) Beyond patriarchy: A class-based explanation of women's subordination. *Antipode* 18(3):311–321

McKittrick K and Woods C (eds) (2007) *Black Geographies and the Politics of Place*. Toronto: Between the Lines

Pilkington A (2011) *Institutional Racism in the Academy: A UK Case Study*. Stoke-on-Trent: Trentham Books

Radcliffe S A (2017) Decolonising geographical knowledges. *Transactions of the Institute of British Geographers* 42(3):329–333

Raghuram P, Noxolo P and Madge C (2014) Rising Asia and postcolonial geography. *Singapore Journal of Tropical Geography* 35(1):119–135

Routledge P (2012) Sensuous solidarities: Emotion, politics, and performance in the Clandestine Insurgent Rebel Clown Army. *Antipode* 44(2):428–452

Shen H (2013) Inequality quantified: Mind the gender gap. Despite improvements, female scientists continue to face discrimination, unequal pay and funding disparities. *Nature* 495(7439):22–24

Sibley D (1998) The problematic nature of exclusion. *Geoforum* 29(2):119–121

Singh G (2011) *Black and Minority Ethnic (BME) Students' Participation in Higher Education: Improving Retention and Success—A Synthesis of Research Evidence*. York: Higher Education Academy

Smith L T (1999) *Decolonizing Methodologies: Research and Indigenous Peoples*. London: Zed

Tickell A (1995) Reflections on "Activism and the academy". *Environment and Planning D: Society and Space* 13(2):235–237

Tolia-Kelly D P (2017) A day in the life of a Geographer: "Lone", black, female. *Area* 49 (3):324–328

Waterstone M (2002) A radical journal of geography or a journal of radical geography? *Antipode* 34(4):662–666

Woods C (1998) *Development Arrested: The Blues and Plantation Power in the Mississippi Delta*. New York: Verso

Woods C (2002) Life after death. *The Professional Geographer* 54(1):62–66

Woods C (2005) Do you know what it means to miss New Orleans? Katrina, trap economics, and the rebirth of the blues. *American Quarterly* 57(4):1005–1018

Pride / Shame

Lynda Johnston

Faculty of Arts and Social Sciences, University of Waikato, Hamilton, Aotearoa New Zealand;
lynda.johnston@waikato.ac.nz

Feelings, emotions and affects of pride and shame shape people and place in profound and mundane ways. Starting with the body, pride and shame serve to connect bodies to places and vice versa. Pride and shame are useful concepts for geographers to pay attention to because they reveal key aspects of everydayness, and at the same time, may force geographers to rethink taken-for-granted interpretations of emotional and affectual geographies. In other words, by thinking critically about pride and shame, geographers are well placed to raise questions about how and in what ways pride and shame constructs place, subjectivities and power.

For feminist, queer, emotional and affectual geography sub-disciplines, the attention to pride and shame comes at a time when the body is frequently placed in the centre of analysis. These critical approaches are not only concerned with understanding places and bodies associated with pride and shame, they are also challenging the view that the human subject is essentially rational, autonomous, unchanging, objective, and somehow free of emotions. The binary constructions of mind/body, masculine/feminine, rationality/emotionality—where mind, masculinity, and rationality have been privileged over body, feminine and emotionality—have been critiqued and deconstructed by feminist geographers. Binary categories are not discreet nor bounded, rather, they inform each other. Spatial expressions of so-called "positive" and "negative" emotions highlight a dynamic relationship between pride and shame. Feminist and queer geographers draw on key theorists such as Sara Ahmed (2004), Elspeth Probyn (2000, 2004, 2005), Sedgwick (1993, 2003), and Tomkins (1995), to prioritise body–space relationships. The resulting geographical research brings to the fore emotionalities of pride and shame, and related dynamically entangled emotions such as of joy and disgust.

In what follows I begin by outlining some of the key theoretical contributors to understandings of pride/shame. This is followed by geographical examples of pride/shame informed research. How pride/shame feels, and what it does, is illustrated by geographies of gay pride parades and festivals; body size and shape; and sport.

Gut Reactions, Deep Blushing

Pride and shame are not separate entities, but rather they are mutually constructing. Shame is often understood as the "mirror image" of pride (Probyn 2000:25) and rather than turning away from shame, scholars have considered its productive affects. Pride and shame, Probyn (2000) argues, is at the centre of

Keywords in Radical Geography: Antipode at 50, First Edition. Edited by the *Antipode* Editorial Collective.
© 2019 The Authors/Antipode Foundation Ltd. Published 2019 by John Wiley & Sons Ltd.

subjectivity. The work of Probyn (2000, 2004, 2005) highlights the productive affects of shame and in doing so raises important questions about pride, the subject, identity and subjectivity. While many scholars have focused on pride, Probyn argues that shame is worth interrogating as it reveals embodied subjectivity. Shame is "the body saying that it cannot fit in although it desperately wants to" (Probyn 2004:345).

The effects of pride and shame are powerful, performative and productive. Probyn (2004:234) notes that shame "works to expose any breaches in the borders between self and others" and hence provides a window to examine who feels they belong in particular places or who does not. Pride—as both a political movement and feeling—is created by shame and vice versa. "Other" bodies are made aware of their "limitations" (and may feel shame) when they are not the "Self", for example, cisgender, male, heterosexual, white, of a certain age, a certain size and shape, etc. Pride movements, such as gay pride, black pride, fat pride, disability pride, and so on, help marginalised people work through their feelings of shame, while at the same time, embrace feelings of pride.

As a gut reaction, shame is something that may be paralysing and negative, but also productive. Feelings of shame may set off "nearly involuntary re-evaluation of one's self and one's actions" (Probyn 2005:78). Yet, these effects are volatile, unstable and unpredictable. The moment of deep blush, for example, triggers a hyper-reflexive moment that turns, Probyn (2005) argues, one inside out—or outside in. To resist shame, bodies may mobilise prideful subjectivities. Therefore, shame simultaneously disrupts and makes subjectivities that align with Self and Other. Embodying pride and shame can be understood as concurrently erasing and defining identities. Political pride movements provide a context from which to consider how pride and shame operate in relationship to performativity, marginalised identities and the politics of belonging.

The cultural politics of emotion have been considered by Sara Ahmed (2004). Like Probyn, Ahmed illustrates the mobility of pride and shame feelings and affects. Ahmed (2004:105) writes: "crucially, the individuation of shame—the way it turns the Self against and towards the Self—can be linked precisely to the inter-corporeality and sociality of shame experiences". Pride and shame are about bodies being close to each other, as well as about an individual's acute sensitivity of one's sense of self. Building on the work of Eve Kosofsky Sedgwick (1993, 2003), Tomkins (1995), Ahmed (2004) and Probyn (2005) acknowledge the lived experiences of pride/shame, insisting on the inter-corporeality and sociality of pride/shame. That is, bodily affects of shame (blushing) are experienced in response to the performativity of the Self and before the gaze of another, which in turn may trigger the self-awareness that shame performs. Pride and shame produce space and vice versa. The next section shows pride/shame geographies in relation to genders and sexualities, body sizes, sport and leisure.

Proud and Shameful Places

A focus on the co-construction of pride and shame for gay pride parades sheds light on the spatial politics of gender and sexual identities, communities, cities, and geographies of (not) belonging (Johnston 2005, 2007; Johnston and Waitt 2015; Waitt and Stapel 2011). Gay pride politics are entwined with spatial politics of shame. Contemporary gay pride parades in many western cities take their starting point from the New York Stonewall riots which began on a June night of 1969 when police raided a gay bar, called the Stonewall Inn, in Greenwich Village, New York City. Three days and two nights of rioting represent the first time in US history when gender and sexually diverse people resisted police harassment. Fighting spurred from resisting shameful feelings and affects, gay pride parades today continue to be emancipatory events (based on feelings of pride) yet also reassert a number of (shameful) hegemonies. They may queer streets and bodies, yet the diversity of identities may not all be included. Gay pride parades and festivals are proud spaces that are also complicit with a range of indignities (the shamed).

Places and spaces are never solely associated with pride, or with shame, rather the two intermingle. The shape and size of bodies, for example fatness, slimness, weight gain and loss, are associated with both pride and shame. Colls and Evans (2009:1012) note that "space and place are becoming increasingly important to the conception and deployment of obesity politics". Fat politics, that is, being out and proud as fat, are also layered with dominant discourses (and feelings of shame) that surround fat bodies (Longhurst 2010). Geographical research conducted on fat women's relationships with different places such as cafes, shopping malls, beaches and work places found that when women are marginalised they feel not only out of place, but shame (Longhurst 2010). Participants also conveyed humour and stories of courage and resistance. Longhurst charted her own emotional geographies of being a fat woman and weight loss, and highlights a number of paradoxical places that are infused with pride and shame (see Longhurst 2014; see also Colls 2006). This research on body size shows that there are powerful and queer ways of thinking about shame.

When bodies are taut and toned, they are usually associated with pride and physical success. Working out, training, and playing sport are activities that are privileged in many places. Pride in sport is intimately connected with sense of self, family and national honour, and most accounts of sport tend to erase shame in favour of pride (Probyn 2000). Tomkins (1995) asserts that shame is unspeakable because of how it is embedded in western societies as "lack" and/or incompleteness. Tomkins (1995:172) states that "shame ... must itself be hidden as an ugly scar is hidden, lest it offend the one who looks". Focusing on men who play competitive country football, Waitt and Clifton (2014) show the spatial dynamics of pride and shame. Their research highlights people's emotional bonds and affective ties of active participation in sport, providing new ways of thinking about football as a site of social engagement. The dynamics of pride and shame provide important clues to understanding the mutual construction of sport, leisure, ruralities and masculinities (see also Waitt and Clifton [2013] for an account of pride/ shame, masculinity and surfing).

The deeply felt dynamics of pride and shame are integral to understandings of body–space relationships. Gay pride parades and festivals, geographies of body sizes and shapes, plus sporting spaces are just some of the pride/shame informed research that geographers are doing. Paying attention to the materialities of bodies, including bodily judgments such as "gut reactions", is another way to understand which bodies are included and/or excluded from place and space. Attention to bodily pride and shame provides opportunities to interrogate normative ideas, performances, subjectivities, power, spaces and places.

References

Ahmed S (2004) *The Cultural Politics of Emotion*. New York: Routledge
Colls R (2006) Outsize/outside: Bodily bignesses and the emotional experiences of British women shopping for clothes. *Gender, Place, and Culture* 13(5):529–545
Colls R and Evans B (2009) Questioning obesity politics. *Antipode* 41(5):1011–1020
Johnston L (2005) *Queering Tourism: Paradoxical Performances of Gay Pride*. London: Routledge
Johnston L (2007) Mobilizing pride/shame: Lesbians, tourism, and parades. *Social and Cultural Geography* 8(1):29–45
Johnston L and Waitt G (2015) The spatial politics of gay pride parades and festivals: Emotional activism. In D Paternotte and M Tremblay (eds) *The Ashgate Research Companion on Lesbian and Gay Activism* (pp 105–119). Farnham: Ashgate
Longhurst R (2010) The disabling affects of fat: The emotional and material geographies of some women who live in Hamilton, New Zealand. In V Chouinard, E Hall and R Wilton (eds) *Towards Enabling Geographies: "Disabled" Bodies and Minds in Society and Space* (pp 199–216). Aldershot: Ashgate
Longhurst R (2014) Queering body size and shape: Performativity, the closet, shame, and orientation. In C Pausé, J Wykes and S Murray (eds) *Queering Fat Embodiment* (pp 13–26). London: Routledge
Probyn E (2000) Sporting bodies: Dynamics of shame and pride. *Body and Society* 6(1):13–28
Probyn E (2004) Everyday shame. *Cultural Studies* 18(2/3):328–349
Probyn E (2005) *Blush: Faces of Shame*. Minneapolis: University of Minnesota Press
Sedgwick E K (1993) Queer performativity: Henry James's *The Art of the Novel*. GLQ: *Journal of Lesbian and Gay Studies* 1(1):1–16
Sedgwick E K (2003) *Touching Feeling: Affect, Pedagogy, Performativity*. Durham: Duke University Press
Tomkins S (1995) *Shame and Its Sisters: A Silvan Tomkins Reader* (eds E K Sedgwick and A Frank). Durham: Durham University Press
Waitt G and Clifton D (2013) Stand up, not out: Bodyboarders, gendered hierarchies, and negotiating the dynamics of pride/shame. *Leisure Studies* 32(5):487–506
Waitt G and Clifton D (2014) Winning and losing: The dynamics of pride and shame in the narratives of men who play competitive country football. *Leisure Studies* 34(3):259–281
Waitt G and Stapel C (2011) "Fornicating on floats"? The cultural politics of the Sydney Mardi Gras Parade beyond the metropolis. *Leisure Studies* 30(2):197–216

Prisons

Matthew L. Mitchelson

Department of Geography & Anthropology, Kennesaw State University, Kennesaw, GA, USA;
mmitch81@kennesaw.edu

Will you take a "speculative leap" with me, and picture a child? A kid of any age will do, and please go with any other details that your mind may conjure. (For example, I am picturing a little boy with curly brown hair and a t-shirt with a dinosaur on it; he is six years old.) Now, importantly, what is the child's name? Let's suppose that you are this child's temporary caregiver now, while their parent is incarcerated. My imaginary kid and your imaginary kid have just joined a cohort of several million very real children with an incarcerated parent. Today, you have spent the morning helping them make a drawing for their parent, and now you will make the drive for an in-person visit. This is a first for the kid. You buckle the child into their seat and you hit the road. And the kid, who has been energetically chattering and busily motoring about all morning, is asleep within minutes. They sleep soundly as you drive for dozens (if not hundreds) of miles. For the child, time has effectively stopped.

The kid awakens, more than a little confused, in the parking lot of a prison. They ask a wonderfully radical question: Where are we? As the following quotations suggest, any response will be quite complicated:

> In most parts of the world, it is taken for granted that whoever is convicted of a serious crime will be sent to prison ... the prison is considered to be an inevitable and permanent feature of our social lives. (Davis 2003:9)

> Henri Lefebvre points out in the opening arguments of *The Production of Space* (1991) that we often use the word "space" ... without being fully conscious of what we mean by it. We have inherited an imagination so deeply ingrained that it is often not actively thought. Based on assumptions no longer recognized as such, it is an imagination with the implacable force of the patently obvious. That is the trouble. (Massey 2005:17)

Our response will not be couched in the terms of inevitability or permanence signalled by Davis. Instead, our answer to the child—who is now looking us in the eye and nervously awaiting our response—will surely have to address the trouble signalled by Massey.

Prisons are the result of a lot of trouble, and they cause a great deal of trouble in their own right. Much of this trouble is spatial in nature (Bonds 2009; Gilmore 2002; Hiemstra and Conlon 2017; Loyd et al. 2012; Moran et al. 2013). Geographic imaginations—including those that we have inherited or passively accepted (as suggested by Massey's quote above)—play a role in all of this trouble, as do material conditions and embodied experiences (Bonds 2013; Brown

Keywords in Radical Geography: Antipode at 50, First Edition. Edited by the *Antipode* Editorial Collective.
© 2019 The Authors/Antipode Foundation Ltd. Published 2019 by John Wiley & Sons Ltd.

and Schept 2017; Gilmore 2007; McKittrick 2012; Peck and Theodore 2008). This entry reflects a dialectic spatial epistemology of prisons, in the hope of inspiring less troublesome space(s) than those we share today. This epistemology is based on two separate tripartite divisions of space: one detailed by Harvey (1973), one by Lefebvre (1991); and their eventual synthesis by Harvey (2006). Geographically (re)imagining prisons through this epistemology, I seek to account for both "the thing" and "the process" at work in imprisonment. Prisons (i.e. buildings) may be the keyword, in this case, but it is their relationality that requires most of our attention.

Why? Because prisons refuse to stay put. Prisons may result from crime and punishment, but this is only part of the story; this can only account for some of the trouble. These remarkably complicated sites are very much on the move, materially and discursively trailing in the wake of criminalisation and dynamically articulated with capital (Hiemstra and Conlon 2017; Loyd 2012). It is not premature to think of the prison as one of the central race-making institutions of our time (Davis and Dent 2001; Gilmore 2002; Wilson 2007). So, crime and punishment necessarily trace back across formations of race, gender, sex, and place (Shabazz 2015)—to the often fatal coupling of power and difference (Gilmore 2002; McKittrick 2012). Along the lines of this retracing, the fixed prison/thing follows and flows with agile force—receiving racial formations, and reshaping the embodiment of race itself because "criminal justice" is a spatial process, and so too are its prisons.

When viewed as a spatial process, it turns out that prisons are asking the world of us. They ask us what kind of means and ends our justice will entail; what kind of communities we will create (and destroy); and, what kind of values our prisons will make manifest. Too often, they are asking us how we choose to value human life and death. By honouring the geographies underlying these questions, we can become more fully conscious of prisons as a space (see Massey above)—more conscious of a prison in a given location, yes, but also of its space of flows and social relations near and far. When we stop accepting prisons as the naturally occurring outcome of crime and punishment, it becomes significantly less difficult to imagine alternative geographies (Davis 2003; Gilmore and Gilmore 2008; Loyd et al. 2012). This piece suggests how less troublesome alternatives to prisons—like any other space—merit attention, in keeping with *Antipode*'s radically persistent reconfiguration of what is possible.

This invitation to reimagine the space of prisons through a "dialectic spatial epistemology" in pursuit of clarity (i.e. regarding prison space) first requires some explanation, because these terms and this approach are admittedly contentious within and beyond the discipline. First, please note that I am generally referring to prisons in the United States, but in terms that I understand to be relevant in other international contexts. Second, I want to ask you to join me on a (conceptual) journey, based on the "speculative leap" proposed by Harvey (2006:281). This journey loosely follows his heuristic matrix; it is a matrix that is neither quantitative nor comprehensive in nature. The intended ontological commitment "within" the matrix is that each of its elements can be "imagined as an internal relation of all the others" (ibid.). One axis is structured by the absolute, relative, and relational elements of space–time first considered by Harvey (1973) in *Social Justice and the City*. Another axis is structured by the experienced, conceptualised, and lived dimensions of space–time explained by Lefebvre (1991) in *The Production of Space*. I will provide further explanation and emphasis for particular terms in what follows, rather than detailing each element here. In sum, the ontological argument here is that prisons can be analysed as productive of (and being produced by) the absolute, relative, and relational spatial process of imprisonment. In this sense, any given prison is many places at once. Rather than a totalising "definition" of prison space, I intend this as an invitation for you to engage with—and work on—different types of spatial questions that centre on the prison (cf. Gilmore 2007:27).

The child tugs at our hand again and reminds us of their question, "Where are we?" (Oh child, where to begin …) Let's start with the basics. What do you, the two of you, see? What differentiates this place from its surrounding environment? Most prisons are simply buildings; the kid's seen plenty of buildings. However, this building is far from simple, because it is premised on confining human beings like their Mom or Dad. A "total institution" (Goffman 1961) like the one in front of you has only one accessible entryway. So, you know where to go—where you can go—in relation to the rest of what you can see. A well marked "guard line" and row after row of razor-wire-lined fencing also tells you where you cannot go. Towers loom above you. Armed officers patrol. How do you translate the overwhelming outward appearance(s) of this place to the child? (Do words like justice and public safety come to mind?)

Your words will create a "representation of space" that will hopefully serve the best interests and needs of the child. So, your conceptualisation in this moment will likely be quite different from those that informed you to drive to this specific prison (as opposed to the other facilities within the prison system). Maps of prisons are often publicly available, as are facility details and driving directions (Google Maps is great, but cell signals are spotty "out here'). These official representations might detail capacities and security levels, for example. In the state where I have done most of my research, you can also "locate" prisoners through an "offender query" online. This dataset includes quite a lot of information: physical descriptions, incarceration details, aliases, and incarceration history. Individuals are coded by name and a unique prison-system-specific identifier that geocodes a life within prisons. In this particular representation of space, prisoners are conceptually "put in their place" for the public (including release details). This piggybacks onto relational representations of space touted by officials who use prisons to "remove violent offenders from society" as if society somehow ended abruptly at the guard line. Returning to the quotes above from Davis and Massey, is it not amazing just how much one can "know" about the geography of a prison or a prisoner without a material encounter?

In the absolute space of this moment, which is holding host to Lefebvre's space of representation (lived space) for you and the child, things are likely quite "messy" (Katz 2001:711). You may be mitigating your own fears and curiosity, with earnest reverence for the monopoly on violence held by the state just overhead; and, simultaneously, trying to provide comfort and reassurance to the kid.

Your senses may be overwhelmed (e.g. some prison grounds are quiet as a pin-drop; the commotion of others sounds as if an entire city has been forced inside). Although prisons are definitively inaccessible sites, we do know (intuitively and empirically) much about what is happening inside of them. You might consider what it is going on with—and what is happening to—the people in this place. Your imagination may run wild, picturing losses of autonomy, dehumanising experiences, and periodic eruptions of violence. And you must be tired, yourself. It took considerable effort to get you two here. It is hard for you when the kid asks if you can get back in the car and go home.

Maybe you pause to reflect on the particular way in which you are here. You and this kid are on a journey. You've come from home, and to there you will return. This is really just a cameo for you. And you are in good company because, with the glaring exception of those who die in prisons, almost everyone goes home (eventually)—more than 90% of prisoners go home; more than half a million prisoners go home each year. And then, before anyone can go home, a funny thing happens. The kid says, "Whoa!", and their mood shifts. Do you know what has caught their attention? It is a fleet of vehicles. The kid loves cars and trucks, and there are a dozen big vans within sight. "Prisoner Transport" is written along their sides. The fleet is sitting there, in that moment, as a testament to the structural mobility inherent to prison space (the state-wide fleet consists of more than 1500 vehicles). This parking lot is one of many locations where fixity is met (and often contradicted) by flow and mobility. Prisons are all about flow(s). If you look in the right places, you can see each of the flows, and how they overlap, avoid, or amplify one another: truckloads of food; kilowatt-hours of electricity; flows of capital; tanks of gasoline; various work and labour; phone calls home. There are many more.

This place, you may come to realise, is being produced and orchestrated from a great distance: from centralised administrative offices, where all of these population flows (including you and the child) have been set into motion through a criminal justice system and bureaucratic protocols that produce a disciplinary knowledge that sets this entire carceral world into motion. You might reflect on a story told to you by a former federal prisoner, who was transferred multiple times across state lines. Once his family (of three) flew halfway across the country, only to learn that he had been transferred back (to their home state). You know that "wheels roll" twice a week as part of the intra-prison-system transfer process in this state. What will you tell the kid now, if you have travelled all of this way and their parent is elsewhere? To whom—and where—will you turn for accountability?

As you are walking inside of the reception hall, the severity of the environment becomes overwhelming, and it is difficult for you to remain mindful of the flows. Now it all feels like fixity. This is no accident. Prisons are imposing, by design, because fixity produces a far more manageable space than does flow. But the child holding your hand reminds you that you are not only "in" this carceral world; you are also helping to facilitate a home (a parent and a child, in this case) in dire times. That homeplace is now part of a larger spatial process in which, as Ruthie Gilmore (2007:16) puts it, "households stretch from neighbourhood to visiting room to courtroom, with a consequent thinning of financial and emotional resources". As the child's caregiver, in a sense, you are walking through the hallway of their second home, metal detectors and all.

As of this writing, we are in grave trouble with our prisons. We are not approximating a holistic justice, the healing of individual and social wounds, or the propagation of meaningful, enduring safety. We are simply off the mark as measured by those standards. Of course all of this is about value (and money value is only one of many values that are potently at work in prisons). Prison abolitionists are currently looking in seemingly distant places for solutions: schools and healthcare systems, we propose, are part of the solution. Reforming the symptom, we suggest, will not cure the cause of such an illness. Time is likely of the essence. When produced as non-dialectical space (i.e. abstract, lethal space), prisons are profoundly dangerous and expansive places (Brenner and Elden 2009; Jones and Popke 2010; Lefebvre 1991). Indeed, prisons are asking the world of us.

But now, the visit is about to begin. The kid asks you, "Where are we now?" They are being so brave.

References

Bonds A (2009) Discipline and devolution: Constructions of poverty, race, and criminality in the politics of rural prison development. *Antipode* 41(3):416–438

Bonds A (2013) Economic development, racialization, and privilege: "Yes in my backyard" prison politics and the reinvention of Madras, Oregon. *Annals of the Association of American Geographers* 103(6):1389–1405

Brenner N and Elden S (2009) Henri Lefebvre on state, space, territory. *International Political Sociology* 3(4):353–377

Brown M and Schept J (2017) New abolition, criminology, and a critical carceral studies. *Punishment and Society* 19(4):440–462

Davis A (2003) *Are Prisons Obsolete?* Toronto: Seven Stories Press

Davis A Y and Dent G (2001) Prison as a border: A conversation on gender, globalization, and punishment. *Signs* 26(4):1235–1241

Gilmore R W (2002) Fatal couplings of power and difference: Notes on racism and geography. *The Professional Geographer* 54(1):15–24

Gilmore R W (2007) *Golden Gulag: Prisons, Surplus, Crisis, and Opposition in Globalizing California*. Berkeley: University of California Press

Gilmore R W and Gilmore C (2008) Resisting the obvious. In M Sorkin (ed) *Indefensible Space: The Architecture of the National Insecurity State* (pp 141–162). New York: Routledge

Goffman E (1961) On the characteristics of total institutions. In id. *Asylums: Essays on the Social Situation of Mental Patients and Other Inmates* (pp 1–124). Chicago: Aldine

Harvey D (1973) *Social Justice and the City*. Baltimore: Johns Hopkins University Press

Harvey D (2006) Space as keyword. In N Castree and D Gregory (eds) *David Harvey: A Critical Reader* (pp 270–294). Oxford: Blackwell

Hiemstra N and Conlon D (2017) Beyond privatization: Bureaucratization and the spatialities of immigration detention expansion. *Territory, Politics, Governance* 5(3):252–268

Jones K T and Popke E J (2010) Re-envisioning the city: Lefebvre, HOPE VI, and the neoliberalization of urban space. *Urban Geography* 31(1):114–133

Katz C (2001) Vagabond capitalism and the necessity of social reproduction. *Antipode* 33(4):709–728

Lefebvre H (1991 [1974]) *The Production of Space* (trans D Nicholson-Smith). Oxford: Blackwell

Loyd J (2012) Race, capitalist crisis, and abolitionist organizing: An interview with Ruth Wilson Gilmore. In J Loyd, M Mitchelson and A Burridge (eds) *Beyond Walls and Cages: Prisons, Borders, and Global Crisis* (pp 42–54). Athens: University of Georgia Press

Loyd J, Mitchelson M and Burridge A (eds) (2012) *Beyond Walls and Cages: Prisons, Borders, and Global Crisis.* Athens: University of Georgia Press

Massey D (2005) *For Space.* Los Angeles: Sage

McKittrick K (2012) On plantations, prisons, and a black sense of place. *Social and Cultural Geography* 12(8):947–963

Moran D, Gill N and Conlon D (eds) (2013) *Carceral Spaces: Mobility and Agency in Imprisonment and Migrant Detention.* Farnham: Ashgate

Peck J and Theodore N (2008) Carceral Chicago: Making the ex-offender employability crisis. *International Journal of Urban and Regional Research* 32(2):251–281

Shabazz R (2015) *Spatializing Blackness: Architectures of Confinement and Black Masculinity in Chicago.* Urbana: University of Illinois Press

Wilson D (2007) *Cities and Race: America's New Black Ghetto.* New York: Routledge

Racial Banishment

Ananya Roy

Luskin School of Public Affairs, University of California, Los Angeles, Los Angeles, CA, USA;
ananya@luskin.ucla.edu

Radical geography is replete with the lexicon of displacement. The conceptual frameworks of displacement, such as gentrification or revanchist urbanism or eviction, have foregrounded the violence of urban transformation. Yet, these frameworks are limited in their capacity to address two key aspects of displacement: the role of the state and the centrality of race. I thus propose a new concept, racial banishment, which emphasises state-instituted violence against racialised bodies and communities. While applicable to the United States, it is a generalisable concept relevant to many other contexts. Banishment is entangled with processes of regulation, segregation and expropriation and it is embedded in the legal geographies of settler-colonialism and racial separation. It often entails "civil death" (Kingston 2005) and indeed even social death. Banishment shifts our attention from displacement to dispossession, especially the dispossession of personhood which underpins racial capitalism.

The present historical conjuncture of urbanism in the United States makes evident the intricate and interlocked processes of racial banishment. First, in metropolitan regions such as the San Francisco Bay Area, black and brown communities are being pushed out of urban cores and relegated to the far margins of urban life. While often described as the suburbanisation of poverty, this is more appropriately understood as "residential resegregation" (Samara 2016). This peripheralisation goes hand in hand with the disproportionate concentration and containment of racialised bodies in urban spaces of impoverishment and surveillance such as Skid Row in Los Angeles. It is also accompanied by the "policing of integration" (Hayat 2016) including systematic efforts by cities to block black and brown residents from settling in suburban locations.

Second, such forms of segregation rest not only on market-driven displacement but also on the public means of criminalisation, specifically what Beckett and Herbert (2010:1) have pinpointed as banishment or "legally imposed spatial exclusion". From civil gang injunctions to sit-lie laws, these forms of banishment target and expel bodies marked as dangerous and disorderly. These bodies are disproportionately black, brown, and poor. Take, for example, crime free leases and nuisance ordinances which serve as the pretext for the evictions of poor women of colour often through the designation of calls to the police as evidence of criminal activity. These are instantiations not only of civil death but also of social death and at times of literal death. As Kurwa (2018) argues, what is at work is the "weaponisation" of municipal ordinances. Such ordinances are the latest iteration of older practices of segregation and quarantine, for example, racially restrictive

Keywords in Radical Geography: Antipode at 50, First Edition. Edited by the *Antipode* Editorial Collective.
© 2019 The Authors/Antipode Foundation Ltd. Published 2019 by John Wiley & Sons Ltd.

covenants and redlining. Their proliferation indicates that banishment is not the movement of racialised bodies from one place to another or what we might call displacement. It is expulsion from everywhere. It is the concerted effort of city governments to block residence, often through police power (Hayat 2016).

Third, racial banishment must be conceptualised as a project of the state. Beckett and Herbert (2010:6) highlight the "central role of the state's coercive power in the exercise of this form of spatial segregation". I situate banishment in what Alexander (2010) has called "the new Jim Crow", the perverse investment in state-institutionalised human caging. Carcerality is not a side-show to racial capitalism; it is a necessary logic. In her path-breaking book on carceral geographies, Gilmore (2007:28) argues that "resolutions of surplus land, capital, labor, and state capacity congealed into prisons". I view racial banishment as the territorial proliferation of such prison logics, manifested in geographies of forced mobility and illegalised presence that stretch far beyond the prison but that are inevitably refracted through the institutions of mass incarceration. I also understand racial banishment to be an instantiation of what McKittrick (2011:951) has designated as "urbicide" or "wilful place annihilation", specifically the "ongoing destruction of a black sense of place in the Americas". The banishment of black, brown, and poor bodies marks not just the disappearance of these residents from urban cores but also the loss of communities and the places and histories they have created.

The study of racial banishment has important implications for radical geography. I have already noted how the concept requires a reckoning with forms of penality and carcerality that are often kept separate from the analysis of capital accumulation and social inequality. Historically, banishment has been a form of punishment that imposes exile, often from the demarcated territory of a city or nation (Alloy 2002; Bleichmar 1999; Borrelli 2003). Implicated in banishment are notions of security and sovereignty. Put another way, banishment as exile is dispossession and in turn such dispossession secures sovereign possession. I interpret racial banishment as the necessary counterpart to what Lipsitz (1998) has called "the possessive investment in whiteness". As banishment relies on the law, so does whiteness as possession. Goldstein (2008:836) thus notes that it is legal reason that has built "proprietary regimes", underpinning settlement as "an entitled and possessive relation to place" and casting Indigenous populations as "supposedly unsettled". Possession, then, Nichols (in Goldstein 2017) argues, must be understood not as preceding dispossession but rather as its effect. "Differential racialisation", Goldstein (2017) emphasises, is necessary for this colonial capacity to possess. These are vitally important insights for radical geography and its interest in the histories of accumulation by dispossession.

Racial banishment also has a distinctive temporality. If it portends death, then following Gilmore (2007:28), such death is inevitably "premature". If it portends punishment, then such punishment is prefigurative, generating criminalisation rather than responding to established crime. These temporalities are constitutive of secure and sovereign territory, of the possession of land and personhood through the dispossession of racial outsiders. As Mitchell (2009:239) argues, the forceful and justifiable removal of individuals and populations from "commonly held spaces and resources" is a "contemporary liberal form of sovereign dispossession" and rests on the designation, in advance, of those who are risk failures. This, she notes, is the making of "pre-black futures".

Finally, the concept of racial banishment raises a set of epistemological and ontological challenges for radical geography. Following Woods (2002), we must ask, is there "life after death"? Woods (2002:62) reminds us that "predictions of the death of impoverished and actively marginalized racial and ethnic communities are premature". How then do we study racial banishment without undertaking what McKittrick (2011:955) has described as the "familiar analytical naturalization of violence, blackness, and death"? As the present historical conjuncture is marked by the revival of practices of banishment, so it is shaped by an exuberant proliferation of resistance. Poor people's movements in cities across the United States are not only fighting against evictions and high rent burdens but also creating new meanings of land, property, and community (Roy 2017). These in turn are bound up with new cartographic and ethnographic methodologies, such as those at the heart of the Anti-Eviction Mapping Project. Responding to Woods's call, this data visualisation and story-telling collective based in the San Francisco Bay Area undertakes "countermapping", such as "collective and public community power maps", to "render visible the landscapes, lives, and sites of resistance and dispossession elided in capitalist, colonial, and liberal topographies" (Maharawal and McElroy 2018:381).

These movements and collectives remind us that the antonym of racial banishment is neither immobility nor integration. It is a radical imagination that refuses the colourblindness of canonical knowledge. That canon includes the insistence on analysing geographies of late capitalism without a theory of racialised dispossession. It includes a vocabulary of neoliberalisation that elides the repeated and current renewal of colonial expropriation. It includes the talk of justice without considering the forms of social death that make entire communities disappear from urban life. The antonym of racial banishment is, as the black radical tradition insists, freedom. These "freedom dreams" (Kelley 2002) do not always make an appearance in the annals of radical geography. But they animate urban struggles around the world, from Los Angeles to Rio de Janeiro, from Cape Town to Chicago. They centre the unfinished work of black reconstruction and the persistent struggle of abolition democracy. A focus on racial banishment thus challenges radical geography to take up the analytical and political centrality of race and to reflect upon, and interrogate, its own possessive investment in whiteness.

References

Alexander M (2010) *The New Jim Crow: Mass Incarceration in the Age of Colorblindness*. New York: New Press

Alloy J S (2002) 158-County Banishment in Georgia: Constitutional implications under the state constitution and the federal right to travel. *Georgia Law Review* 36:1083–1108

Beckett K and Herbert S (2010) Penal boundaries: Banishment and the expansion of punishment. *Law and Social Inquiry* 35(1):1–38

Bleichmar J (1999) Deportation as punishment: A historical analysis of the British practice of banishment and its impact on modern constitutional law. *Georgetown Immigration Law Journal* 14:115–162

Borrelli M (2003) Banishment: The constitutional and public policy arguments against this revived ancient punishment. *Suffolk University Law Review* 36:469–486

Gilmore R W (2007) *Golden Gulag: Prisons, Surplus, Crisis, and Opposition in Globalizing California*. Berkeley: University of California Press

Goldstein A (2008) Where the nation takes place: Proprietary regimes, antistatism, and US settler colonialism. *South Atlantic Quarterly* 107(4):833–861

Goldstein A (2017) "The Ground Not Given: Colonial Dispositions of Land, Race, and Hunger." Paper presented to the "Race and Capitalism: Global Territories, Transnational Histories" conference, Institute on Inequality and Democracy, UCLA Luskin

Hayat N B (2016) Section 8 is the new n-word: Policing integration in the age of black mobility. *Journal of Law and Policy* 51:61–93

Kelley R D G (2002) *Freedom Dreams: The Black Radical Imagination*. Boston: Beacon

Kingston R (2005) The unmaking of citizens: Banishment and the modern citizenship regime in France. *Citizenship Studies* 9(1):23–40

Kurwa R (2018) Paper presented to the "Freedom is a Place: Land, Rent, and Housing" workshop, Institute on Inequality and Democracy, UCLA Luskin

Lipsitz G (1998) *The Possessive Investment in Whiteness: How White People Profit from Identity Politics*. Philadelphia: Temple University Press

Maharawal M and McElroy E (2018) The Anti-Eviction Mapping Project: Counter mapping and oral history toward Bay Area housing justice. *Annals of the American Association of Geographers* 108(2):380–389

McKittrick K (2011) On plantations, prisons, and a black sense of place. *Social and Cultural Geography* 12(8):947–963

Mitchell K (2009) Pre-black futures. *Antipode* 41(s1):239–261

Roy A (2017) Dis/possessive collectivism: On property and personhood at city's end. *Geoforum* 80:A1–A11

Samara T R (2016) *Race, Inequality, and the Resegregation of the Bay Area*. Oakland: Urban Habitat. http://urbanhabitat.org/sites/default/files/UH%20Policy%20Brief2016.pdf (last accessed 16 August 2018)

Woods C (2002) Life after death. *The Professional Geographer* 54(1):62–66

Radical Globalisation

Ipsita Chatterjee

Department of Geography and the Environment, University of North Texas, Denton, TX, USA;
ipsita.chatterjee@unt.edu

Globalisation is difficult to conceptualise because it has always existed in some form—people have always travelled, goods have moved, money has circulated, and information flowed. What is so different about the post-1980s that calls for a new concept? Sociologists call this new phase "the intensification of worldwide social relations" (Giddens 1990:64), "a compression of the world and the intensification of consciousness as a whole" (Robertson 1992:8). Economists call it the removal of barriers to free trade and a closer integration of national economies into an international economy (Bhagwati 2004:440; Stiglitz 2003:ix). While neoclassical economists like Bhagwati are ecstatic about the emancipatory potentials of free trade globalisation, development economists like Stiglitz emphasise the need to proceed with caution. Geographers like Harvey (1989:98) call it "time–space compression" where a transition to a post-Fordist flexible accumulation regime annihilates space and time to create conditions for a postmodern hyper-consumptive, hyper-speculative, and hyper-exploitative world. Anthropologists like Appadurai (2001) claim that it is not all homogenisation of tastes and preferences, but rather globalisation creates new scapes of ideas, images, ethnicity, and finance that alter the global map of neat container nations. Social movement theorists, on the other hand, claim that globalisation produces opportunities for ordinary people to organise (Anti-WTO, anti-corporate, Zapatismo) and resist various oppressions (Starr 2000). In other words, although the world has always been global, the current process of globalisation is more intense, almost all pervasive, and because of information technology revolutions and economic deregulation, it is as if globalisation has acquired a new aura.

In the early years of the 2000s, I was trying to work on an MPhil dissertation on Indian national identity, and it became impossible to read anything without stumbling into "globalisation". It is as if while I slept, aliens took over geography, disrupted the neat little nations and divided the world into North and South, or integrated it into a "village". The intellectual world I inhabited was simmering with ideas about how nations have become redundant and the world was a global village. What was I going to do about my now redundant thesis topic! The dissertation of course was completed after adding chapters on the leakiness and porosity of "nation", which could not, however, resolve some of the theoretical tensions that had seeped into concepts like "local"/"global". "Globalisation" had me completely, it brimmed over with concept-metaphors (ethnoscapes, mediascapes, time–space compression), touched various academic niches (culture, politics, economics, geography), was the subject of introspection from the left and the right. I was determined to delve into it. As I researched, it became clear that like any other concept, it was not innocent. Eisenstein (2005) and Fraser (1995), for example, were cautioning that the glitter of globalisation was seducing women into temporary work, call-centre jobs, and sweatshop slavery. Trading the paddy field for industrial homework in the city is an illusion of freedom, and treating this transition as emancipatory could put feminism in liaison with corporate capital. Harvey (2003:140) was demonstrating that globalisation is actually new style imperialism that operates through "accumulation by dispossession". Peet (2003) explained that while globalisation allowed the working class, women and Indigenous people to come together, neoliberal globalisation, on the other hand, is exploitative and imperialistic. Spivak (1999:364) advised that "to think globality is to think the politics of globality. How are the loose outlines of popular politics inscribed?" The "globe girdling" (Spivak 1999:390) politics is loose indeed, because while it may strengthen activism against race-gender-class exploitation, it may become the conduit for expatriate funds that foment religious fundamentalism in the guise of multiculturalism, or pitch First World labour against Third World labour. How to think of a process that is not new, but is a bit different nonetheless (World Bank and IMF associated structural adjustment), impacts many aspects of living and academic/non-academic thinking about living, a process that most contemporary commentators comment upon, but don't agree upon, something they say can be at once good and bad. Therefore, how to think the politics of globality so that it is anti-imperialistic, anti-elitist, anti-racist, anti-sexist, anti-homophobic, pro-poor? How to think that politics and express it conceptually? For it is in this thought-concept, in the learning to learn (Spivak 1999) that "globalisation" can be radicalised.

In tightening our thinking about our "globe girdling" politics, we can position ourselves in critique of capital revealing how it works to exploit labour, racial minorities, indigenous minorities, women, the environment, and other marginal groups. In revealing these many exploitations, therefore, we align with the oppressed and hence transform a duplicitous concept into one that is potent with possibilities for social and environmental justice. In thinking about a progressive politics of globalisation, we can thus create a thought-concept that is implicitly and explicitly anti-exploitative; such a concept is radical globalisation.

Unfortunately, much of globalisation literature is not radical, it does not think about its politics, or perhaps it is political in how it thinks about it, and therefore presents a dichotomous view of the world into modern (global, good!) and tradition (local, bad!). Friedman's (1989) popular *The Lexus and the Olive Tree* is an example—the new global world order seen as a clash between the "western" forces of free market capitalism (Lexus) and forces that oppose commodity culture (olive tree). In Friedman's dual view of the globalising world, power to consume is the true representative of modernity, and the spread of fast food and information technology is equivalent to emancipation. Any opposition to this version of globalisation is seen as suspicious, un-modern, anti-enlightenment, parochial, and provincial. In a similar dual cast of character, but through a more critical lens, Barber (2000) calls globalisation a clash between McWorld and Jihad forces. McWorld is an outward acting expansive force of corporate culture, while "Jihad" is the inward acting force of tribalism, both can coexist within the same society. Bhagwati (2002) laments that globalisation has become the site for clash between globalisation and anti-globalisation protests. Critiquing the post-war communists, revolutionaries, and socialists, who according to Bhagwati understand globalisation as the new phase of capitalism and hence direct the trauma of their "vanished dream" (the collapse of the Soviet Union) towards it. This dejected generation has raised a new army of young detractors who now demonstrate on the streets of Seattle, Prague, Genoa. Capitalism steered by corporations has, according to Bhagwati (2002), the power to do tremendous good in their community, just as non-profit organisations do to theirs. China and India, argue Bhagwati and Srinivasan (2002), have achieved faster economic growth on account of trade liberalisation that has infused entrepreneurialism and increased consumption. Huntington's (2000) almost paranoid portrayal of the "West" under imminent attack by the "Rest" veers on the same script of western dominance. In a fast globalising world, the threats are no longer national but civilisational. Globalisation is inherently bellicose, and the "West" must be astute in exploiting the identity politics of the "Rest". The politics of thinking about globality in these approaches reaffirm orientalist imaginations of the Middle Eastern cauldron painted simplistically in terms of a "Judeo-Christian West" and the "Islamic world". Geo-economically, they give hope to post-colonial countries that global economic integration can be a short cut to super-growth. Because the politics of thinking globality is racist/capitalist, the concepts that emerge from this politics preserve the status quo (West/free-trade/good, Rest/absence of free-trade/not-so-good; Christian/good/civilised, Muslim/terrorist/uncivilised).

As opposed to the dualistic models, Appaduarai (2001) understands globalisation as an assemblage of scapes: ethnoscape, technoscape, ideoscape, mediascape, and finacescapes. This fluid assemblage of culture, economy and politics cannot be easily dissected into neat analytical boxes. Appadurai's disruption of dualism is useful as it allows concepts to reflect the messiness of globalisation, which is simultaneously economic, political, and cultural. However, his "scapes" say little about power, oppression, or exploitation, they risk complicity with capital, fundamentalism, patriarchy, and heteronormativity. If we are to invest in a progressive politics of globalisation, we must rethink the "scapes" and push them further so that they can help us learn radical globalisation better.

Many on the left understand globalisation as the new free market policy regime, or as neoliberalism, that is inherently pro-market, hence pro-capital, and therefore anti-people and anti-labour (Harvey 2003; Klein 2007; Peet 2003; Smith 2005). The shift towards trade and financial liberalisation, encouragement of foreign direct investment (FDI), reducing public expenditures have created a corporate-led globalisation. This corporate-led globalisation, or neoliberal globalisation, allows capital to globalise freely, increasing the power of those classes, groups, ethnicities, castes, and communities that have traditionally controlled capital. While Peet calls it neoliberal hegemony, Harvey calls it "new imperialism" achieved through "accumulation by dispossession" (2003:137). "Neoliberalism", "accumulation by dispossession", "new imperialism", "neoliberal hegemony" allow us to imagine a politics of globality that is anti-capitalist, and therefore they are potent with radical possibilities, however they need to explicitly connect capital's alliance with patriarchy, racism, cultural imperialism, and heteronormativity in order to comprehensively radicalise globalisation.

Since my MPhil days, my work has been a conceptual struggle to pry, wrest, and usurp "globalisation" from racist geopolitical theories and intellectual free marketers that say so much about "clash of civilisations", the "war on terror", the "west under attack", "spread of democracy", altruism of corporations, and yet say nothing, for example, about occupation of Palestine as a global imperial project, say nothing about Islamophobic hysteria against Muslims and other immigrants, misadventures of American neocolonialism and production of terrorism, say nothing about the exploitation of labour in free zones. How to come towards a radical globalisation so as to talk about all these without conceptually prioritising one or the other? How to prevent globalisation's imprisonment into economic (e.g. neoliberal), political (e.g. war on terror), cultural (e.g. Islamophobia, anti-immigrant hysteria), and spatial (global, local; North, South) containers so that we can interrogate exploitative global reality, which is simultaneously neoliberal, neo-fundamentalist, neo-racist, and neo-sexist? A radical onslaught that aligns with the oppressed cannot challenge one axis of exploitation feeling vaguely sorry for all the other forms. And, if globalisation as a process is truly all-encompassing, more trenchant than ever before, it must be possible to conceptualise its exploitative impacts in all-encompassing, but non-totalising ways. Conceptual radicalisation is praxis, because theory is not a second step *after* material processes have happened, mental constructs and reality are simultaneously material, and hence radical globalisation *is* praxis and can be truly transformative. Marx's (1976) all-encompassing way was to talk about globalisation as capitalism, and synthesise exploitation of labour in its very kernel, the commodity. How can globalisation be conceptualised radically so that we understand exploitation in its entirety? How to synthesise global exploitation in non-reductionist, contextual and yet all-encompassing ways so as to comprehensively dismantle the status quo?

In thinking the politics of globality, and (un)learning to learn globalisation, I have found Marx's dialectical approach useful. Because the dialectical approach conceives of reality in its full range of material interactions (Harvey 1996; Marx 1993; Ollman 2003), it allows us to conceptualise labour, women, blackness, indignity, coloniality as imbricated. Because dialectics does not compartmentalise reality, and abstractions are always concrete, there is no "culture" or "economy" to begin our conversation from, nor is there a "global" and separate "local", nor McWorlds and Jihads, or the West and the Rest. A dialectical approach to globalisation can help us think about a politics of globality that is anti-imperialistic, anti-elitist, anti-racist, anti-sexist, anti-homophobic, pro-poor, and therefore (un) produce globalisation as radical globalisation. After all, the alt-right movement is all-pervasive—it ingeniously weaves an anti-immigrant stance with anti-Muslim hysteria, with anti-women fervour to emerge into pan-European and pan-American movements like Young European Alliance for Hope (YEAH), Patriotic Europeans Against the Islamization of the West (PEGIDA), and Youth for Western Civilization (YWC) (Hafez 2014; Kallis 2015). Based on a globalising identity politics, they materialise exploitation by simultaneously attacking the migrant, the labourer, the Muslim, the Mexican, and the woman all at once. Countering this requires a radical (dialectical) stance that exhumes globalisation from its bourgeois banality, extricates it from its conceptual sterility, and ruptures dualities. Radical globalisation means transcending our conceptual container-boxes to speak of exploitative reality simultaneously as markets, mergers, financial crisis, drone strikes, displaced slums, occupied Palestine, trafficked bodies, deposed migrants, and stigmatised Muslims. Culture and economy, politics and society, class and gender, religion and race, base and superstructure co-constitute the "whole" of exploitation. Our radical concept must be trenchant in de-stabilising exploitation as a whole, only then can globalisation be emancipatory.

References

Appadurai A (2001) Disjuncture and difference in the global cultural economy. In M Featherstone (ed) *Global Culture* (pp 295–310). London: Sage

Barber B R (2000) Jihad vs. McWorld. In P O'Meara, H D Mehlinger and M Krain (eds) *Globalization and the Challenges of a New Century* (pp 23–33). Bloomington: Indiana University Press

Bhagwati J (2002) Coping with anti-globalization: A trilogy of discontents. *Foreign Affairs* January/February

Bhagwati J (2004) Anti-globalization: Why? *Journal of Policy Modeling* 26(4):439–463

Bhagvati J and Srinivasan T N (2002) Trade and poverty in the poorer countries. *American Economic Review* 92(2):180–183

Eisenstein H (2005) A dangerous liaison? Feminism and corporate globalization. *Science and Society* 69(3):487–518

Fraser N (1995) From redistribution to recognition? Dilemmas of justice in a "post-socialist" age. *New Left Review* 212:68–93

Friedman T L (1989) *The Lexus and the Olive Tree*. New York: Farrar, Straus and Giroux

Giddens A (1990) *The Consequences of Modernity*. Cambridge: Polity

Hafez F (2014) Shifting borders: Islamophobia as common ground for building pan-European right-wing unity. *Patterns of Prejudice* 48(5):479–499

Harvey D (1989) *The Condition of Postmodernity*. Oxford: Blackwell

Harvey D (1996) *Justice, Nature, and the Geography of Difference*. Oxford: Blackwell

Harvey D (2003) *The New Imperialism*. New York: Oxford University Press

Huntington S P (2000) The clash of civilizations? In P O'Meara, H D Mehlinger and M Krain (eds) *Globalization and the Challenges of a New Century* (pp 3–23). Bloomington: Indiana University Press

Kallis A (2015) Islamophobia in Europe: The radical right and the mainstream. *Insight Turkey* 17(4):27–37

Klein N (2007) *The Shock Doctrine*. New York: Metropolitan Books

Marx K (1976 [1867]) *Capital, Volume 1*. London: Pelican

Marx K (1993 [1858]) *Grundrisse*. London: Penguin

Ollman B (2003) *Dance of the Dialectic: Steps in Marx's Method*. Urbana: University of Illinois Press

Peet R (2003) *Unholy Trinity*. New York: Zed Books

Robertson R (1992) *Globalization*. London: Sage

Starr A (2000) *Naming the Enemy*. New York: Zed Books

Stiglitz J E (2003) *Globalization and Its Discontents*. London: Norton

Smith N (2005) The endgame of globalization. *Political Geography* 25(1):1–14

Spivak G C (1999) *A Critique of Postcolonial Reason*. Cambridge: Harvard University Press

Radical Vulnerability

Richa Nagar

Department of Gender, Women, and Sexuality Studies, University of Minnesota, Minneapolis, MN, USA;
nagar@umn.edu

Roozbeh Shirazi

Department of Organizational Leadership, Policy, and Development, University of Minnesota, Minneapolis, MN, USA;
shir0035@umn.edu

Hungering Collectively for Justice

[The movement] … has been teaching us that transformation does not happen simply by bringing into the same space bodies who desire change. The real movement begins when the bodies enter into intense embodied journeys in order to wrestle with incommensurable gaps in lived experiences and in theoretical and political positions through radical vulnerability. (Sangtin Kisan Mazdoor Sangathan [SKMS], in Nagar in journeys with SKMS and Parakh 2019)

When 35 saathis [comrades] of SKMS, most of them dalit kisans [farmers] and maz-doors [laborers], first sat down to engage collectively about the struggles of transgen-der people with Meera Sanghamitra, an activist whom we had invited to the Sangathan's three-day long meeting, several people in our group had never previously uttered the word, "transgender". There were long silences as well as an open sharing of dangerous stereotypes and stories of encounters defined by fear, confusion, and mistrust. By the end of the meeting, however, everyone was identifying not only deep connections, but also the impossibility of separating the threads that entangle the struggles of transgender communities in India generally, and those of dalit kisans and mazdoors in Uttar Pradesh, more specifically. Saathis reflected on the ways in which our families, societies, and religions often refuse to understand or respect the truths of our bodies and desires, push us into suffocating relationships and institutions, encage us in everyday spaces where it becomes impossible to breathe fully with dignity, sometimes in ways where every dimension of a person's existence—from their right to livelihood, housing, and health care to basic expressions of the body and soul—are brutally crushed. Saathis noted that the only way to fight against this injustice is to stand with one another while also being forever attentive to the convergences and divergences in each of our locations, paths, and journeys. (Notes from SKMS meeting in Lucknow, 2–5 July 2018)

We write at a time when there is an onslaught of state violence, right-wing extremism, financial austerity measures, neoliberal development, racialised polic-ing and surveillance against, among others, the poor, undocumented, minori-tised, and displaced—people wrestling with the violence of precarity in its many forms. Though these conditions are by no means new, what makes them feel unprecedented are the ways in which these forms of violence appear to operate in efficient transnational concert with each other. At the same time, these violent times have birthed multiple movements and lines of dissent, each with distinctive dialects and tactics of resistance. These dissenting modes of speaking, theorising, and undoing of militarised racism, capitalism, heteropatriarchy, and other forms of oppression, have contributed enormously towards sharpening broader understandings of how the injuries inflicted by these systems are differentially circulated and experienced. As well, these modes of resistance have highlighted where and how these systemic powers may overlap and intertwine with one another. Not surprisingly, these same processes have also catalysed new roles and identities in and around struggles for justice—roles and identities that inevitably constitute epistemic and ontological divisions and hierarchies, including through such conceptual categories as "community", "ally", and "persons of color" (Shirazi 2017).

Even as we are often swept along with contemporary political currents, we are faced with the difficult work of grappling with, and translating (imperfectly) the place-specific vernaculars for analysing and undoing oppression. The second epigraph at the beginning of this essay suggests, however, that these vernaculars also set forth complex politics of incommensurability that tend to fix or ontologise difference along such lines as race, ethnicity, caste, class, gender, sexuality, and rural versus urban. Yet, the multiple border crossings that define human lives, histories, knowledges, and bodies, defy neat typologies of difference; as such, we must confront the peculiar dilemmas of being committed to deepening just possibilities of being human while also sorting through ontological and epistemological frameworks or practices that may be incommensurable with those very goals.

It is from this specific dislocation that we ask, what ways of being otherwise are possible? What does it mean to consciously forge a collectivity by becoming a "blended but fractured we" across multiple axes of power and difference (Sangtin Writers and Nagar 2006)? What possibilities might we unlock by travelling together to forge a shared vision of justice and what kind of sacrifices might such co-travelling entail? What kind of collectivity is worth building, at what cost, and to whom in the collective? It is on such turbulent terrain—deeply grounded in, and responsive to, space, place, and time—that radical vulnerability works as the ethical basis of building collectivity *as* situated solidarity, one that actively embraces the ongoing and uneven risks and potentials of forging shared visions and journeys in ever-unfolding political struggle, that is inevitably marked by both critical convergences and divergences in the positions and aspirations of those who commit to walking together.

In articulating and advocating for such collectivity through radical vulnerability, however, we do not seek to excise or erase the divides that mark our bodies and movements as raced, gendered, casted, or classed. Far from reducing collectivity to sameness or simple consensus, our material, intellectual, and embodied differences inform the possibilities and challenges of forming such togetherness. The rips and fissures that we live and embody emerge from our inherited histories and geographies. Rather than being forgotten or vanquished, it is these fractures that ground the work of building and being otherwise in the profound inequality and injustices of our lived conditions: they are ever present reminders that risk is never fairly distributed, and that different people have different capacities for experimentation and making mistakes. At the same time, we insist on a need for an ontology of togetherness that dares to imagine past these divides, without ever dismissing them.

Entangled Concepts: Radical Vulnerability, Situated Solidarities, and Hungry Translations

The co-constitutive concepts of radical vulnerability, situated solidarities, and hungry translations are rooted in evolving journeys of saathis of Sangtin Kisan Mazdoor Sangathan or SKMS, a movement of 8000 small kisans and mazdoors in Sitapur District of Uttar Pradesh in India that emerged from the battles summarised in *Playing with Fire* (Sangtin Writers and Nagar 2006) and *Muddying the Waters* (Nagar 2014). Embraced as a "blended but fractured we", situated solidarity is a foundational tenet of SKMS. Arguing against the common sense that mastery of certain theories and methods can enable well-meaning researchers or practitioners to bring justice to those they pronounce to be the "marginalized", *Muddying the Waters* translates the methodology of Sangtin Writers and SKMS as "radical vulnerability"—an essential requirement for journeying together as co-authors in a political landscape disfigured by non-stop epistemic violence. This co-authorship hinges on sharing authority through a situated solidarity that derives its meanings from its groundedness in place, time, and struggle and that continuously strives for radical love through the dissolution of egos.

Hungry Translations (Nagar 2018; Nagar in journeys with SKMS and Parakh 2019) further extends this concept of radically vulnerable co-authorship as a relation of translation—one that is always hungry for journeying together in search of justice while also recognising that there will always be faults in both the relation as well as in the search. Moreover, it is in the impossibility of arriving at completion and of knowing the destination with certitude that the hope for the journey resides. The possibility of justice, too, is contingent on the shared need to continue journeying together; for, the end of such need is the end of the ethical relation that is hungry translation. Hungry translation, then, is a forever evolving entanglement—a relation of radical vulnerability and openness, a politically aware submission to one another to become a "we" that also struggles to overcome intense mistrust, bitterness, hatred, or suspicion at times. As an embodied mode of co-living that co-constitutes its own aesthetics and politics, it is as frequently marked by tears of joy and frustration as it is by silences emanating from trust and distrust. In this co-living or situated solidarity, spiritual activism becomes one with political activism, and theory and epistemology become inseparable from pedagogy (see Nagar in journeys with SKMS and Parakh 2019).

When Does Vulnerability Become Radical?

Radical vulnerability cannot be an individual choice. Vulnerability becomes radical only when it becomes a collectively embraced mode in search of the shared creative power it has the potential to enable. This collective creativity—which is never fully attained and always in progress—emerges slowly as we learn to let go of the threads of stories that we have inherited and that have made us, and as those threads get entangled with the words and worlds of our saathis or co-travellers. In such letting go, narratives about our childhoods and our ancestors, our relationships and our losses, our fears and our dreams become collectively owned narratives that enable a hitherto unknown awareness of spatialities and temporalities.

Far from being marked by endless generosity and openness, then, radical vulnerability is about grappling with the simultaneous co-existence of inhumanity and humanity within each one of us. It is about reminding ourselves and one another of the violent histories and geographies that we inherit and embody despite our desires to disown them. Confronting these realities and contradictions implies working through the specific moments in which resistance to collectivity emerges as well as the significant lessons that those moments may carry about context-specific meanings of ethics, aesthetics, and justice. Such grappling with the simultaneous presence of the protagonist and antagonist in all of us is necessary for just translations—that is, for modes of retelling that agitate against the structures and epistemes invested in guarding the binary of "the emancipators" and "those in need of emancipation" (see Bargi 2014).

This kind of hunger in a translation cannot be demanded or achieved through mechanical protocols. It can only emanate from an intense relationality and co-ownership of dreams among those who occupy different locations in predominant epistemic and sociopolitical hierarchies; it involves cultivating what María José Méndez (2017) poetically terms as "a disposition to listening to incommensurable worlds where rivers tell stories and call upon us". Such relationality inspires a situated solidarity where our minds and bodies, our hearts and tongues can always be open to diverse ways of knowing and co-creating in a world where the humans alongside the land, forests, rivers, and non-human animals can become our ever present teachers and interlocutors (Barla and Bettelyoun 2017). An intentional and shared search to continue learning how to breathe and flow relationally, while also making space for refusals that may resist that flow, is what makes a translation or retelling hungry. Such hungry translations can dare to fight and interrupt projects that seek to educate, modernise, or emancipate certain categories of bodies—including "the rural poor" and "women"—through globalised vernaculars that reproduce a landscape of intellectual enfranchisement and dispossession.

Radical Vulnerability as Surrender

To trust, to put faith in something or someone then, is a form of surrender: a surrender similar to love. Surrender does not mean loss of power, or acquiescence to being dominated. Rather, it recognises that the logics of our contemporary political moment—notably, living through self-interest and self-protection—which in turn rely on the notion of a self-reliant subjectivity, cannot take us very far within a framework of justice. Surrendering, then, represents an affirmation of the generative political possibilities of vulnerability, loss, and the unknown itself. Radical vulnerability embraces these risks and possibilities. It recognises that there is something to be gained through our shared refusals and through voluntarily sur-rendering modes of being that diminish us or that imprison us in narrowly defined identity or community that fixes us in (already known) difference.

Radical vulnerability, then, seeks to reimagine the temporalities and meanings of knowledge-making partnerships by surrendering to a politics of co-travelling and co-authorship, politics that are accompanied by difficult refusals. Such radical vulnerability cannot be an individual pursuit. Indeed, it is meaningless without collectivity. Yet, this collectivity does not seek to erase the singular by subsuming everything in a larger whole; rather, the singular relearns to breathe and grow dif-ferently in the plural. While this praxis may be reminiscent of Butler's (2004) idea of vulnerability as a necessary condition for an ethical relationship, it builds from lessons learned in and through the ongoing work of fiercely alive collectives that define the conditions for solidarities across borders. This praxis of radical vulnera-bility opens up the possibility of a togetherness "without guarantees" (Hall 1996:45): it does not seek to know prior to the journey where the shared paths will lead us but it commits to walking together with the co-travellers over the long haul in the struggles and dreams that we all have chosen to weave, unweave, and reweave together. In foregoing the very category of a "subject" in the form of a singular, autonomous self, and in actively co-constituting an inter-subjective space, such a praxis does not look for corporeal or moral protection of one individual from another. It recognises that each of us is limited by our loca-tions and languages, by our pasts and presents, by our desires and complicities (Sangtin Kisan Mazdoor Sangathan 2018). Within this contingent space of inter-subjectivity, practicing radical vulnerability entails an ontological and epistemolog-ical shift. One in which the willing surrender of the subject—or the willingness to reimagine subjectivity as/with/through collectivity—is the precursor to embarking on a journey of ever-unfolding collectivity. As a mode of togetherness, it at once acknowledges and moves beyond our available memberships and ascribed identi-ties to bring new possibilities into view.

Agitating "the Social" and "the Political"

A relationality embedded in radical vulnerability strives to internalise that our self is intensely co-constituted and entangled with the other. Whatever we learn, whatever we come to be, becomes deeply contingent on what each one of us is prepared to give to the collective journey that seeks to unite the I/we with the you/they. By disrupting such categories as writer, educator, activist, artist, farmer, and labourer, these ever evolving solidarities enable the formation of multiple interpretive communities so that people in the so-called margins cannot become raw materials or suppliers of stories. Each co-traveller wrestles with unlearning, relearning, and negotiating which stories can cross which borders, in which form, when and with what intentionality. Such co-authoring moves across contexts and socio-political idioms and vocabularies; across forms and genres; and it recognises how different forms of labour that constitute protest must shift with every staging according to context and audience, each time pushing for new forms of co-con-stitutive (re)theorising, (re)strategising, (re)calling, and (re)telling.

In this translational praxis, the meanings of the political cannot be learned in a straightforward journey. Rather, this praxis is a complex choreography between, among, and across multiply located discursive sites. For the one who participates in this praxis as a writer, a responsible grappling with this dance asks that the page become a stage where the saathis or co-travellers located in each site can come alive as co-critics and co-performers who co-own authority, insights, and courage. Together, we shape and refine the narratives that emerge in the dance. Theatrical and political positions as well as storylines and rules of narration must co-evolve as every encounter or movement in one site leads to a new round of revisions in the next retelling. This storytelling—where words, effects, and affects are always in continuous and critical creative motion, and purposeful and collectively owned revision—fundamentally complicates our received ideas about "the expert". It can unsettle the cultural and material economies of intellectual and material enfranchisement and disenfranchisement that are embedded or entangled in dominant conceptualisations of expertise and the accompanying authori-(si)ng practices.

For those of us located in the academy, a desire to partake in, and contribute to, such hungry translations requires that we do not merely travel to the Othered worlds that form the basis of our knowledge claims. Rather this desire comes with the responsibility to embed ourselves in the relationships and hopes that form our entwined worlds, so that that which has been Othered in dominant imaginaries may emerge differently in our consciousness and conscience and in our ways of being. By enabling a continuous and deeply difficult process of unlearning and relearning, such a process can become a politics without guarantees that is dedicated to continuous becoming. This openness to continuous unfolding of politics is key in disrupting pre-formed proposals about "the social" as well as formulations that compartmentalise research, writing, art, engagement, and activism.

Reminders/Caveats

For those who remain sceptical of the idea of radical vulnerability because we are all vulnerable in unequal ways, or because the institutions we breathe in are hostile or dismissive of such a mode of being, there are several points to consider: to begin with, such scepticism is essential for radical vulnerability and hungry translations to continue evolving. Relatedly, if hungry translation is the only relation through which one can search for justice with those one has encountered as one's Others—and if that journey must continue without arriving—then it is solely by offering pieces of oneself that one can hope to enact hungry translation. Moreover, it is the continuation of the hunger in the relationship that gives each co-traveller the courage to continue offering new layers of themselves to one another, without hoping for (re)solutions.

Second, in this continuous cycle, where radical vulnerability and hungry translation (re)make one another in search for justice, the terms of the relationship—or the nature of offerings—cannot be demanded in accordance with any predetermined formula or recipe: the journey of this entanglement must remain an organic process that unfolds serendipitously to realise a shared goal; one in which not everyone may choose to remain together until the next halt or turn. Third, even though such a journey asks for sacrifices, the stories of those who have walked together in this mode attest to one common truth: that the joys and lessons of moving and creating together in a radically vulnerable mode are often deeper than the sacrifices made by individual travellers. Last but not least, there will never be a field manual to "correctly" undertake this journey, but only an invitation to step into a realm of ontological and epistemological possibilities and risks. One always must be prepared for the impossibility of achieving radical vulnerability as well as hungry translation. After all, the labour and poetics of forging togetherness across difference through radical vulnerability, situated solidarities, and hungry translation can only realise their transformative potential as politics without guarantees.

Acknowledgements

This essay draws centrally on the discussion in *Hungry Translations: Relearning the World Through Radical Vulnerability* (Nagar in journeys with SKMS and Parakh 2019). We are grateful to David Faust and Vinay Gidwani for their comments on earlier versions of this essay.

References

Bargi D (2014) The dilemma of an upwardly mobile, English speaking, Bengali Dalit woman. *Roundtable India* 20 January. https://roundtableindia.co.in/index.php?option=c om_content&view=article&id=7175:the-dilemma-of-an-upwardly-mobile-english-spea king-bengali-dalit-woman&catid=119:feature&Itemid=132 (last accessed 16 May 2018)

Barla D and Bettelyoun C F (2017) "Indigeneity, Colonialisms, and Justice: A Conversation between Dayamani Barla and Cante Suta-Francis Bettelyoun." In Tarun Kumar's film *Our Land Won't Become Your Downtown! Conversations Across Indigeneity*. https://www.youtu be.com/watch?v=ZcLdYDfMMWk&feature=youtu.be (last accessed 20 August 2018)

Butler J (2004) *Precarious Life: The Powers of Mourning and Violence*. New York: Verso

Hall S (1996) The problem of ideology: Marxism without guarantees. In D Morley and K-H Chen (eds) *Stuart Hall: Critical Dialogues in Cultural Studies* (pp 25–46). New York: Routledge

Méndez M J (2018) "The river told me": Rethinking intersectionality from the world of Berta Cáceres. *Capitalism Nature Socialism* 29(1):7–24

Nagar R (2014) *Muddying the Waters: Coauthoring Feminisms across Scholarship and Activism*. Champaign: University of Illinois Press

Nagar R (2018) Hungry translations: The world through radical vulnerability. *Antipode* https://doi.org/10.1111/anti.12399

Nagar R in journeys with SKMS and Parakh Theatre (2019) *Hungry Translations: Relearning the World Through Radical Vulnerability*. Champaign: University of Illinois Press

Sangtin Kisan Mazdoor Sangathan (2018) Ek aur Manthan: Jati, Dharm, aur Pahchan ki Rajneeti. *Hamara Safar* 13(1):7–10

Sangtin Writers and Nagar R (2006) *Playing with Fire: Feminist Thought and Activism through Seven Lives in India*. New Delhi: Zubaan Books / Minneapolis: University of Minnesota Press

Shirazi R (2017) How much of this is new? Thoughts on how we got here, solidarity, and research in the current moment. *Anthropology and Education Quarterly* 48(4):354–361

Rift

Katherine McKittrick

Department of Gender Studies, Queen's University, Kingston, ON, Canada;
k.mckittrick@queensu.ca

The intent of this keyword was, originally, to read the "rift" against itself and displace its geological meaning (the permanency of splits and large-scale faulting in large rock plates, for example), and instead think about how the term signals conceptual and theoretical openings in relation to race, geography, liberation. The idea was, more specifically, to focus on conceptual and theoretical openings (cracks, fissures, splits) and show how black intellectual thought offers new ways of theorising oppression, humanity, and resistance. Much of my work is on black methodologies and the ways black intellectuals—scholars and non-scholars—not only challenge disciplinary silos but produce work that always expresses some kind of radical interdisciplinarity. This interdisciplinarity work understands liberation as a struggle that is ongoing but never resolved; this work understands that threading together multiple sources is necessary to reimagining our worlds; this work signals the ways in which black people have always shared ideas with each other and with non-black people, too, uneasily and generously, in order to struggle against prevailing systems of knowledge that are beholden to and profit from racial and sexual violence. In this sense, the rift is a conceptual opening that recognises subversion and disruption as a form of black thought; this thought seeks out and engenders the intellectual praxis of living this world and seeking liberation by challenging prevailing systems of knowledge that devalue black people, their ideas, and their worlds. Seeking liberation, in a world that does not and cannot imagine or support black liberation, disrupts (cracks, pulls apart) oppressive infrastructures and architectures and archives and social systems. Seeking liberation allows us to gather and learn about ways of liberation that are already here (Gilmore 2017; Simone 2019). Black intellectual thought cannot be contained by the normative intellectual structures that shape our worlds. Normative intellectual structures are fields of study that are imagined and produced as totally discreet. Normative intellectual structures understand geography and literature and physics and geology to be absolutely different. Of course, there is overlap. We know there is overlap. But noticing or building overlap is discouraged.

Even so, even though we all know there is overlap, those who research and write in black studies insist on unsettling how we know and what we know by reorganising and reinventing ways of knowing. This interdisciplinary research is not just about thinking across knowledge systems and fields; it is thinking across knowledge systems and fields with the purpose of recalibrating who and what we are and what they think we are. In the field of black studies this is urgent precisely because disciplinary thinking—the fields of study—are animated by colonial-plantocratic race thinking. I will not go into a long explanation here, but the ongoing refrain that "we" (black people) "were never supposed to be here" (the academy) summarises not only how racism and displacement frames academic life, but also how the figure of the black stands in as the oppressed and violated and empty object of study across a range of research inquiries. We can look to the work of Nick Mitchell to wade our way through the thorny racial-political economies of academic infrastructures (Mitchell forthcoming).

For those of us working in black studies and who engage the discipline of geography, we think across fields of study while also recognising how the discipline of geography is empowered by ongoing practices of dispossession. There has been a lot of amazing work written on this (Mahtani 2014; Peake and Schein 2000; Rose 1993). The short hand is that patriarchal, colonial, and imperial legacies continue to inform the discipline of geography; the theoretical and methodological purpose of the discipline is twinned with exploration and conquest and European masculinist ways of knowing. This kind of analytics is unkind to marginalised communities; the unkindness is demonstrated by and through hiring practices, admissions, student funding, mentorship, non-mentorship, failing students, overworking faculty of colour, delegitimising black students and their ideas publicly and privately, delegitimising black faculty and their ideas publicly and privately, teaching racist materials, fiercely defending key thinkers and key literatures, stacking editorial boards, and more. This is something we see and experience in other disciplines, but in geography it is always eerily clear and always sharply in focus, for the discipline—as a knowledge system—seems to know that it must replicate a fictive spatial-colonial-essence or die (cf. Katz 1996).

For those of us working in black studies and who engage the discipline of geography we seek to know and imagine the world without the mandate for conquest. The task is not to refuse geography but to see how, where, why, and when the production of space enables liberation. In this way, those of us working in black studies and who engage the discipline of geography look for, demand, and create, as LaToya Eaves notes, a geographic perspective that "disrupt[s] methodological normativities, challenge[s] geographic containment, and decenter[s] assumptions and pathologies" (2017:93). While the discipline of geography expresses and enacts both discursive and real racial violences, I do not want to cast it as solely an object of complaint. This is because we make geography what it is; the discipline is made and upheld by its practitioners. The discipline of geography is not an object; it is inhabited and energised and animated by its disciplinarians and all of us, globally, as we make sense of our various geographic worlds. If we make geography what it is—space, place, and the discipline—then it is alterable. If this is so, perhaps we can think not simply about geography oppressing us but rather how it is an oppressive system of knowledge that incites rebellion. Indeed, for me, the more interesting and urgent project is to think about what geography enables in terms of liberation.

When I sat down to write this paper, to think about how geography might learn something from the rifts black people make in prevailing systems of knowledge, I quickly realised I was, at least in part, betraying the intellectual project of black studies. Of course, yes, yes, geography can learn from black studies! Most non-black studies should engage black studies and unlearn racism and other forms of oppression and learn that the political work of the ex-slave archipelago is tied to the "individual and collective narrative arcs that persistently tend toward freedom ... [and] the radical process of finding anywhere—if not everywhere—in political practice and analytical habit, lived expressions (including opacities) of unbounded participatory openness" (Gilmore 2017:237). So, yes, geography can learn from black studies. But the project of black studies, as I see it, is one of relationality and conversation and uncomfortable debate; this means we cannot simply sit with and point to the violence of disciplinary thinking and the coloniality of geography. We must also engage the discipline, rigorously, so as to undo it and put it into conversation with other ways of knowing and, at the same time, notice the radical black reimagining and reinvention of geographic processes.

As I sat with rifts I became increasingly agitated because I kept thinking in metaphor. What I have learned from black studies is that thinking in metaphors is delimiting and constraining. I came to this understanding by studying Sylvia Wynter's bios-mythois in concert with the 1993 piece by Neil Smith and Cindi Katz, "Grounding Metaphor" (Smith and Katz 1993). The piece has allowed me to think about the interdisciplinary underpinnings of black studies beyond an additive model but also contains within it, I think, lessons for black studies. "Grounding Metaphor" thinks about how the language of space, inside and outside geography, must interlace the material *and* the metaphoric. What Smith and Katz ask us to think about are the ways different kinds and types of geographic terms —space, location, position, mapping, and so on—are often utilised without attending to the politics that underwrite these terms. Specifically, the material, concrete, and grounded work of physical-material space goes missing from the spatial concept itself. This reifies the "taken-for-grantedness" of space and positions it as an empty container, thus naturalising uneven geographies and their attendant social inequities. Leaning heavily on metaphoric concepts risks fixing social identities in place because it ostensibly puts forth a "floating world of ideas" that are simply hovering around us (Smith and Katz 1993:80). This kind of outlook actually removes social actors from the production of space. Leaning on metaphors—and here my struggle with rifts comes into sharp view—disregards how space, place, and rifts are socially produced. The metaphor I began with, the rift as "conceptual opening", forecloses the social and geologic conditions through which the fault is produced. Importantly, "Grounding Metaphor" does not disregard metaphor, but instead asks that we take seriously how spatial metaphors are necessarily illuminating other ways of imagining geographic politics. This essay offers a very meaningful lesson. Smith and Katz ask that we think of space as simultaneously material and metaphoric. I run with this to notice that this simultaneity gives us a framework to think relationally across disciplines while also taking seriously the violence of geographic thought and practice.

The lesson is especially important for those who work in the fields of black studies and black geographies, because our analytical sites, and our selfhood, are often reduced to metaphor, analogy, trope, and symbol. To borrow from Hortense Spillers, black people are, in many instances, conceived through "mythical prepossession" (2003:203). We are purely metaphor. In terms of geography, our sense of place is, as I have argued elsewhere, pre-conceptualised as dead and dying and extends outward, from that death, toward extinction (McKittrick 2011). The dead spaces are inextricably linked to the dehumanising scripts—they require one another: one cannot have welfare queen without the objectionable infrastructures that surround her. At the same time, black geographies are often described through metaphor: fleeting, fugitive, underground, non-existent, tomb, womb, aquatic. Often, but not always, these metaphors are delinked from their material underpinnings, which means the question of racial violence is analogised. The redoubled workings of death and metaphor—everything and everywhere we are—flattens out black life. I believe strongly we cannot politicise ourselves, collectively, unless we address how the racist trope and absolute space are co-relatedly working against black life (see also McKittrick forthcoming).

Conceptualising black geographies as both material and metaphoric illuminates how racism authenticates marginalisation. At the same time, this framework pushes us to enliven the production of space. The intellectual task is not to simply describe how absolute space harms black people; it is to get in touch with the materiality of our analytical worlds. What happens to the rift if is not just a metaphor for a "conceptual opening"? What happens, at least for me, is the hard task of working out how environmental disaster, ecology, and geology, are tied to black studies, while also not leaning heavily on geologic or other metaphors to work out the connections. I do not find this is to be an easy exercise. In her essay "Anthropogenesis", Kathryn Yusoff writes about how geology and narrative are fused. She writes, "[g]eology is always involved in ontological questions precisely because of its empirical function in generating origin stories" (2015:5). She also thinks about how the human temporally frames geologic meaning, thus foreclosing human and non-human connections. Yusoff (2015:17) notes we often fail to understand the capaciousness of the planet—the shuffling of pebbles and erratic boulders, the pleasure of snails, bacterial ingestions over millennia. With this conceptual move, she thickens the rift beyond its metaphoric function and creates a space to think about black geographies in relation to retemporalised geologic worlds. Here, the work of Sylvia Wynter—especially her extension of Aimé Césaire's "science of the word" and her bios-logos conception of the human—provides an entry point to observe how the natural sciences (geologies, ecologies, and physiologies), human activity (origin stories) and psychic activity (emotionality) are not only entwined but emerging, simultaneously, as knotted knowledge systems that can read our planetary futures outside market time. Wynter's work with geology—specifically the millennial old writing on walls that linguistically refuse a biocentric origin story—illustrates how this kind of thinking might be useful for black studies (McKittrick and Wynter 2015). In a sense, Wynter reimagines geology, to observe how rock-writing in Southern Africa shapes, inscribes, and historicises the natural environment and *simultaneously* interrupts (rifts, pulls apart)—narratives of evolution that posit black people as less-than-human. Without summarising her monumental project here, I observe simply that her black geologic history infers how the physical environment—in this case aggregate minerals that form rock surfaces—are coupled with an enunciation of black life that is outside normative human-geology inquiries. The coupling of geology and blackness *is* the rift. With this in mind, we can also think, imagine, track, black geologies elsewhere. Here, the plantation and plantation labour show that racial violence and the objectification of the enslaved are necessarily tied to a massive global economic system, wherein extraction-exploitation-accumulation-dispossession is bound to black people's familiarity and intimate knowledge of geological surroundings: "He looked at her again and nodded toward a rock that stuck out of the ground ..." (Morrison 1987:106). The rock sticking out of the ground cannot be delinked from objectification, plantation extraction, imperialism and black rebellion. This kind of knotted interdisciplinary black studies worldview also takes seriously the ways in which plantation slavery and colonialism—extractive economies—underwrite ecocides of racial capitalism. There is much at stake, then, when imagining and reimagining black geologies. The rift is not just a conceptual opening, a metaphoric analogy that sits within the perimeters of academic debate; it must be, to borrow from Césaire, read as a "revolutionary image, ... [a] distant image, the image that overthrows all the laws of thought" (1990:li). The rift must, in other words, not just take us somewhere new, or old, conceptually, it must provide the conditions for us to think carefully about how the work of liberation is tied to the uneasy work of getting in touch with the materiality of our analytical worlds.

References

Césaire A (1990 [1982]) Poetry and knowledge. In *id. Lyric and Dramatic Poetry, 1946–1982* (trans C Eshleman and A Smith) (pp xli–lvi). Charlottesville: University of Virginia Press

Eaves L (2017) Black geographic possibilities: On a queer black South. *Southeastern Geographer* 57(1):80–95

Gilmore R W (2017) Abolition geography and the problem of innocence. In G T Johnson and A Lubin (eds) *Futures of Black Radicalism* (pp 226–240). New York: Verso

Katz C (1996) Towards minor theory. *Environment and Planning D: Society and Space* 14(4):487–499

Mahtani M (2014) Toxic geographies: Absences in critical race thought and practice. *Social and Cultural Geography* 15(4):359–367

McKittrick K (2011) On plantations, prisons, and a black sense of place. *Social and Cultural Geography* 12(8):947–963

McKittrick K (forthcoming) *Dear Science and Other Stories*. Durham: Duke University Press

McKittrick K and Wynter S (2015) Unparalleled catastrophe of our species? Or, to give humanness a different future: Conversations. In K McKittrick (ed) *On Being Human as Praxis* (pp 1–89). Durham: Duke University Press

Mitchell N (forthcoming) *Disciplinary Matters: Black Studies, Women's Studies, and the Neoliberal University*

Morrison T (1987) *Beloved*. New York: Plume

Peake L and Schein R (2000) Racing geography into the new millennium. Studies of "race" and North American geographies. *Social and Cultural Geography* 1(2):133–142

Rose G (1993) *Feminism and Geography: The Limits of Geographical Knowledge*. Minneapolis: University of Minnesota Press

Simone A M (2019) *Improvised Lives: Rhythms of Endurance in an Urban South*. Cambridge: Polity

Smith N and Katz C (1993) Grounding metaphor: Towards a spatialized politics. In M Keith and S Pile (eds) *Place and the Politics of Identity* (pp 67–83). New York: Routledge

Spillers H J (2003) Mama's baby, Papa's maybe: An American grammar book. In *id. Black, White, and in Color: Essays on American Literature and Culture* (pp 203–229). Chicago: University of Chicago Press

Yusoff K (2015) Anthropogenesis: Origins and endings in the Anthropocene. *Theory, Culture, and Society* 33(2):3–28

Seeing

Brett Christophers

Department of Social and Economic Geography, Uppsala University, Uppsala, Sweden;
brett.christophers@kultgeog.uu.se

The history of radical geography is in significant part a history of the struggle—political as much as intellectual—for improved ways of seeing the world. Ways of seeing that help us better perceive, understand, and empathise. Ways of seeing that which is veiled by capitalism's inherent fetishisms and reifications or that which is actively rendered *in*visible: power, oppression, exploitation, injustice. For as David Harvey, so central to the germination and flowering of radical geography in its formative years, observed, prior to the radical turn, the discipline of geography's dominant ways of seeing the world rendered it wholly inadequate as a progressive diagnostic or directive social-scientific discourse. A revolution, or radicalisation, in geography was required precisely because there was "a clear disparity between the sophisticated theoretical and methodological framework which we are using and our ability to say anything really meaningful about events as they unfold around us" (Harvey 1972:6). Tethered to spatial science, geography, as a way of seeing, was bankrupt, defunct.

Radical geography has fundamentally changed that. Through the development of radical modes of geographic enquiry, especially although not only in the pages of *Antipode*, geographers have fitted themselves (and others) with the lenses necessary to see the world anew, and thus to speak about it, *pace* Harvey, meaningfully. Think, for example, of the work of someone like the geographer-artist Trevor Paglen, whose determination that the unseen—such as covert or hidden military operations and facilities—be seen has led him to develop startlingly original, not to mention brave, visual methodologies and modes of representation (e.g. Paglen 2009). If radical geography is about seeing power, and about seeing power moreover that in an age of mass surveillance is itself predicated on its own very different ocular modalities, then Paglen's work is arguably geography at its most radical. And more of it is required. Data releases such as the Panama and Paradise Papers and by groups such as WikiLeaks provide repeated reminders that much about the world that concerns radical geography stubbornly resists being brought to light.

For many radical geographers, of course, simply seeing evidence of injustice and exploitation, and even understanding how they arise and are perpetuated, is not nearly enough. Harvey (1972:10), for example, said at the outset that scholarship depicting "more evidence of man's patent inhumanity to man", such as "daily injustices to the populace of the ghetto", is insufficient. More recently, J.K. Gibson-Graham (2008:619) has made a similar point, claiming that the primary objective of radical geography is not merely "to extend knowledge by confirming

Keywords in Radical Geography: Antipode at 50, First Edition. Edited by the *Antipode* Editorial Collective.
© 2019 The Authors/Antipode Foundation Ltd. Published 2019 by John Wiley & Sons Ltd.

what we already know, that the world is a place of domination and oppression". The point, rather, is to change it (Castree et al. 2010).

One of the key insights of Gibson-Graham's work, however, has been that seeing and changing are fundamentally interrelated. Not only can one not hope to change the world—at least not for the better—if one does not see its problems for what they are. But seeing, and seeing well, can actively abet, perhaps even "perform", positive transformation. The progressive potential in radical geography lies, Gibson-Graham says, in "help[ing] us see openings, to provide a space of freedom and possibility" (2008:619). In her own work, perhaps the best example of this transformative power of radical envisionings relates to "diverse economies" and the so-called iceberg model of the economy. Gibson-Graham (2006) argues that we can change the world for the better by promoting and producing alternative—semi- or non-capitalist—economic arrangements: diverse economies not mediated by exploitation and the profit calculus. Many other anti-capitalists, needless to say, have made similar entreaties. But what distinguishes Gibson-Graham's intervention is the particular role she accords seeing. To maximise the likelihood of producing and sustaining diverse economies, she says, it is vital to recognise that they are already there, only not widely seen. They are below the surface of visibility. Mainstream economic discourse only sees and legitimates "pure" capitalism, the tip of the iceberg above the waterline; the rest of the economy, in the form of alternative economic arrangements, lies beneath. It is there. But it is neither widely seen, nor validated, nor therefore as robust or extensive as it could be. Formally seeing existing diverse economic forms, Gibson-Graham claims, can help corroborate their viability and worth (lifting them above the waterline), and in the process increase the likelihood of their maintenance and proliferation. If, in radical geography, all we see is capitalism—in the shape of a pervasive and unrelenting hegemon—then, says Gibson-Graham, we constrain our ability to imagine, validate, and generate alternatives.

None of this is to say, though, that radical geographers have ever been in agreement about *how* exactly we should or do see the world. In fact, not only is it the case that radical geography has long been characterised by considerable discomfort and disagreement about ways of seeing, but these particular scopic fault-lines are in important respects representative—even constitutive—of wider differences and divergences. Some radical geographers, with Harvey prominent among them, have consistently aspired to produce "strong", generalisable, macro-scale explanations of the social world and its historical-geographical development. Meanwhile, others, including notably Gibson-Graham, have questioned this totalising impulse, arguing that we can (and should) only aspire to more modest claims. Our theories of the world should be "weaker"—not weaker in the sense of usefulness or transformative potential, but, rather, of explanatory ambition and confidence (Gibson-Graham 2008). We should reject the conceit and futility of mastery in favour of "minor" theory (Katz 1995). And, crucially, we should do so partly in view of our restricted and compromised ability to see. There is, for instance, no God's-eye, omniscient, view-from-nowhere: all envisionings of the world, and hence all claims about it, are partial in the sense of being neither complete nor disinterested. Everything we say about the world and its core transformations—the transformations of, today, globalisation, neoliberalisation and financialisation—is dependent upon the perspective from which we look at it (Christophers 2017).

Harvey, to whom much of the critique of mastery and its scopic vanity has been directed, would likely argue, with some justification, that the Marxism to which he appeals has never in reality been predicated on the totalising perspective often imputed to it. It actually was and is, rather, a quintessential example of the type of "situated knowledge" famously championed by Donna Haraway (1988) in her striving for a more realistic—that is, more situated—version of objectivity than that on offer in natural and indeed much of social science. As Harvey (1972) observed in his call to fellow geographers to substitute Marxism for spatial science, one of the most significant ways in which Marx radicalised the political economy of Adam Smith and David Ricardo was *precisely* to situate it—to look at capitalism from a particular, hitherto unfamiliar, vantage-point. And this move was revolutionary not only in the sense of being novel, but of being threatening to the bourgeoisie. Marxism was "clearly dangerous" insofar as it provided

> the key to understanding capitalist production *from the point of view of those not in control of the means of production* and consequently the categories, concepts, relationships, and methods which had the potential to form a paradigm were an enormous threat to the power structure of the capitalist world.   (Harvey 1972:5, emphasis added).

This was not the "god trick of seeing everything from nowhere" (Haraway 1988:581); Marx sought to see *something*—exploitation—from *somewhere*.

And yet to the extent that schisms within radical geography are rooted in differences around seeing, the latter are not reducible to questions of situatedness (or otherwise). J.K. Gibson-Graham and Cindi Katz do not criticise the scopic modalities of radical geography's "master" theorists only for their pretensions— real or perceived—to universality and disembodiment. Gibson-Graham's and Katz's critiques are based on a much wider set of concerns with the ocular regime underwriting rationalist, modernist thinking, and which they associate with much of the "strong" theorising in radical geography. Their concerns mirror those that Martin Jay (1993) has catalogued in his discussion of the denigration of vision in late-20th century French thought, ranging from Foucault's critique of the medical gaze and panoptic surveillance, through Debord's nightmarish account of the society of the spectacle, to Irigaray's identification—via the imbrication of ocularcentrism with phallocentrism—of the centrality of vision to patriarchy. Thus in her critique of the work of another of radical geography's master (and male) theorists, Derek Gregory, Katz (1995:166) maintains that for all his gesturing toward "different ways of reading, writing, talking"—toward, in short, a different type of working and theorising—Gregory "literally does not see it given the focus of his lens on the distant (more luminous?) horizon … where only the largest objects are visible".

So how might radical geographers of the current generation navigate the treacherous waters of "scoping" the worlds they seek to understand and transform? Are there safe, even iceberg-free routes between the Scylla of the God's-eye view and, at the other extreme, the Charybdis of abandoning the project of accurately envisioning the world—and perhaps lapsing into some disabling variant of *non*-representational theory (Thrift 2008)—on the grounds that all visual representation is compromised? Jay thinks so. He remains "unrepentantly beholden to the ideal of illumination" (1993:17). So too does Haraway. She is adamant that the ambition to see the world well, and even, in a fashion, "objectively", does *not* have to "signify a leap out of the marked body and into a conquering gaze from nowhere". Indeed her project, if we can call it that, is, as she says, to "reclaim the sensory system" from those who would—and have—put it to unfeasible or unseemly ends. Insisting on the ineluctability of the visual, but also, simultaneously, on "the particularity and embodiment of all vision", allows us, Haraway submits, "to construct a usable, but not an innocent, doctrine of objectivity" (1988:581–582).

To this end, Jay (1993) calls for a plurality of what he calls "scopic regimes"; and this is arguably precisely what we find in radical geography today. Some practitioners, to be sure, seek out the totalising, detached vantage-point; others insist on a language of embodiment and proximity instead of distance. And while the tension between these two regimes—and the theoretical frameworks they inform and underwrite—is sometimes combative, even antagonistic, it is just as often receptive and engaged. Perhaps an engaged pluralism (cf. Barnes and Sheppard 2010) of scopic regimes is the best radical geography can hope for. And perhaps, if this engaged pluralism can be successfully put into practice, radical geography in toto can aspire to what Jean Starobinski called a "complete critique"—one that demands distance *and* intimacy/proximity because truth lies in neither pole alone but "in the movement that passes indefatigably from one to the other" (quoted in Jay 1993:19–20), or in what Susan Buck-Morss (1991) called the *dialectics* of seeing. The future of radical geography is likely to consist in significant part in the struggle to see the world, collectively, in this dialectical way.

References

Barnes T J and Sheppard E (2010) "Nothing includes everything": Towards engaged pluralism in Anglophone economic geography. *Progress in Human Geography* 34(2):193–214

Buck-Morss S (1991) *The Dialectics of Seeing: Walter Benjamin and the Arcades Project.* Cambridge: MIT Press

Castree N, Chatterton P, Heynen N, Larner W and Wright M W (eds) (2010) *The Point is to Change it: Geographies of Hope and Survival in an Age of Crisis.* Oxford: Wiley-Blackwell

Christophers B (2017) Seeing financialization? Stylized facts and the economy multiple. *Geoforum* 85:259–268

Gibson-Graham J K (2006) *A Postcapitalist Politics.* Minneapolis: University of Minnesota Press

Gibson-Graham J K (2008) Diverse economies: Performative practices for "other worlds". *Progress in Human Geography* 32(5):613–632

Haraway D (1988) Situated knowledges: The science question in feminism and the privilege of partial perspective. *Feminist Studies* 14(3):575–599

Harvey D (1972) Revolutionary and counter revolutionary theory in geography and the problem of ghetto formation. *Antipode* 4(2):1–13

Jay M (1993) *Downcast Eyes: The Denigration of Vision in 20th Century French Thought.* Berkeley: University of California Press

Katz C (1995) Major/minor: Theory, nature, and politics. *Annals of the Association of American Geographers* 85(1):164–168

Paglen T (2009) *Blank Spots On the Map: The Dark Geography of the Pentagon's Secret World.* New York: Penguin

Thrift N (2008) *Non-Representational Theory: Space, Politics, Affect.* London: Routledge

The Anthropo(Obs)cene

Erik Swyngedouw

Department of Geography, University of Manchester, Manchester, UK;
erik.swyngedouw@manchester.ac.uk

Politicising the Anthropo(Obs)cene

The "Anthropocene" is now commonly mobilised by geologists, Earth Systems scientists, and scholars from the humanities and social sciences as the name to denote the new geological era during which humans have arguably acquired planetary geo-physical agency. While recognising a wide-ranging and often contentious debate (see e.g. Castree 2014a, 2014b, 2014c; Hamilton et al. 2015), I argue that the Anthropocene is a deeply depoliticising notion that off-stages political possibilities. This off-staging unfolds through the creation of "Anthropo-Scenes", the *mise-en-scène* of a particular set of narratives that are by no means homogeneous, but which broadly share the effect of silencing certain voices and forms of acting (Bonneuil and Fressoz 2016). The notion of the Anthropo(Obs)-cene then, is a tactic to *both attest to and undermine* the performativity of the depoliticising stories of "the Anthropocene" (see Swyngedouw and Ernstson 2018).

Earth scientists, who coined the term "Anthropocene" (see Crutzen 2002), now overwhelmingly understand the earth as a complex, non-linear, and indeterminate system with multiple feedback loops and heterogeneous dynamics in which (some) human activities are integral parts of these terraforming processes. The capitalist forms of uneven and combined physico-geo-social transformation are now generally recognised as key drivers of anthropogenic climate change and other deep-time socio-environmental transformations that gave the Anthropocene its name (Moore 2016). Both human and non-human futures are irrevocably bound up in this intimate and intensifying metabolic—but highly contentious—symbiosis. The configuration of this relationship has now been elevated to the dignity of global public concern as deteriorating socio-ecological conditions might jeopardise the continuation of civilisation as we know it.

Indeed, a global intellectual and professional technocracy has spurred a frantic search for "smart", "sustainable", "resilient", and/or "adaptive" socio-ecological management and seeks out the socio-ecological qualities of eco-development, retrofitting, inclusive governance, the making of new inter-species eco-topes, geo-engineering, and technologically innovative—but fundamentally market-conforming—eco-design in the making of a "good" Anthropocene. These techno-managerial dispositifs that search for eco-prophylactic remedies for the predicament we are in have entered the standard vocabulary of both governmental and private actors, and are presumably capable of saving both city and planet, while assuring that civilisation as we know it can continue for a little longer. Under the banner

Keywords in Radical Geography: Antipode at 50, First Edition. Edited by the *Antipode* Editorial Collective.
© 2019 The Authors/Antipode Foundation Ltd. Published 2019 by John Wiley & Sons Ltd.

of radical techno-managerial restructuring, the focus is now squarely on how to "change" so that nothing really has to change!

A More-Than-Human Ontology?

The proliferation of prophylactic socio-technical assemblages to make our socio-ecological metabolism "sustainable" and "resilient" coincided with the emergence of a radical ontological shift articulated around non-linearity, complexity, contingency, "risk", and "uncertainty". In addition, theorists from both the social sciences and the humanities mobilised these new earthly cosmologies to propose new materialist perspectives and more-than-human ontologies that point towards grasping earthly matters in more symmetrical human/non-human, if not post-human, constellations. This symmetrical relational ontology, variously referred to as more-than-human, post-human, or object-oriented ontology, fuels the possibility of formulating a new cosmology, a new and more symmetrical ordering of socio-natural relations (see e.g. Coole and Frost 2010; Harman 2016; Morton 2013). Nonetheless and despite its radical presumptions, we contend that these cosmologies also open up the spectre, albeit by no means necessarily so, for deepening particular capitalist forms of human–non-human entanglements and can be corralled to sustain the possibility for a hyper-accel-erationist eco-modernist vision and practice in which science, design, geo-engineering, terraforming technologies, and big capital join to save both earth and earthlings (Neyrat 2016). In the process, the matter of ecology is fundamentally de-politicised.

The geo-sciences and, in particular, Earth System experts discern indeed in the advent of the Anthropocene the possibility, if not necessity, for a careful "adaptive" and "resilient" massaging of the totality of the Earth System. The recognition of the earth as an intricately intertwined, but indeterminate, socio-natural constellation opens up the possibility that the earth, with loving supervision, intelligent crafting, reflexive techno-natural nurturing and ethical manicuring, can be terraformed in manners that sustain a deepening of the eco-modernising and eco-capitalist process. As Bruce Braun (2015) insisted in his dissection of the historiographies of the new materialisms, the parallel between non-deterministic geo-science, "resilience" studies, and the varieties of new materialisms associated with a more-than-human ontology within neoliberalism are not difficult to discern. Indeed, in this staging of the "good" Anthropocene, the new symmetrical relational ontology can function as a philosophical quilt for sustaining and advocating accelerationist hyper-modernising manifestos (Neyrat 2014). To save the world and ourselves, we need not less capitalism, but a deeper, a more intense and radically reflexive form, one that works to terraform earth in a mutually benign and ethically caring co-constitution. Covering up the multiple contradictions of capitalist eco-modernisation, the apparently revolutionary new material ontologies offer new storylines, new symbolisations of the earth's past and future that can be corralled to help perform the ideological groundwork required. In the next section, we shall show how this perspective enters the field of politics, the governing of things and people in common in troublingly de-politicising manners.

The De-politicised Politics of the Anthropocene as Immuno-biopolitical Fantasy

As suggested above, some Anthroposcenic narratives provide for an apparently immunological prophylactic against the threat of a hitherto presumably irredeemably external and revengeful nature. In what ways can the mainstreaming of critical and radical new ontologies whose explicit objective was and is the unsettling of modernist cosmologies be understood? Roberto Esposito's analysis of bio-political governmentality, enhanced by Frédéric Neyrat's psychoanalytical interpretation, may begin to shed some light on this (Esposito 2008, 2011; Neyrat 2010). Esposito's main claim expands on Michel Foucault's notion of biopolitical governmentality as the quintessential form of modern liberal state governance by demonstrating how this biopolitical frame today is increasingly sutured by an immunological drive, a mission to seal off objects of government (the population) from possibly harmful intruders and recalcitrant or destabilising outsiders who threaten the bio-happiness, if not sheer survival, of the population, and guarantees that life can continue to be lived. Immuno-biopolitics are clearly at work, for example, in hegemonic Western practices around immigration, health, or international terrorism. Is it not also the case that many of the sustainability, "resilience", "smart" technologies, and adaptive eco-managerial policies and practices are precisely aimed at re-enforcing the immunological prowess of the immune system of the body politic against recalcitrant, if not threatening, outsiders (like CO_2, waste, bacteria, refugees, viruses, Jihadis, ozone, financial crises, and the like) *so that* life as we know it can continue?

Alain Brossat (2003) calls this a fantasy of *immunitary democracy*. This is a dangerous fantasy, as the immunitary logic entails nothing else than the destruction of community, of politics. Necessarily, this immunitary logic creates the continuous production of the exposed and the exiled (the non-immunised —the dying ones) as the flipside of the immunised body, and leads to de-politicisation. As Esposito argues further, the immunological biopolitical dispositif turns indeed into a thanatopolitics, of who should live or die; it turns into making life and making die (Mbembe 2003). In the excessive acting of the immunological drive, the dispositif turns against what it should protect. It becomes self-destructive in a process of auto-immunisation. The construction of eco-bubbles and "sustainable" enclaves for the privileged produces simultaneously the unprotected exiles and deepening socio-ecological destruction elsewhere. This is eco-gentrification elevated to new heights. In other words, the processes that secure life in some places end up threatening its very continuation elsewhere, at all geographical scales. This infernal dialectic, Neyrat argues, is predicated upon re-doubling the fantasy of absolute immunisation, that is the fantasy that despite the fact we (the immunised) know very well we shall die, we act and organise things as if life will go on forever (Neyrat et al.

2014). The symmetrical human–non-human ontology on which the Anthropo-(Obs)cene rests promises to cut through the unbearable deadlock between immuno- and thanato-politics without really having to alter the trajectory of socio-ecological change. It is the process that makes sure that we can go on living without staring the Real of eventual (ex-)termination in the eye. It is the hysterical position that guarantees that death remains obscure and distant, an obscene impossibility.

Re-Centring the Political

While the controversies over the Anthropocene are mobilised in all manner of ways, suggesting indeed a politicisation of the stuff of things, the "political" cannot and should not be grounded on the eventual truth of the Anthropocene. There is no code, injunction, ontology in the Anthropocenic narratives that can or should found a new political or politicising ecology. The ultimate de-politicising gesture resides precisely in letting the naming of a geo-social epoch and a contingent "truth" of nature decide our politics, thereby disavowing that the "our" or "the human" does not exist. It is yet again a failing attempt to found a new politics on a contested truth of nature. What is required is to assume fully the trauma that the decision is ours and ours only to make.

It is indeed surprising that post-foundational political thought is rarely articulated with more-than-human ontologies of the stuff of matter. Indeed, the post-foundational intellectual landscape that brought into conversation complexity theory and the new materialisms, and claims to open up radical new possibilities, is symptomatically silent of the post-foundational political thought that emerged alongside and in a comparable context. Jacques Rancière (1998), for example, understands the political as the interruptive staging of equality by the "part of no-part". The political appears when those that are not counted within the count of the situation (the excluded, the mute, the exposed, and exiled) make themselves heard and seen—that is, perceptible and countable—in staging equality. For these thinkers, the political emerges symptomatically as an immanent practice of appearance—as Hannah Arendt would put it, or an event in Alain Badiou's terminology (Badiou 2007), that interrupts a given relational configuration or constellation. This performative perspective of politics needs no grounding in any current or historical order or logic, based on say nature, the non-human, ecology, race, class, abilities, or gender. The political is a public aesthetic affair understood as the ability to disrupt, disturb, and reconfigure what is perceptible, sensible, and countable (Rancière 2004).

Indeed, a wide range of political theorists, despite their often radically opposing views, share this search for renewing political thought in a post-foundational ontological landscape characterised by inconsistency, radical heterogeneity and incalculable immanence. Badiou (2008), for example, insists that the attempts to re-found the political philosophically are in fact an integral part of what he diagnoses as a pervasive process of depoliticisation. For him, "ecology is the new opium of the masses". A re-emergence of the political, he insists, resides in fidelity, manifested in militant acting, to egalitarian political events that might open a political truth procedure. Turning a politically progressive event into a political truth procedure requires the emergence of political subjects that maintain a fidelity to the inaugural egalitarian event, aspire to its generalisation and coming into being through sustained actions and militant organisation. It is a fidelity to the practical possibility of the coming community, but without ultimate ontological guarantee in history, theory, technology, nature, ecology, the Party, or the State. Yet it is one that slowly and relentlessly carves out a new socio-physical and socio-ecological reality, often in the face of the most formidable repression and violence. This requires sustained action, painstaking organisation, and the lengthy process of radical egalitarian transformation. Above all, it necessitates embracing the trauma of freedom and abandoning the fear of failing as fail we shall; more-than-human unpredictable and uncaring behaviour guarantees that.

References

Badiou A (2007) *Being and Event*. London: Bloomsbury

Badiou A (2008) Live Badiou—Interview with Alain Badiou, Paris, 2007. In O Feltham (ed) *Alain Badiou: Live Theory* (pp 136–139). London: Continuum

Bonneuil C and Fressoz J B (2016) *The Shock of the Anthropocene*. London: Verso

Braun B (2015) New materialisms and neoliberal natures. *Antipode* 47(1):1–14

Brossat A (2003) *La Démocratie Immunitaire*. Paris: La Dispute

Castree N (2014a) The Anthropocene and geography I: The back story. *Geography Compass* 8(7):436–449

Castree N (2014b) Geography and the Anthropocene II: Current contributions. *Geography Compass* 8(7):450–463

Castree N (2014c) The Anthropocene and geography III: Future directions. *Geography Compass* 8(7):464–476

Coole D H and Frost S (eds) (2010) *New Materialisms: Ontology, Agency, and Politics*. Durham: Duke University Press

Crutzen P J (2002) Geology of mankind. *Nature* https://doi.org/10.1038/415023a

Esposito R (2008) *Bios: Biopolitics and Philosophy*. Minneapolis: University of Minnesota Press

Esposito R (2011) *Immunitas: The Protection and Negation of Life*. Cambridge: Polity

Hamilton C, Bonneuil C and Gemenne F (eds) (2015) *The Anthropocene and the Global Environmental Crisis*. London: Earthscan

Harman G (2016) *Immaterialism: Objects and Social Theory*. Cambridge: Polity

Mbembe A (2003) Necropolitics. *Public Culture* 15(1):11–40

Moore J W (ed) (2016) *Anthropocene or Capitalocene? Nature, History, and the Crisis of Capitalism*. Oakland: PM Press

Morton T (2013) *Hyperobjects-Philosophy and Ecology after the End of the World*. Minneapolis: University of Minnesota Press

Neyrat F (2010) The birth of immunopolitics. *Parrhesia* 10:31–38

Neyrat F (2014) Critique du géo-constructivisme Anthropocène & géo-ingénierie. *Multitudes* 56. http://www.multitudes.net/critique-du-geo-constructivisme-anthropocene-geo-ingenierie/ (last accessed 22 August 2018)

Neyrat F (2016) *La Part Inconstructible de la Terre*. Paris: Editions du Seuil

Neyrat F, Johnson E and Johnson D (2014) On the political unconscious of the Anthropocene: Frédéric Neyrat, interviewed by Elizabeth Johnson and David Johnson. *SocietyandSpace.org* 20 March. http://societyandspace.org/2014/03/20/on-8/ (last accessed 22 August 2018)

Rancière J (1998) *Dissensus*. Minneapolis: Minnesota University Press

Rancière J (2004) *The Politics of Aesthetics: The Distribution of the Sensible.* London: Continuum
Swyngedouw E and Ernstson H (2018) Interrupting the Anthropo-obScene: Immuno-biopo-
    litics and depoliticising ontologies in the Anthropocene. *Theory, Culture, and Society*
    https://doi.org/10.1177/0263276418757314

The Common

Miriam Tola

Media and Screen Studies, Northeastern University, Boston, MA, USA;
m.tola@northeastern.edu

Ugo Rossi

Dipartimento Interateneo di Scienze, Progetto e Politiche del Territorio,
Università diTorino, Turin, Italy;
ugo.rossi@unito.it

Over the last three decades or so, radical scholarship has identified the continuous and ongoing enclosure of the commons as critical for the expansion of global capitalism (Caffentzis 2016; Federici 2004; Linebaugh 2008). Largely focusing on capitalist apparatuses of dispossession and expropriation (Rossi 2013), this line of inquiry has shed light on the colonising logic of neoliberal globalisation. Taking a different perspective, authors working within the autonomist-Marxist strand of "radical Italian thought" (Hardt and Virno 2006) have drawn attention to the productive dimension of what, using the singular form, they define "the common" and its role in the context of cognitive capitalism. In revisiting the concept of "the common", this entry assesses its significance as an intellectual and political category, particularly against the backdrop of the current moment of political authoritarianism and intensified racial, gendered and environmental violence.

From the Commons to the Common

Underpinning the autonomist conception of the common is the notion of "living labour", the creative human activity that Marx identified with the collective potential of workers' bodies (Marx 1973:361). In the *Grundrisse*, Marx reflects on the relationship between living labour and the "dead labour" crystallised in technology and machines. He suggests that capitalism increasingly depends on the "general intellect", that is, the accumulation of general social knowledge that capital objectifies in the form of machinery. Autonomist Marxists have applied the concept of general intellect to the analysis of contemporary post-Fordism, a regime of accumulation that connects disparate places and forms of increasingly flexible and precarious labour. Creatively drawing on Marx, they contend that general social knowledge cannot exist independently from "the interaction of a plurality of living subjects" (Virno 2006:193). If the general intellect results from the interrelation of living labour and the fixed capital of machinery, the driving force of the productive process is collective human intelligence that, although incorporated into technologies, is continuously renewed through the interactions of a heterogeneous multitude of workers. This idea that the post-Fordist regime

Keywords in Radical Geography: Antipode at 50, First Edition. Edited by the *Antipode* Editorial Collective.
© 2019 The Authors/Antipode Foundation Ltd. Published 2019 by John Wiley & Sons Ltd.

of production extracts value from social cooperation is variously explored in autonomist circles.

For Paolo Virno (2015), contemporary capitalism produces value by mobilising those potentialities of speech and relationality that set human animals apart from the rest of the living. These shared human abilities are put to work as living labour in the circuits of post-Fordist accumulation. Language, sensory perceptions and specific motor skills constitute the common of humanity, the "biological invariant", but are performed differently by the multitude of human beings. Michael Hardt and Antonio Negri's extensive work on the common diverges from Virno's focus on the bio-linguistic abilities of the human species. But like Virno, they insist on identifying the common with living labour, the productive social cooperation that capitalism incessantly expropriates (Hardt and Negri 2009). Simultaneously, they argue, the common comprises the earth we share and the natural resources on which we depend. If the "natural common" is described in terms of scarcity and limits, the "social common" of ideas, code, and images is limitless and reproducible, governed by a logic of abundance and proliferation (Hardt 2010). In acknowledging the distinction between the natural and the social as fraught with problems, Hardt and Negri (2009) gesture towards an "ecology of the common" that would result from the mutual transformation between human and non-human beings. This point, however, remains underdeveloped in their work.

Sandro Mezzadra and Brett Neilson (2013) complicate the notion of natural common by placing emphasis on the intricate web of human activities behind its production and reproduction. This means, for example, that one cannot think of water as a natural common without taking into account the socio-material infrastructures that organise its distribution and usage. But this, for Mezzadra and Neilson, entails focusing on living labour as a transformative force, "the moment of excess that characterizes the common with regard to the commons" (2013:297). Writing with Verónica Gago, Mezzadra has further developed this point in the analysis of extractivist economies in Latin America (Gago and Mezzadra 2017). They propose to expand the concepts of extractivism and the common to shift focus from the intensified dispossession of raw materials in Latin America to broader operations of capitalism in which territories are linked to the dynamics of finance. The circuits of extractivism, driven primarily by finance, target not just "inert materials" (Gago and Mezzadra 2017:579) but also the networks of social cooperation, the labour and life of populations. They connect forms of dispossession and exploitation, rural and urban contexts, sites of extraction in Latin America with digital economies of Europe and the United States. Such reworking of extractivism productively complicates the tendency to frame the common through the binary between natural common and social common. Yet, it problematically frames nature as inert, a repository of resources that are mobilised in the operations of capitalism. In this respect, autonomist thought inherits from Marx what Maurizio Lazzarato (2004:4) defines as an "ontology of subject/object" for which "the constitution of the world is thought in terms of production".

Feminist and Decolonial Challenges

Scholars in feminist and decolonial studies have exposed several blind spots in autonomist accounts of the common. Here we focus on three interrelated points that both challenge and contribute to the politics of the common. First, the identification of the common with the creative force of living labour carries with it a universalising impetus that misses the ways in which ongoing histories of racialised, gendered, and environmental violence have increased the potentialities of particular bodies and enormously reduced those of others. This is particularly clear in Virno's reliance on the notion of species. His identification of the common with the cognitive and linguistic abilities of human animals fails to consider the historical imbrications between the dominant species discourse and colonial racism (Tola 2017). Second, Indigenous studies scholars have addressed the political implications of universalising claims to the common. When autonomist theorists and activists invoke the generalised reappropriation of the common, they tend to ignore specific histories of dispossession and risk construing Indigenous claims to land and sovereignty as regressive and exclusionary (Byrd 2011). Even more, they overlook Indigenous insights into the relationship between people and land, humans and non-humans, and their value for political movements struggling for more just worlds (Coulthard 2014). Thinking about the common, then, requires attention to ongoing forms of colonial dispossession, including the dispossession of ways of relating with the material world that do not entail a binary distinction between active subjects and inert resources. This leads us to the third blind spot.

As a growing body of scholarship shows, while the Italian-autonomist emphasis on immaterial labour and collective intelligence may be useful for investigating aspects of cognitive capitalism (Pasquinelli 2014; Terranova 2014; Vercellone 2015), this strand of literature still has to engage environmental questions and the ecological dimension of global capitalism. Silvia Federici (2010:286–287) notes that Italian radical theorists overlook the fact that "online communication/ production depends on economic activities—mining, microchip and rare earth production—that, as presently organised, are extremely destructive, socially and ecologically". If critical theorisations of cognitive capitalism within Italian Radical Thought tend to focus on how capitalism captures the proliferating social common in technology-intensive economies (Roggero 2010), too little attention has been devoted to its ecological costs and the destructive impact on populations and territories. Radical geographers have provided important input for addressing this point. Sara Nelson and Bruce Braun (2017) point out that the acceleration of planetary change that goes under the name of the Anthropocene challenges the political imaginary of autonomist Marxism, specifically, its reliance on the expansion of human productive capacities. What is needed, they suggest, is a critical assessment of the dichotomy between nature and culture that underpins autonomist articulations of the common. Central, in this respect, is the reconsideration of living labour in ways that, as Elizabeth Johnson (2017) notes, account for the manner in which non-human life is enrolled in the informational order of cognitive capitalism. Taken together, these engagements with the common point toward fruitful directions for theoretical and activist work.

Concluding Remarks

As we have seen, feminist and decolonial critics have questioned the universalist narrative that underpins Italian-autonomist accounts of the common. Such universalistic perspective is not limited to Italian autonomists but can be found in other Marxist-inspired views of the neoliberal expropriation of the commons that, as some critics have argued, gloss over questions of race (Dawson 2016; Wang 2018). Despite these lacunae, Italian-autonomist authors have had the merit of showing how the subsumption of the common is a process inherent to the expanded extractivism of contemporary capitalism, rather than the product of an external colonising force, as in theories of the commons that emanate from ontologies of dispossession and expropriation.

A decade after the systemic crisis of the late 2000s, today's economic-political conjuncture appears characterised by the peculiar combination of a resilient neoliberal governmentality with a monopoly-based platform capitalism that exploits collective intelligence in a potentially infinite number of socio-physical settings across the global North and the global South. A renewed engagement with the common, therefore, has to deal with the challenges posed by this advanced globalisation in the capitalist process of extraction, appropriation and reproduction of common wealth.

References

Byrd J (2011) *The Transit of Empire: Indigenous Critiques of Colonialism.* Minneapolis: University of Minnesota Press

Caffentzis G (2016) Commons. In K Fritsch, C O'Connor and A K Thompson (eds) *Keywords for Radicals: The Contested Vocabulary of Late-Capitalist Struggle* (pp 93–101). Chico: AK Press

Coulthard G (2014) *Red Skin, White Masks: Rejecting the Colonial Politics of Recognition.* Minneapolis: University of Minnesota Press

Dawson M C (2016) Hidden in plain sight: A note on legitimation crises and the racial order. *Critical Historical Studies* 3(1):143–161

Federici S (2004) *Caliban and the Witch: Women, the Body, and Primitive Accumulation.* New York: Autonomedia

Federici S (2010) Feminism and the politics of the commons in an era of primitive accumulation. In Team Colours Collective (ed) *Uses of a Whirlwind: Movement, Movements, and Contemporary Radical Currents in the United States* (pp 274–294). Oakland: AK Press

Gago V and Mezzadra S (2017) A critique of the extractive operations of capital: Toward an expanded concept of extractivism. *Rethinking Marxism* 29(4):574–591

Hardt M (2010) Two faces of apocalypse: A letter from Copenhagen. *Polygraph* 22:265–274

Hardt M and Negri A (2009) *Commonwealth.* Cambridge: Harvard University Press

Hardt M and Virno P (eds) (2006) *Radical Thought in Italy: A Potential Politics.* Minneapolis: University of Minnesota Press

Johnson E (2017) At the limits of species being: Sensing the Anthropocene. *South Atlantic Quarterly* 116(2):275–292

Lazzarato M (2004) *La politica dell'evento.* Cosenza: Rubbettino

Linebaugh P (2008) *The Magna Carta Manifesto: Liberties and Commons for All.* Berkeley: University of California Press

Marx K (1973 [1858]) *Grundrisse: Foundations of the Critique of Political Economy.* New York: Penguin

Mezzadra S and Neilson B (2013) *Border as Method, or, the Multiplication of Labor*. Durham: Duke University Press

Nelson S and Braun B (2017) Autonomia in the Anthropocene: New challenges to radical politics. *South Atlantic Quarterly* 116(2):223–235

Pasquinelli M (2014) Italian *Operaismo* and the information machine. *Theory, Culture, and Society* 32(3):49–68

Roggero G (2010) Five theses on the common. *Rethinking Marxism* 22(3):357–373

Rossi U (2013) On the varying ontologies of capitalism: Embeddedness, dispossession, subsumption. *Progress in Human Geography* 37(3):348–365

Terranova T (2014) Red stack attack: Algorithms, capital, and the automation of the common. In A Avanessian and R Mackay (eds) *#Accelerate#: The Accelerationist Reader* (pp 379–399). Falmouth: Urbanomic

Tola M (2017) Species, nature, and the politics of the common: From Virno to Simondon. *South Atlantic Quarterly* 116(2):237–255

Vercellone C (2015) From the crisis to the "welfare of the common" as a new mode of production. *Theory, Culture, and Society* 32(7/8):85–99

Virno P (2006) Virtuosity and revolution: The political theory of exodus. In M Hardt and P Virno (eds) *Radical Thought in Italy: A Potential Politics* (pp 189–210). Minneapolis: University of Minnesota Press

Virno P (2015) *When the Word Becomes Flash*. South Pasadena: Semiotext(e)

Wang J (2018) *Carceral Capitalism*. South Pasadena: Semiotext(e)

The Union of Socialist Geographers

Eric Sheppard

Department of Geography, University of California, Los Angeles, Los Angeles, CA, USA;
esheppard@geog.ucla.edu

Linda Peake

The City Institute, York University, Toronto, ON, Canada;
lpeake@yorku.ca

Active between 1974 and 1982, the Union of Socialist Geographers (henceforth USG[1]) was a Canadian–US centred academic/activist collaborative venture in socialist geography, operating independently from the mainstream professional associations (the Canadian Association of Geographers, henceforth CAG, and the then-named Association of American Geographers, henceforth AAG). Its origins were in the Geographical Expeditions organised in the late 1960s and early 1970s by the first generation of self-styled radical geographers in Detroit, Toronto, Vancouver, and Sydney (Australia). By 1973, the Vancouver Geographical Expedition, headquartered out of Simon Fraser University, was the largest cluster of activist radical geographers in North America, involving more than 40 graduate and undergraduate students, supported by Michael Eliot-Hurst (the Chair of SFU's Geography Department) and junior and visiting faculty. In 1974, members of this group, driving to the CAG annual meeting in Toronto in vans organised by Eliot-Hurst, convened with members of Bill Bunge's Toronto Geographical Expedition at the Expedition house on Brunswick Avenue and participated in a four-day meeting attended also by other radical geographers (including Clark Akatiff, David Harvey, Richard Peet, Wilbur Zelinsky and Bill Blaut). The idea to create a USG emerged during this encounter. Its founding statement, composed by Clark Akatiff, signed by 29 white men and four white women and dated 28 May 1974, presented its agreed on mission:

> The purpose of our union is to work for the radical restructuring of our societies in accord with the principles of social justice. As geographers and people, we will contribute to this process in two complementary ways:
>
> 1. Organising and working for radical change in our communities; and
> 2. Developing geographic theory to contribute to revolutionary struggle.
>
> Thus we subscribe to the principle: from each according to their ability, to each according to their need. We declare that the development of a humane, non-alienating society requires, as its most fundamental step, socialisation of the ownership of the means of production. (Akatiff 1974)

Keywords in Radical Geography: Antipode at 50, First Edition. Edited by the *Antipode* Editorial Collective.
© 2019 The Authors/Antipode Foundation Ltd. Published 2019 by John Wiley & Sons Ltd.

Between that meeting and its dissolution in 1982, the USG primarily engaged in two kinds of activities: publication and distribution of the *USG Newsletter*, and the organisation of its own academic meetings. Membership was declared open to anyone subscribing to its principles and willing to pay its $5 annual dues. Communication with the membership was initially envisioned as occurring through existing outlets (the AAG Newsletter, *Transition*—a journal published by the Society of Socially and Ecologically Responsible Geographers, and *Antipode*), but the *Newsletter* came to replace this.

The USG, paralleling the structure of labour unions, had both a national organisation and local affiliates (the latter having far more autonomy than in a typical labour union). The national organisation collected dues, oversaw production of the *Newsletter*, and held annual meetings. USG annual meetings were organised to occur alongside but outside the annual meetings of the CAG and the AAG, alternating between the two (a pattern that persisted throughout its lifetime). This saved on nationwide travel to conferences, but the USG annual meetings also were used to plan radical interventions into the mainstream AAG and CAG events, putting on socialist geography sessions, packing business meetings to organise anti-war motions and the like, and deputising members to harangue sessions and speakers seen as particularly inimical to socialism. These gatherings occurred annually from 1975 to 1982 (Peake 2019), occasionally meeting twice in a given year alongside both the AAG and CAG.

Vancouver created the first local, closely followed by a short-lived one in Toronto building on the USG's founding event. Others emerged through connections with, and the mobility of, USG members, particularly out of Vancouver. A Montreal local was created in 1976, there was a short-lived group at Michigan State University (until the faculty behind the Detroit Geographical Expedition were forced out of the academy), an east coast local centred on Clark University, and a long-standing Midwest local around the University of Minnesota, the University of Wisconsin, Madison, and Valparaiso University. Locals also were formed outside North America, in New South Wales (Australia), Denmark, Ireland, and the (self-named) British Isles. Other more loosely affiliated but independent groups formed at Queen's University and Laval University (Canada). By 1979 interest in the USG had spread as far as Africa, Latin America, and Asia, especially India, including to non-geographers and non-academic geographers.[2] Peake (2019) has identified some 190 individuals, a list that is almost certainly incomplete. These locals also created their own local USG conferences and conference sessions. There were some 16 of these: six in the US Midwest, two in Vancouver and in Worcester (MA), and one each in Toronto, Kingston (Ontario), the US East Coast, Dublin, and Manchester and Lancaster (UK).

The *USG Newsletter* comprised a total of eight volumes, expanding and declining in size over time in parallel with the fortunes of the USG itself, and stretching beyond North America to draw others into the organisation.[3] The first five issues were edited by the Vancouver local, which meant writing entries, soliciting contributions from others, collecting members' subscriptions, and copying and mailing issues to those members. Thereafter, *Newsletter* production skipped from place to place. Volume 2 was coordinated by east coast USG locals at Johns Hopkins (Baltimore, USA) and McGill (Montreal, Canada), and Volume 3 was produced by locals at the University of Toronto, the University of Minnesota, and McGill. Volume 4 stretched beyond the continent to include issues edited by collectives in Sydney (Australia) and London (UK). The organisation and distribution of Volumes 5 and 6 were carried out at Minnesota (including issues edited from McGill and Queens University in Canada) and London (UK), and Volume 8 was edited by Chrys Rodrigue at CSU Northridge (California). As production of the *Newsletter* circulated, so did the origins of its content. Increasingly a focus emerged on issues relevant to the region where an issue was produced. The various issues thus provide insight into the interests at that time of radical geographers located in the places where they were produced.

Almost every issue included minutes from USG meetings, reports on other meetings and fieldtrips, course syllabi and information on forthcoming activities, while also cajoling the membership to contribute articles and send membership dues. Beyond this, the *Newsletter* increasingly evolved into a venue for academic articles. Aligned with what was an emergent consensus, until the mid-1980s, that radical geography should be Marxist geography, articles on Marxism were common, but from the beginning the *Newsletter* was also a space to discuss the relations between geography and feminism, anarchism, indigeneity, homosexuality, Third World movements, "nature", and race (to a much lesser extent). It became an important venue for radical geography graduate students to whet their academic writing skills, in an era when a substantial grey literature existed beyond (and was excluded from) the refereed journals published by presses and academic associations. This meant that the *Newsletter* was more heterodox in the radical traditions that it embraced than *Antipode*. Academic articles came to dominate journal issues, culminating in the last issue of Volume 6 (#2) which weighed in at 82 pages. Thereafter, the wind absconded from the *Newsletter*'s sails: the sole issue of Volume 8 (December 1982) was four pages long, consisting of a reflective essay by Chrys Rodrigue (then at California State University, Northridge) on the decision to revert the mission of the *Newsletter* to that of reporting on the actions of the USG's activities, supplemented by a membership renewal form.

The USG also produced other publications: an 87-page *Studies on Imperialism* for the 1976 New York City AAG meeting, the *Canadian American Geographical Expedition Papers* by Bill Bunge and Ron Horvath, and the 1983 *Society and Nature: Socialist Perspectives on the Relationship Between Human and Physical Geography* (a forerunner for what now is known as critical physical geography). A planned radical geography textbook never materialised.

The demise of the USG was triggered by four factors: exhaustion of those willing to take on the organisational tasks associated with meetings, the newsletter and dues collection; increasing difficulty in collecting dues from members; the move of some original USG members onto activism elsewhere and into faculty positions in and beyond Geography (an ironic consequence of how USG activities enabled its members to gain footholds in the academy); and the decision to work inside as well as outside the system. After the AAG decided to allow its members to create their own specialty groups within the Association (on the basis of 100 members signing a petition to form such a group), there was active debate at USG annual general meetings about whether socialist geographers should take up this opportunity. As Peake (2019) notes, the "crux of the debate was whether incorporation within the AAG would blunt radical geography's radicalism". Neil Smith recalls:[4]

> We wanted organisation for the USG but were resistant in our post-60s way to heavy organisation, and this made the USG a bit haphazard. I think many of us felt that for all its wonderful energy, the USG was not likely to become more central, indeed had already served its best role, and, far from ideal as it was, we couldn't afford not to be involved in the SGSG [Socialist Geography Specialty Group of the AAG].

This debate began in 1979, a petition was filed in 1980 to form the SGSG, which was accepted by the AAG in 1981. The SGSG's initial Board was made up entirely of USG members (Eric Sheppard, Mickey Lauria, Ruth Fincher, Gerry Hale and Richard Hansis), and its mission (Socialist Geography Specialty Group 1980) was stated as:

> To examine geographical phenomena critically, questioning the implications of geographical research for the well-being of social classes. To investigate the issue of radical change toward a more collective society.

Energy was rapidly ebbing from the USG, which held its last annual meeting in April 1982 alongside the San Antonio (TX) AAG meeting and published its last, rump newsletter in December of that year.

While lasting just eight years, the USG was a transformative force for radical geography in North America (and beyond). It enabled direct collaboration across the 49th parallel that divides Canada and the US in the name of socialist geography (unlike the national-scale organisation of much of academic geography, radical and otherwise, today). It created space for more-than-Marxist radical geography and radical geographers (such as formation of a radical feminist reading group in 1975). This space was rapidly occupied after the mid-1980s, sometimes under the label of critical geography, generating such journals as *Society and Space*, *Gender, Place, and Culture*, and the online *ACME: An International Journal for Critical Geographies*, alongside *Antipode*. Even if its locals had limited longer-term impact on the places where they operated (with the possible exception of Minneapolis; Lauria et al. 2019), the USG was a space where activism and the academy co-existed; a significant number of its members made activism central to their post-USG activities. Finally, it was an incubator for an emergent generation of radical geographers, including ourselves, to enter (and colonise) mainstream Anglophone human geography.

Endnotes
[1] For a full account, see Peake (2019).
[2] See Ed Vandervelde's note in Volume 4 (#3) of the *USG Newsletter* (p. 33).
[3] The *Newsletter*, except for Volume 7 (for which we have been unable to locate a copy), is available in digital form at AntipodeFoundation.org: https://antipodefoundation.org/2017/06/28/usg-newsletter-archive/ (last accessed 16 August 2018).
[4] In an email to the authors on 22 March 2011.

References

Akatiff C (1974) "Union of Socialist Geographers: Preamble and Minutes." Manuscript available from the author

Lauria M, Higgins B, Bouman M, Mathewson K, Barnes T J and Sheppard E (2019) Radical geography in the Upper Midwest. In T J Barnes and E Sheppard (eds) *Spatial Histories of Radical Geography: North America and Beyond.* Oxford: Wiley-Blackwell (forthcoming)

Peake L (2019) The life and times of the Union of Socialist Geographers. In T J Barnes and E Sheppard (eds) *Spatial Histories of Radical Geography: North America and Beyond.* Oxford: Wiley-Blackwell (forthcoming)

Socialist Geography Specialty Group (1980) "Annual Report, 1980." Manuscript available from the authors

"Value"

George Henderson

Department of Geography, Environment and Society, University of Minnesota, Minneapolis, MN, USA;
hende057@umn.edu

Three Parts In Motion / To Be Read In Every Order

1 (or 2 or 3)

My yearly mistake. Once every year I teach a course on the urban geography of the United States. Early in the semester I assign a book by Thomas Sugrue, *The Origins of the Urban Crisis* (2005), in order to share with students Sugrue's incontrovertible, empirical proof that capitalism, inequality, and injustice are fully compatible. They get it. In the very next module, we treat the question "What is capitalism anyway?", a question that, oddly, does not concern Sugrue. And here is where things get tricky. I share with students a dictionary definition of "capitalism", and show them that with this particular definition, there is no way to distinguish between, say, an antebellum plantation worked by enslaved Africans and a New England textile factory employing (newly proletarianised) female wage labourers. These are two value-producing systems that, for all their ties to each other, students nonetheless *know* have fundamental differences. I introduce them to Marx's concept of the mode of production as at least a more incisive approach than the dictionary's, even if it is not the last word. My goal is just to ask that they see conceptually what they already know to be true, even though this particular perspective, Marxism, is something they have been taught to be suspicious of. My goal is for them to grapple with the *concept* of the difference they already know to have existed. I always think this approach is rational, or at least as sensible a way as any other to introduce the class to Marx, who through his various interpreters, has been so influential in urban geography. Who would not be convinced that there is at least something *interesting* here? Yet every year there are a couple of students who stand up and walk out; every year a group of varying size who does not show up again until the syllabus seems to indicate that Marx has been dealt with and safer ground returns. So that is my yearly mistake, to adopt what is to my mind a rational approach to my seemingly simple goal.

I want, after all, to appeal to students' sense of reason, given the US's history of anti-Marxism. I want to construct an argument based on evidence. But what does this imply about my assumptions of what people are actually like? So, here is my second mistake, which I'll put like this: I assume that these paying students who *do* follow me through my abbreviated tour of Marx's stamp on urban geography, do so because my approach is indeed reasonable. But this rationality is surely entangled with other, affective, even fetishistic, processes: just what is the commodity students have bought when they have paid for the course—is it the message or a messenger? And just how is value circulating through the class setting where each week the next portion of the commodity (the course), itself about value-creating systems, is delivered? On the one hand, I have shaped with my labour the commodity they have bought, honing my presentation over the years until now it's more confidently performative than it used to be. There are more emphatic gestures, there is more grain in my voice, more eye contact, a nice shirt, and all-around aesthetic amperage. On the other hand, against the pre-sumptive innocence of my own self-possession, I am the very means for a larger process of social production and reproduction in which dividends are paid to teachers, especially males, with white-skin privilege, a process that is itself the legacy of the manufacture of "white". What exactly is the commodity these students (most of them also with white-skin privilege) have bought then? What exactly are the ideas they have been sold and on what basis would they or wouldn't they entertain them? Where else might these ideas be encountered and in what other forms?

2 (or 1 or 3)

In the Marxist strain in geography, with some exception, there is a place for concepts where affect gets its due and concepts reserved for reason/rationality. In the former we might group ideology, consciousness, fetishism, species being. To the latter most especially belongs value. Value is the place of accounting—accounting for labour time, for connection between labour time and the flux of prices, for labour discipline, for economic rationalisation. It is put to work theoretically to effect certain distinctions—between the economic and non-economic, the pro-ductive and non-productive, capitalism and not-capitalism, the abstract and concrete, the law (of value) and incidentals, base and superstructure, essence and appearance. Recall Marx's first step. To exchange commodities is to render "abstract" the concrete labour specific to the production of the different com-modities exchanged. This abstraction, accomplished *only if* commodities are exchanged, implicitly is a matter of valuing labour time (and, soon in Marx's anal-ysis, socially necessary labour time) over the concrete materialities of the labour process and the materials worked upon. This abstraction presumes a mechanism —money, in fact, another kind of "universal" commodity—for its own representa-tion, and is *presumed to be* the underlying mechanism through which use values are distributed so that concrete labour may reassert its relevance in the many untold acts of consumption. Value may be seen as making use of immeasurable affective qualities (e.g. immaterial labour, emotional labour, affective labour, etc.) but at the end of the day those qualities are subsumed into the reductive rational-ity of value, the "law of value" even. Value seems therefore always to be a site of a reduction and always on the side of capital.

Value is rendered as the idea that it doesn't matter what we feel or believe about it. It just is the core truth about capital, because it exposes ideology as ide-ology, fetishism as fetishism, proper consciousness as proper consciousness. The modalities of reason and unreason are closer than some might think, however, and the coupling of value and capital is to my mind an under-reading. Value is an imaginary that thrusts us, by means of logic and illogic, into the realm of future possibilities, of what is not (yet) practical. Because if we follow the various currents of how Marx works with the value framework, value ("inside" or "outside" capitalism) only ever is the full round of activities through which societies produce and reproduce themselves, think themselves, feel themselves, change themselves, and alter the world in so doing. Understanding what capitalism is requires a value framework, and so too does thinking about what post-capitalist value could be like. The Marx who writes of the irrationality of the capitalist drive to produce surplus, is, for example, the same Marx who writes of the necessity to produce surplus in the future society of "associated labour". The Marx whose materialist approach emphasises the compulsion in capitalist society to reproduce its social relations is the same one who insists on a degree of indeterminacy, of excess, regarding the question of whether tomorrow (or next year or what have you) needs to be the same as today. The Marx who uses his reasoning capacities to demolish bourgeois notions of value is the same Marx—never forgetting that human beings are sensory beings—who constantly engages in rhetorical flourishes to get a laugh or a rise out of his readers.

3 (or 1 or 2)

The art gallery is cave-like when you first peer through the doors. Enter, and hundreds and hundreds of white shirt sleeves, virtually filling the ceiling space, hang like stalactites low above your head. Pause. Look left, pause again, see dozens and dozens of outlined chalk figures filling a long, dark wall. What?

Something strange and something familiar is here, something old and new, sacred and profane. What is the feeling of being in this space? Because it wants to be felt before it wants to speak, to confound before it teaches. What is the feeling? The sensual excitement of number, quantity, pattern; mere shirts, repeated beyond number until they become abstract, enveloping form. What is the feeling? The feeling of feeling, per se. *Shroud*, an installation created by the Minneapolis artist and professor of art, Rachel Breen, is, on its face, a critical commemoration of two disasters: the 1911 Triangle Shirtwaist Factory fire in New York City and the 2013 Rana Plaza factory collapse in Dhaka, Bangladesh. The disaster in New York killed 146 textile workers, mostly young immigrant women. 1134 died at Rana Plaza, textile workers also, mostly women, working under no less inhumane conditions. *Shroud*, smartly numerical, is made of 1280 white shirts sewn together at the hem into a single enormous garment and hung upside down, so the cuffs might nearly graze your forehead (Figure 1). *Shroud* is an imagined garment for the notional collective worker and the cooperative spirit of labour, a "free gift" to capital, says Marx. (We can't be responsible if it dies, says the factory owner.) The chalk drawings along the wall are recognisable as outlines of shirt patterns and pattern remnants, the positive and negative spaces of the materials worked up by human labour into shirts. Figures of a labour process whose final product looms weightily above.

But *Shroud* doesn't just commemorate the death of workers with the "dead labour" of the commodity. It is built on new labour time, too. It took five months

Figure 1: *Shroud,* installation by Rachel Breen, Carleton College, Northfield, Minnesota, 2018 (photo by author)

for Rachel Breen to find enough white shirts. During weekly pilgrimages to Goodwill she bought them by the pound. She dumped them into the back of the car, hauled them up the steps to her house, filling her living room … dining room … study. Then two months of new collective labour to sew the hems together. The shroud is "Made in Minneapolis" from shirts "Made in Honduras", "in Cambodia", "in Mexico" and, inevitably, "in Bangladesh". You must imagine small gatherings of (women) friends workshopping the shroud, and workshopping their very friendship. Free gift to?

What a project getting the shroud out of the house and down to Northfield, Minnesota, an hour south. Another project getting it into the gallery. And yet another fixing it to the ceiling grids. The chalk drawing is another project still. You run paper through a sewing machine, "drawing" on it with an unthreaded needle, turning it into a stencil for powdered chalk to go through. It takes a lot of very particular labour for the shroud and the stencil to signify, scold, and point to the possibilities of something beyond, the subsuming abstractions of value that circle the globe. All this within the very workshop of the history and geography of value: the shroud and the stencil hang in a gallery at Carleton College, one of the crowning achievements of the "immaterial", "information" economy of greater Minnesota.

"The wealth of those societies where capitalism prevails", Marx writes at the beginning of *Capital* (1887), "presents itself as an immense accumulation of commodities". Sure enough, in *Shroud,* you see a slice of that immense accumulation and you see it with startling intensity. The accumulation here is not the emporium Marx suggests, but is instead a mass of a single kind of thing. Yet nowhere does this single kind of thing normally accumulate. *Shroud* is not a demurring piece of art, then. It surpasses what it mimics. Consider: it is not an example of a commodity, like Marx's famous coats to be exchanged in some number for some linen. *Shroud* stands apart, wrested from market exchange and withdrawn from ordinary consumption. It points toward the world of circulating commodities, which generally "shrouds" (obscures, veils) labour, but refuses absorption into it. It is a radical hoard. This is the shock of it; that these white shirts are not going anywhere and are not, if normal rules applied, even supposed to be here. This illogic and un-reason, so central to the magical realist beauty of the shroud, stands in resolute contrast to facts (shared in other parts of the gallery installation) that have become horrifically banal: the unenforced workplace standards, the exploitation of poor women, and the power geometries of globalised production. Juxtaposed to the accumulation of such data—there seems no end to these and related facts—hangs the shroud. *Shroud* commemorates the dead, leveraging the commodity to do so. Think of it, all this dead labour brought back to life by a quite different model of convivial work and a quite different vision of what conviviality is for. Think of what you can do with the "fetishism of commodities", as Marx called capitalism's inevitable characteristic: haul it into a cave and refigure it so the present can be understood as prehistory of the future.

References

Marx K (1887 [1867]) *Capital, Volume 1* (ed F Engels; trans S Moore and E Aveling). Moscow: Progress
Sugrue T J (2005) *The Origins of the Urban Crisis: Race and Inequality in Postwar Detroit* (new edn). Princeton: Princeton University Press

Wiggle Room

Jessica Dempsey and Geraldine Pratt

Department of Geography, University of British Columbia, Vancouver, BC, Canada;
jessica.dempsey@geog.ubc.ca, gerry.pratt@geog.ubc.ca

Commitments to social struggles drive so much of our scholarly work. And yet so much of our theory and practice intimate failure. Critics such as Nancy Fraser (2009:99) have noted, for instance, the "elective affinities" between some of the powerful critiques made by second-wave feminists, the rollback of the welfare state, and the expansion of feminised, racialised precarious employment: "In a fine instance of the cunning of history, utopian desires found a second life as feeling currents that legitimated the transition to a new form of capitalism: post-Fordist, transnational, neoliberal". As another example, strong persuasive critiques of the whiteness of most postsecondary institutions in North America, Australia and England have led to the creation of seemingly progressive equity and diversity policies. And yet in her ethnography of diversity policy in practice, Sara Ahmed (2012) brings us close to the "brick wall" encountered by those who attempt to implement such policy, the radical disparity between high-minded words and action, and the absorption of progressive policy into the status quo of white supremacy.

All of this takes place on the material terrain of "late liberalism", which keeps relations of accumulation in place through the production and governance of difference and markets, across human and non-human forces of existence. This governance is improvised and takes shape differently in different contexts "according to local materials and conditions" (Povinelli 2016:173–174). Socialist, black feminist and Indigenous and postcolonial theorists (and others) teach us that these patterned arrangements of accumulation involve persistent devaluations of some places, peoples, bodies and ways of life. Reversing all-too-familiar criticisms of feminist, queer and anti-racist struggles, Ahmed (2015) insists that capitalism too is an identity politics in which "the few become the universe/universal". Such a universal is "handy" for accumulation, she writes, because "it makes others into the hands, helping hands, those who have to help reproduce the very system that reproduces their own subordination, or risk becoming unhandy hands, who are grasping at something that is not theirs".

Those helping hands can and do make their own practices but for the non-dominant, politics typically takes place in cramped spaces of deprivation, contradiction and compromise: for instance, when Indigenous communities in settler colonial societies, searching for alternative ways of generating income from their lands, are dependent on extractive capital to fund these ventures, or when Philippine-based migrant organisations wiggle through the seeming inconsistency of simultaneously opposing the Philippine state's labour export policy and assisting

Keywords in Radical Geography: Antipode at 50, First Edition. Edited by the *Antipode* Editorial Collective.
© 2019 The Authors/Antipode Foundation Ltd. Published 2019 by John Wiley & Sons Ltd.

migration to other countries. "We are stuck with the problem of living despite economic and ecological ruination" (Tsing 2015:19). Anna Tsing's underlying claim about what she calls salvage capitalism is that acts of translation across various social and political spaces make commensurate for capital utterly divergent forms of work and nature, such that capitalism is able to salvage value produced without, alongside or outside of capitalist control. She makes this argument in relation to independent matsutake mushroom pickers in Oregon, who treat the forest "as if" it were an "extensive commons", and are nonetheless patched into a highly profitable global market for the most valuable mushroom in the world. This is a new twist on an old story, of capitalism subsuming older ways or other forms of life. However, rummaging for survival, wiggling around obstacles—or simply enduring—might also be openings to something new. Among the mushroom pickers in Oregon, there is pervasive talk of freedom in relation to their activity. This freedom, Tsing writes, has nothing to do with the kinds of freedom imagined by economists or within political liberalism. It is "performative, communally varied, and effervescent. It has something to do with the rowdy cosmopolitanism of the place ... I think it only exists in relation to ghosts" (2015:76), hauntings from many times and places: including Indigenous dispossession from the Oregon forests and wars throughout southeast Asia.

Feminist, radical, and critical geographers (and more—activists, advocates, movements, unions, artists) are trying to grasp the contours of cramped spaces of late liberalism and the possibilities of that something new. Mime-like, feeling their hands around brick walls seeking and sensing spaces or cracks, paying close attention. Radical geographical practice is often about finding "wiggle room".

Why Wiggle Room?

Wiggle room is at once a recognition of these cramped spaces, the material constraints, the structuring logics of patriarchy, racism, the incessant drive to accumulation, and the something more. The room we are thinking of is not a space *outside* of hegemonic power relations. Rather, by wiggle room we emphasise the always present "traces" (a la Derrida and Spivak) and possibilities for being otherwise within those relations. It's not as much about a space as making space. Rather than the binary of inside–outside, for us wiggle room evokes repeated actions, a certain verbiness—*movement*. The emphasis is on the sweaty, laborious, restless, creation (and re-creation) of dominant power relations/norms/codes that are in constant "parasitic" engagement with similarly ceaseless resisting, evading, playing, enduring, inventing, refusing, strategising, and living according to "infinite other, *non*- or *a*-capitalist affective capacities and potentials" (Gidwani 2008:195). These potentials might work as elective affinities that are absorbed within, support and expand existing social and economic relations or they might lead somewhere else.

Wiggle room is a sensibility that recognises that the context of the theoretical or political intervention affects how we theorise. Asked whether, in a private familial household, just a partner—or a partner and children, appropriate a woman's surplus labour, J.K. Gibson-Graham respond that the question cannot be answered outside a particular theoretical intervention: "Answers to these questions fall into place only when we know what is at stake" (1996:214). Wiggling involves holding onto and living within the indeterminacy of not knowing when a strategy or approach or refusal is being absorbed or making gains. Wiggling accepts that there are different paths towards the same orientation, and that universal, unifying, homogenising theory not only replays Eurocentricism but limits strategy and political options. Wiggling is a wilful practice of movement. As a killjoy feminist, Sara Ahmed (2015) writes of the wiggling work of feminism, driven by anger: "Anger against objects or events, directed against this or that, moves feminism into a bigger critique". Anger has moved feminism beyond its focus on women and gender, and this ungrounding is not a "failure of feminist activism, but is indicative of its capacity to move, or to become a movement". In radical geographical work, such wiggling involves carefully mapping cramped spaces and their histories, to understand the structuring logics, but also the disarticulations and contingencies and room for movement. Below are some promising approaches to wiggling towards other futures.

Incomplete and Failed Capitalist Colonial Formations

For us, Vinay Gidwani's theorising of capitalism exemplifies a wiggling sensibility. Drawing from Marx, Althusser, and postcolonial critiques of historicism (among others), Gidwani theorises capitalism "as an incomplete totality constantly striving for self-adequacy" (2008:xxiii); he insists on narrating capitalism as a social formation with a structuring logic (its relentless drive to expand and accumulate) but one that is "continuously interrupted by other logics" (2008:xxiv). Read through Gidwani, capital "draws its force by attempting to divert or attach itself to other kinds of energy or logic—cultural, political, non-human—whose contributions, like those of history's subalterns, are erased from conventional accounts" (2008:xix). Through this picturing, even where capital does succeed in harnessing such energies—as with labour—it is never all-encompassing: "*there is a living, creative potential in labour that is irreducible;* that persistently survives objectification by capitalist social relations" (2008:225). This resonates with Audra Simpson's theorisation of the settler-colonial project as ongoing, violent, and deeply "strangulating" but never complete, even a failure: "it fails at what it is supposed to do: eliminate Indigenous people; take all their land; absorb them into a white, property-owning body politic" (2014:8). At the centre of her book, *Mohawk Interruptus*, are the Mohawks of Kahnawa:ke who continue to refuse "the 'gifts' of American and Canadian citizenship" (2014:7), insist on Haudenosaunee governance, and do "all they can to live a political life robustly, with dignity as Nationals" (2014:3)— Nationals of the Mohawk nation. As a result of the unending labour of refusal of the Mohawks of Kahnawa:ke, Simpson concludes: "there is more than one political show in town" (2014:11), a kind of sovereignty within sovereignty that creates "terrific tension", posing "serious jurisdictional and normative challenges to each other" (2014:10). Shiri Pasternak (2016) too shows the Canadian state's "permanent disability" to effectively take control over the Algonquins of Barriere lake, positing imperfect, sutured together, often threadbare geographies of territorial control as the norm, as opposed to an imaginary of fully formed, air-tight state jurisdiction. These representations reclaim a wilful persistent agency by tracing how labour and Indigenous People shape, scupper, and interrupt empire.

Remaindered Life and Hetero-spatiotemporalities

In her dissatisfaction with reductive readings of disposable, redundant or superfluous populations, Neferti Tadiar (2009, 2013) reaches for what "falls away" from feminist, marxist and other accounts, for overlooked modes of social experience, social cooperation and forms of life-making that elude existing narratives, with the possibility that this opens up "other genealogies for understanding those remaindered ways of *living* in the world that move and generate that world in ways we would otherwise be unable to take into political account" (2013:43). She conceives of Filipino migrant domestic workers, for example, through less bounded ontologies of self and sociality than are legible in narratives of racialised and exploited workers. Framed within pre-colonial and colonial relations of spirit mediumship, "they act in a very practical but also otherworldly sense as forms of human media—technologies of reproduction rather than full-fledged sovereign (self-determining, self-owning) individual objects" (2015:152). Imagining the world as radically heterogeneous, with multiple ontologies of self, emerging out of and situated in other histories and temporalities, other social practices and other social logics, potentially opens unanticipated histories and plural normative horizons. It might allow us to:

> write over the given and privileged narratives of citizenship other narratives of human connections that draw sustenance from dreamed up pasts and futures where collectivities are defined neither by the rituals of citizenship nor by the nightmare of "tradition" that "modernity" creates. (Chakrabarty 2000:46)

Singular historical narratives can foreclose the claims of the disadvantaged or dispossessed in many different ways and retrieving "crowded histories" (Collins 2017) may be one way of creating wiggle room.

Diverse Economies

J.K. Gibson-Graham have worked with feminist and poststructural critiques to reimagine capitalism as a set of practices rather than a systematic concentration of power in a closed system, practices that coexist with other non-capitalist economic activities. In their early work on coal-mining towns in Australia (Gibson-Graham 1996), they theorised the familial household as a site of non-capitalist economic activity (rather than a condition of existence for the social reproduction of the capitalist economy). This gave them the wiggle room to theorise ways that the lives of coal miners' wives, within the feudal non-capitalist relations of the household, had been made intolerable by their husband's seven-day shift roster, so as to position women's organising against the mines' management within their own rather than their husbands' class relations. Along with this analytical reframing of capitalism is their insistence on a disposition towards hopefulness. In line with Eve Sedgewick's characterisation of much critical theory as paranoid, they opt for a reparative approach to theory, for theory "to help us see openings, to help us to find happiness" (Gibson-Graham 2006:7) and to foster "affinities and even affiliations" among those oriented in the same direction "even if pursuing different paths" (2006:8). Working with much the same body of theory and in a similar affective register, David Seitz (2017) draws on Melanie Klein's distinction between paranoid and depressive psychic structures. As opposed to a paranoid position in which the possibility of suffering forecloses attachment, a depressive position admits the complexity of the world (as a contradictory amalgam of good and bad objects) to allow attachment to "good enough" objects and the psychic resources to wiggle on. For Gibson-Graham wiggling on with good enough objects tends to happen through local practices of non-capitalist economic experimentation (e.g. community gardens, community currencies, the social economy), potentially inventing new embodied practices of politics and economic relations. For Seitz, it can take place within the imperfect politics of a queer church in local, national and transnational struggles.

Decolonising and Queering Everyday: Wiggling in the Everyday

Drawing from queer Indigenous scholars and their own praxis, Sarah Hunt and Cindy Holmes animate and verbify theory. They insist on decolonis*ing* and queer*ing* as central practices to unsettling "White settler colonialism and the colonial and gender categories it relies on" (2015:156). Their theorising of resistance brings into focus the "everydayness" of decolonising, centring the "daily acts of embodying and living Indigeneity, honoring long-standing relationships with the land and with one another" (2015:157). Hunt and Holmes argue that decolonising politics is found not only in the public displays of blockades or rallies, but in the intimate and interpersonal, between friends and within families, across differences. Friendships are one route towards allyship, a route that while pleasurable is not "easy" or smooth, requiring "negotiations of tensions of confronting colonialism and White privilege" (2015:161) that arise when attempting to build coalitions between socially dominant and marginalised groups. Their essay narrates relationships that are complicated, power-laden; but, they insist, are also the basis of decolonising and queering. Any movement to confront structuring forces is forged in relationships—personal, work, organisational—that matter on their own terms as Hunt and Holmes insist, but are also the foundation of collective organising:

> Friendship is also a source of power. Oppression maintains itself by putting us in competition with one another, making us distrust and distance ourselves from each other … Recovery of friendship gives us back the power of trust, equality, connection, value, and respect, and gives us the ability to become conscious, heal, organize and act together. (Bishop 2015:102)

And so—in friendship—here's to wiggling on, together.

References

Ahmed S (2012) *On Being Included: Racism and Diversity in Institutional Life*. Durham: Duke University Press

Ahmed S (2015) It is not the time for a party. *feministkilljoys* 13 May. https://feministkilljoys.com/2015/05/13/it-is-not-the-time-for-a-party/ (last accessed 17 August 2018)

Bishop A (2015) *Becoming an Ally*. Victoria: Fernwood

Chakrabarty D (2000) *Provincializing Europe: Postcolonial Thought and Historical Difference*. Princeton: Princeton University Press

Collins E (2017) "Of Crowded Histories and Urban Theory". Unpublished paper presented at annual meetings of the American Association of Geographers

Fraser N (2009) Feminism, capitalism, and the cunning of history. *New Left Review* 56:97–117

Gibson-Graham J K (1996) *The End of Capitalism (As We Knew It): A Feminist Critique of Political Economy*. Oxford: Blackwell

Gibson-Graham J K (2006) *A Postcapitalist Politics*. Minneapolis: University of Minnesota Press

Gidwani V (2008) *Capital, Interrupted: Agrarian Development and the Politics of Work in India*. Minneapolis: University of Minnesota Press

Hunt S and Holmes C (2015) Everyday decolonization: Living a decolonizing queer politics. *Journal of Lesbian Studies* 19(2):154–172

Pasternak S (2016) *Grounded Authority: The Algonquins of Barriere Lake*. Minneapolis: University of Minnesota Press

Povinelli E (2016) *Geontologies: A Requiem to Late Liberalism*. Durham: Duke University Press

Seitz D (2017) *A House of Prayer for All People: Contesting Citizenship in a Queer Church*. Minneapolis: University of Minnesota Press

Simpson A (2014) *Mohawk Interruptus (Political Life Across the Borders of Settler States)*. Durham: Duke University Press

Tadiar N X M (2009) *Things Fall Away: Philippine Historical Experience and the Makings of Globalization*. Durham: Duke University Press

Tadiar N X M (2013) Life-times of disposability within global neoliberalism. *Social Text* 31(2):19–48

Tadiar N X M (2015) Decolonization, "race", and remaindered life under empire. *Qui Parle: Critical Humanities and Social Sciences* 23(2):135–160

Tsing A (2015) *The Mushroom at the End of the World: On the Possibilities of Life in Capitalist Ruins*. Princeton: Princeton University Press